ELECTRON·OPTICS

P. GRIVET

Professor at the University of Paris

With the Collaboration of

M.-Y. BERNARD, F. BERTEIN, R. CASTAING,
M. GAUZIT AND A. SEPTIER

ELECTRON OPTICS

SECOND ENGLISH EDITION

IN TWO PARTS

Translated by P. W. HAWKES
Revised afresh for this Edition by A. SEPTIER

Part 1 OPTICS

PERGAMON PRESS

OXFORD · NEW YORK
TORONTO · SYDNEY · BRAUNSCHWEIG

Pergamon Press Ltd., Headington Hill Hall, Oxford
Pergamon Press Inc., Maxwell House, Fairview Park, Elmsford,
New York 10523
Pergamon of Canada Ltd., 207 Queen's Quay West, Toronto 1
Pergamon Press (Aust.) Pty. Ltd., 19a Boundary Street,
Rushcutters Bay, N.S.W. 2011, Australia
Vieweg & Sohn GmbH, Burgplatz 1, Braunschweig

First English edition 1965
Second (revised) English edition 1972
Library of Congress Catalog Card No. 72-132964

This book is a translation, with much revision and
addition, of *Optique Électronique*, originally
published in French by Bordas Éditeur, Paris

Printed in Great Britain by A. Wheaton & Co., Exeter
08 016226 6

CONTENTS

PART I. OPTICS

CONTENTS OF PART 2

PART 2. INSTRUMENTS

PREFACE TO THE SECOND ENGLISH EDITION

WE HAVE tried to improve this new edition of *Electron Optics* for the benefit of two categories of reader. First of all, we have endeavoured to make it more useful to the graduating student. With this in mind we have, for example, added some introductory sections describing selected but nevertheless classical methods of calculating electric and magnetic fields; these indications are intended as a guide for the beginner who sometimes feels lost in the abundant specialized literature.

On the other hand, we thought it essential to keep the research worker (even the beginner) informed of the important recent progress in the field: this seems to us an essential prerequisite for a book which aims at helping scientists to form their scientific taste and choose their own path in research.

This proved to be a fascinating but difficult task because activity in the field of electron optics has been both important and successful during recent years. The familiar magnetic objective has been significantly perfected, mainly by its discoverer, Professor Ruska, while the practical introduction of superconducting lenses gives some realistic hopes of a breakthrough in the domains of high resolution and high energy: today, several careful measurements prove that a resolving power of 4 Å is already achieved and that one can distinguish lines spaced 1 or 2 Å apart: the traces of Bragg planes are clearly distinguishable and these results give good reason to hope that in a not too distant future the individual atoms of most solids will be made visible by high-voltage electron optics.

On the theoretical side, the phenomenon of interference between two electron beams has gained in interest, as it may be usefully compared to the superconducting interferometer, a more mysterious but also more practical instrument making use of the stability of Cooper electron pairs and producing beats in the time domain instead of fringes in space. For this reason, greater care has been taken to bridge the gap between refractive index theory and the theory of the phase shift for electron waves.

The new appearance of this second edition is again mainly due to the efforts of Dr. Septier and the new text was also very ably translated into English by the English specialist of electron optics, Dr. Hawkes. I am glad to thank them for their renewed efforts on behalf of this book, in the name of all our team. The success of the first edition induces me to express our gratitude to the founder of this team during the hard times of the forties, Academician Dr. Ponte. We hope that the qualities of this book reflect in no negligible part the precious lessons of his two teachers during the twenties, Professor L. de Broglie and Sir John (J. J.) Thomson: their creative enthusiasm still had a beneficial impact on us 40 years later.

PIERRE GRIVET

PREFACE TO THE FIRST ENGLISH EDITION

WITH the appearance of this new edition in English it is a pleasure to look back at the original French version of the book; the latter is not so very old, and yet how important have been the advances which have been made since its publication, in 1955 and 1958. The most fundamental sphere of application of electron optics, electron lenses, seemed at that time to have been fully explored, but nevertheless, the subject has a fresh appeal to young research workers. The methods of calculation have developed beyond recognition, for nowadays no-one hesitates to invoke the aid of an electronic computer, so that the atmosphere of slightly timorous respect in which the differential equations to be solved used to be enveloped has practically disappeared. The importance of these electronic computers has been shown by the major discoveries which have resulted from their use, the behaviour of the quadrupole lens, for example. The need to consider this new type of lens was imposed by the requirements of nuclear physicists; none of the ordinary methods was suitable for studying these lenses, but even so, their properties and their surprising qualities are now well known after less than five years' study—ordinary lenses, with their much simpler structure, had to wait almost two decades before being thoroughly understood.

Other pleasant surprises await us in the instrumental field. To begin with, the electron microscope is the most perfected instrument, having bettered the performance which the theoreticians were reasonably hoping for around 1949. In their caution, these latter were doubtless too severe in their estimates of the limiting resolving power which could be attained, for all over the world today, industrial instruments of various nationalities— English, German, Japanese, French and Russian—are in current use and reach and even slightly exceed the most optimistic of the estimates of the theoreticians of ten years ago.

Still more astonishing are the results which have been obtained with this microscope. Ten years ago, it was still thought that the electron microscope was capable of giving only a rather frozen view of reality, unnaturally stiff like the view which we are given of the life of prehistoric organisms by paleontology. But at Toulouse, Professor Dupouy has already begun to study living biological cells with the aid of very high voltages; and even in more ordinary laboratories, we have no longer to be satisfied with an examination of a replica or desiccated skeleton of reality. Proper sections can now be cut which show the real structure of the dead cell. The success is still more striking where solids are concerned; with an electron microscope we can fasten on to the very life of a metal, by filming, for example, the evolution of the dislocations when the metal is heated; this can be performed directly on a very thin layer cut from a block of the metal being studied.

Metallurgy provides us with yet another example of a great success. The metallurgists have now whole-heartedly adopted the electron probe; with its aid, they are able to perform extremely detailed and accurate metallurgical analyses.

This analyser was invented in France in 1949, and since 1954 Castaing has made a highly perfected form available to the metallurgists. It has taken six years for metallurgists all over the world to become convinced of the excellence of this method, and 1960 marks the beginnings of world-wide employment of this micro-analyser. This example demonstrates how useful it always is to struggle for the propagation of scientific ideas and methods, even when they have received the most brilliant experimental confirmation in the laboratory.

In these present times it is always necessary to make a very great effort towards the diffusion of scientific thought and the renewal of experimental methods and to disseminate them among a wide public, as, in the last resort, it is this public which reaps all the benefit. It is for this reason that the senior author of the French edition feels justified in his delight at seeing that this book is going to present the ideas of the French school of electron optics to an English-speaking public, to a public whose language is that of the country in which the "electron" was born.

It is a pleasure to thank the many colleagues, both at home and abroad, who have graciously contributed to this work by helping us to clarify some difficult points, or by providing us with illustrations taken from their own personal work. From among them, we should like to single out Professors Castaing, Dupouy, Fert, Lallemand, Möllenstedt, Müller, Nozières and Pernoux, and Doctors Duchesne, Haine, Mulvey, Stohr and Wlerick.

We are also grateful to various industrial firms who furnished us with certain documents, and in particular to C.S.F. and O.P.L. (Paris) and Zeiss (Oberkochen).

This book, in its present form, owes much to the painstaking work of Dr. Septier, who brought it up to date. Dr. Hawkes has been more than a translator, and his profound knowledge of Electron Optics is present, if not apparent, in many parts of this book. Last, but not least, I should like to express my personal gratitude, and that of most of my colleagues in this enterprise, to Professor Marton; without his continual encouragement during the past fifteen years, much of the research and teaching included within these pages would perhaps never have been performed.

PIERRE GRIVET

PART 1

OPTICS

CHAPTER 1

FUNDAMENTAL FEATURES AND BASIC TECHNIQUES

1.1 THE HISTORICAL BACKGROUND

Towards the end of the nincteenth century, physicists directed a large proportion of their efforts towards elucidating the properties of cathode rays which, by virtue of their rectilinear propagation, exhibited a close affinity with light rays. An example is furnished by the celebrated experiment in which Crookes introduced a metal cross into the path of the rays and threw its shadow onto a fluorescent screen. That there was, however, a fundamental difference between light rays and these cathode rays composed of electrons in motion, was demonstrated by Pluecker and Goldstein who showed that the latter are deflected by magnets and by charged bodies.

Later, with the revolutionary new ideas which followed the discovery of the wave nature of the electron by L. de Broglie, it became apparent that this deviation could be regarded as an optical property; further, the exploration of the vast new sphere of phenomena, now classed generically as "electron diffraction", could begin.

The creation of this "physical optics" of electrons led to a more profound examination of particle mechanics, and the analogy between the dynamics of a material point and geometrical optics, which had been formulated in theory by Hamilton in 1827, soon provided the foundations for the new subject of "electron optics". This "corpuscular" optics—valid for charged particles other than electrons : ions for example—consists in studying the laws which govern the behaviour of electrons from their source, through the "object", to the image which they trace out upon either a fluorescent screen or a photographic plate.

After the first lenses had been built, by H. Busch (1927) and by Davisson and Calbick (1931), the new optics developed rapidly, considerably aided as it was by the important advances in "radio valve" techniques which were taking place.

In fact, many of the difficulties were becoming less important: in particular, to obtain a vacuum within a capacious envelope containing large metal surfaces had become a straightforward procedure and fluorescent powders were now standard industrial products. Technology, too, provided fascinating scientific problems to be solved, and from the beginning, electron optics has been used successfully both to unravel the mysteries of oxide-coated cathodes and secondary emission, and to satisfy the requirements of television engineers.

To-day, electron optics is a branch of science which has achieved maturity, a branch represented by a rich diversity of instruments—cathode ray tubes, electron microscopes and electronic image converters for example. Throughout the whole of this vast field of applied electron optics, however, it remains true that there are two alternative ways of controlling a beam of cathode rays. The beam passes through the aperture either of an "electrostatic lens", in which the rays are deviated by the electric field produced by a system of annular electrodes, or of a "magnetic lens", where annular magnets or current-carrying coils surround the beam.

Two quite different methods of exploring the theory of electrons in electric and magnetic fields are possible.

If, from the beginning, we examine the analogy between the trajectory of an electron and the track of a light ray, it is possible to define a refractive index, a function of position $n(r, z)$, in such a way that the path of the particle is identical with that which a light ray, obeying the usual rules of refraction, would trace out in a medium with refractive index n. Once n has been established, it becomes a simple matter to tabulate the properties of electron lenses by a direct transposition from those of their optical models; this is not, however, physically particularly meaningful in the case of the magnetic lens. In his book *Optique Electronique et Corpusculaire* (1950), de Broglie has displayed both the power of this method, and the symmetry which can be established between optics and electromagnetic theory.

Alternatively, the elementary classical formalism of the laws of mechanics and electromagnetism can be used, but this involves rejection of the powerful range of optical methods. However, although the investigation is made less concise, it does relate the optical properties more directly to the concrete features of the electrical experiment. Even if we deliberately restrict the extent of our vision, we do nevertheless avoid the abstract mathematical reasoning which is necessary to relate elementary geometrical optics to Hamilton's theory. For this reason, we shall—to begin with—employ the trajectory method.

1.2 OUTLINE OF A DEMONSTRATION IN ELECTRON OPTICS

The formation of an image by an electron lens can be demonstrated with the apparatus shown schematically in Fig. 1. An electron gun, which comprises an oxide-coated cathode and an accelerating grid C_1 held at a potential Φ_1, is placed within an evacuated envelope. The cathode is held at a temperature of about 800°C by an incandescent heating element and electrons, in consequence of their thermal motion, are ejected with a small velocity which corresponds for the majority to an energy of less than 0·1 eV.

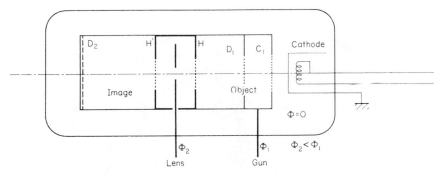

FIG. 1. Diagram of the basic demonstration.

The velocity v_1 at which they pass through the grid is given by the conservation of energy relation

$$\tfrac{1}{2}m\,v_1^2 = e\,\Phi_1.$$

The initial velocity is negligible in comparison with the velocity produced by Φ_1, which is usually of the order of thousands of volts. The beam then illuminates the object, which is a grating cut in the diaphragm D_1, passes through the lens L, and finally forms an image on the plate D_2, which is coated with some fluorescent substance. The potential of the lens L is negative with respect to the potentials of D_1 and D_2, which are held at the same potential as the outer electrodes H and H', the purpose of which is to localize the effect of the lens. The extent of the lens field is of the order of magnitude of the diameter of the holes in the grids; outside (between the object and the grid H and between the grid H' and the screen), the field is zero and the potential has the constant value Φ_1.

1.3 THE OBJECT

Electronic devices have been adapted to enable a great variety of objects to be examined; however, despite outward appearances, the optical properties of these instruments can always be traced back to those of the apparatus shown in Fig. 1, an electron gun in conjunction with a grating, the properties of which we shall now examine in more detail.

1.3.1 Opaque Object

A grating is a characteristic example of an object which is genuinely opaque to fast electrons; the bars of a grid are at least some dozens of microns thick, and even though they be made of some light metal, aluminium or beryllium for example, the electrons will be stopped and their energy

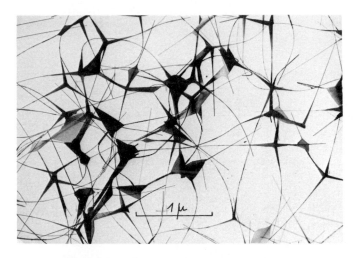

FIG. 2. The electron image of an opaque object (zinc oxide); magnification 20,000.

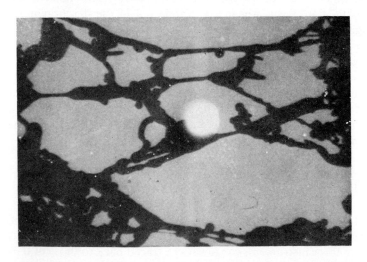

FIG. 3. The image of an opaque object (zinc oxide) "illuminated" by Li⁺ ions
(Gauzit, 1951).

converted into heat, as Lenard showed towards the end of the last century, by a series of very careful measurements (see the article by Bothe in Geiger and Scheel's *Handbuch der Physik* **24**). Such objects—powders, fibres, colloids, for example— are thus genuinely opaque in the usual sense of the word, and it is natural that they should give "black and white" images. They are, in fact, the only types of object which can at present be observed in ion microscopy, as such heavy particles are halted by films of matter only a few millimicrons thick (see Figs. 2 and 3).

1.3.2 Thin Objects for Transmission Microscopy

This category, to-day extremely wide—microscopic observation of such objects is an everyday occurrence—produces images of which the black regions represent only an *apparent opacity*. These objects have in fact the form of very thin films, only a few millimicrons thick (between 10 and 50 mμ for electrons accelerated to 60 kV), and such thin layers are effectively transparent to electrons, which pass through them without any appreciable loss of speed. Each electron does undergo, however, a slight deviation—it is "elastically scattered" through an angle θ. A very fine incident pencil is therefore scattered into a cone of a few degrees at the apex, on leaving the thin film; the scattering angles are in fact those which are observed directly in electron diffraction patterns, and lie between a few hundredths and a few tenths of a radian, whereas the initial angular aperture of a beam from a typical electron gun scarcely exceeds a thousandth of a radian. For a given substance, the mean scattering angle θ is proportional to the film thickness; for a heterogeneous specimen, θ is essentially proportional to the mass traversed by the electrons. The effect of the microscope is to convert the variation of the mean scattering angle between different regions of the specimen into an apparent variation of opacity. A diaphragm with a small opening (a few hundredths of a millimetre for a focal length f of a few millimetres) is placed in the focal plane of the objective lens. Rays which have left the object at an angle θ to the optic axis intersect the focal plane at a distance $r = \theta f$ from the axis; the radius ϱ of the hole in the diaphragm is chosen so that only those rays which were scattered through an angle α or less can pass through ($\varrho = f \cdot \alpha$), while those scattered through angles θ greater than α are intercepted ($\theta > \alpha$ implies $\theta f > \varrho$). The presence of a "contrast diaphragm" subtracts a certain number of electrons from each pencil, an action which becomes progressively more important as the scattering becomes more marked, which occurs when the thickness—or, more generally, the quantity of matter traversed—is increased (or again, for a given thickness, as the atomic weight of the constituent elements is made greater). The apparent opacity which appears in the image is therefore proportional to the quantity of matter through which the electrons have passed, and is thus a direct visual representation of variations of thickness

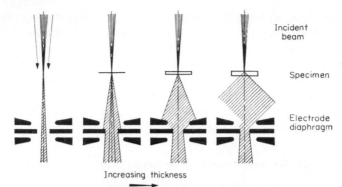

FIG. 4. The principle behind the formation of contrast by a transparent object
(for clarity, the rays are drawn as rectilinear).

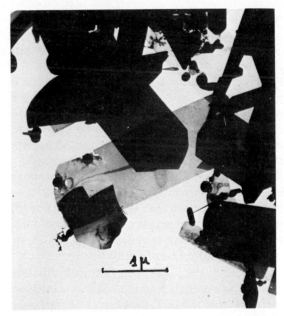

FIG. 5a. Contrast in the image of a transparent object (molybdenum oxide);
$G = 20,000$ (Gauzit).

or of structure of the object. Figure 4 shows schematically this imaging
mechanism; for simplicity's sake, the contrast diaphragm is represented as
lying between the object and the objective lens, while the electrons follow
rectilinear paths, an arrangement which has been successfully used experimentally (Hall, 1948).

Figure 5a is a photograph of the image of a transparent object. The contrast of particular regions of the object can be accentuated by covering
them with a thin layer of "heavy" metal (gold or uranium), by the shadow

technique, or by a method similar to the phase contrast techniques of classical optics (Locquin, 1953).

The complex action of the object can, therefore, be summarized by ascribing to it a fictitious transparency, characteristic of the structure at each point. Although clearly only a first approximation, this description proves to be quite adequate as a foundation for the theory of the instruments customarily used to examine such objects. It is, too, extremely difficult to acquire more detailed knowledge about the interaction between a thin object and an electron beam; recent work is only just beginning to yield a few definite results, which will be mentioned later (see also the publications of Marton, 1946; Gabor, 1945; and Boersch, 1947).

1.3.3 Self-luminous Objects

These objects behave as if each point of their surfaces were emitting a fine pencil of electrons ($\alpha = 10^{-4}$ radians, for example); as geometrical optics alone is necessary to describe the mechanism which leads to this conclusion, we shall postpone its description to the paragraph devoted to immersion objectives. Meanwhile, we simply summarize the principle underlying the phenomena which produce in the image features characteristic of the structure of the object, that is, which produce variations in the rate of emission of electrons at different points on the source. Self-luminous objects can be classified according to the type of emission which they produce.

Thermionic electron emission: the object is heated to a temperature at which it emits electrons spontaneously (at least 1300°C for pure metals).

Photoelectric emission: the object is illuminated by an auxiliary light source (usually a source of ultra-violet radiation).

Secondary emission: the object is bombarded with an auxiliary beam of electrons or ions.

Thermionic emission is the most familiar of these, and we shall take it as an example, but experimental evidence shows that in the two other cases the phenomena are qualitatively similar. Within a heated metal, the electrons acquire a thermal excitation, distributed at random among the individual particles, which is ranged about a mean value according to the Maxwellian probability distribution. The electrons are held within the solid by a potential barrier, known as the "work function", whose height depends not only upon the nature of the substance but also upon the condition of its surface. The current density which is emitted depends, therefore, upon three factors, the natural conditions, specimen preparation, and the temperature. By natural conditions, we means such properties as the nature of the metal and the nature of the crystal face at the point being considered, since a slight variation in work function produces a large variation in the emission intensity. The manner in which the specimen has been prepared is reflected in the cleanliness and relief of the surface, the nature of such impurities as may accidentally have been absorbed at the surface, and by the nature of layers known as "activation layers" with which the surface may become covered as a result of some particular preparation to which the specimen has been subjected. Finally, increasing the specimen temperature increases the energy of the thermal excitation.

FIG. 5b. The image of a self-emissive object (the surface of recrystallized niobium); $G = 300$ (Septier, 1951).

The factors which help to provide contrast in the image are relief and variations in crystal orientation. We usually try to work under conditions of controlled cleanliness, choosing whichever are most favourable for revealing the surface structure of the crystal; this structure can be seen clearly with the refractory metals, for example, when they have been cleaned by prolonged heating in a vacuum at high temperature (1400 to 2000°C); with the less refractory metals, the surface structure is more elusive—it is necessary to hold them at a low temperature (800°), after first cleaning the surface, and then covering it with an activation layer of one of the alkaline earths (usually barium or strontium).

These processes are described in detail in the literature of emission electron microscopy (Gauzit and Septier, 1950, 1951). Figure 5b shows the image of such a cathode.

Objects which emit positive ions can be studied equally easily, and again provide images with contrast, under certain conditions; in the case of ions, however, the origin of this contrast is still unknown.

1.4 THE FLUORESCENT SCREEN

As electrons are invisible, a *fluorescent screen* is necessary to make the image perceptible; the screen fulfils the double function of transforming into light as large a part of the energy of the incident particles as possible, and of ensuring that the flow of electrons is towards the positive pole of the source of the accelerating potential.

1.4.1 Fluorescence

The first function of the screen, that of fluorescing under electron bombardment, is satisfied nowadays by a whole range of substances which differ in the colour of the light emitted, in the duration of the emission of light, in resistance to electron or ion bombardment or in the efficiency with which they convert electron energy into light. So many phosphors are now available that whatever may be the combination of properties necessary for some particular application, it is usually possible to obtain it. Here, we shall simply mention the range of substances which are usually employed in conjunction with an electron optical bench: calcium tungstate, which gives a blue light and can support extreme bombardment and ionic discharge, but is rather insensitive; zinc silicate, or willemite, which fluoresces green, and zinc and cadmium sulphide, which give a white or a yellow light, depending upon which activating element is used, are considerably less robust although much more sensitive. High quality specimens, distributed over a metal support and used at high tension (60,000 volts) will transform up to 10 per cent of the incident electron energy into light, and an efficiency of three or four per cent is standard. For further details, the reader is referred to the books by Kröger (1948), Garlick (1949), and Leverentz (1950).

The light intensity emitted by the screen obeys a law of the form:

$$B = A f(i)(V - V_0)^n$$

Under normal conditions (the current density i not larger than a few $\mu A\ cm^{-2}$), B is proportional to i, but not to the accelerating potential which affects B according to $(V - V_0)^n$, where the index n varies between one and three according to the substance used, and V_0, depending upon the condition of the screen, is of the order of hundreds of volts. For will-emite, $n \sim 2$. With larger current densities, B approaches a constant value; for zinc silicate, this saturation appears beyond about 10 A cm^{-2}, and for zinc sulphide activated with silver or with copper, beyond 100 A cm^{-2}. These saturation conditions occur whenever the electrons are concentrated into a spot as nearly as possible a geometrical point, which is the usual situation in television sets and cathode ray tubes. It is worth mentioning that fluorescent substances provide an extremely sensitive detector of radiation. On the final screen of an electron microscope, for example, with a direct magnification of 50,000, the current density rarely exceeds 10^{-10} A cm^{-2}, but nevertheless, the screen is sufficiently luminous for the eye to be capable of distinguishing clearly the detail in the image, once it has become accustomed to the darkness, and for the focus to be adjusted with precision (the situation known as "scotopic" vision). In other applications, television for example, the spot emits a blinding light. At such a high brilliance, however, the temperature of the fluorescent screen rises, and its quality deteriorates correspondingly. It is necessary to provide a cooling mechanism for the screen, and this is done by using a copper support, cooled by a current of air or of water.

Nevertheless, with very intense beams of electrons, the sensitive layer is destroyed gradually by the impact of negative ions which are produced at the cathode and by ionization of residual gas in the tube.

The high sensitivity of fluorescent substances explains why it has been for a long time possible to dispense with electron lenses when studying the properties of electrons. "Electron rays", stopped down by small holes, a tenth of a millimetre in diameter, cut in a lead screen, still possess sufficient energy to produce very bright spots, and by such simple means as these, quite brilliant diffraction patterns can be obtained.

For the last few years, the question of the sensitivity and the robustness of fluorescent screens to the arrival of powerfully accelerated ions has arisen, the ions now being considered no longer as stray destructive particles, but as the "illuminating" particles in ion microscopy. We now know how to produce sources of monoenergetic ions sufficiently intense to give visible images, although the image quality improves markedly as the intensity is reduced, as this renders the numerous parasitic effects less harmful.

All the fluorescent substances commercially available have been systematically investigated, and classified according to their sensitivity and their resistance to bombardment (Gauzit, 1954). Certain of the zinc sulphides, and willemite, have proved to be superior to calcium tungstate which nevertheless remains the most robust, with a lifetime of several hours at a current density of about 10^{-12} A cm^{-2}.

The object, however, is strongly affected by an ion beam, even though the intensity of the latter be comparable with the values of beam intensity which are usual in electron microscopy.

1.4.2 The Screen as an Electron Sink

A fluorescent screen comprises an active layer formed on a passive support. The active layer is sufficiently thin and is composed of a substance sufficiently translucent for the layer to be very transparent—the image, therefore, can be observed from either side of the screen, either by "reflexion" looking at the side on which the electrons actually impinge, or by transmitted light, provided that the support is transparent.

The first method is used in electron microscopes, and the support is then a metal plate whose high optical reflecting power enables some of the light which is emitted in directions away from the observer to be used. The majority of the incident charge is conducted away across the fluorescent film to the conducting support; secondary emission effects play a minor part, a phenomenon which will be described in more detail later.

If, on the other hand, we choose to observe the rear face of the fluorescent layer through a transparent support, which is usual in cathode ray oscillographs, the incident current circuit is closed by secondary electron emission. The screen is effectively insulated, as the conductivity of the fluorescent layer and the glass is too poor for there to be an appreciable current towards the conducting ring at their edge. Fortunately, however, provided the accelerating potential lies between 100 and about 10,000 V, each incident electron ejects at least one secondary electron as it arrives at the screen, and frequently many (four or five, say, or even as many as nine if the nature of the layer and the magnitude of the potential are favourable); these secondary electrons are ejected at low speeds, in the direction opposite to the incident electrons, and are collected either on a metal electrode, or on a band of colloidal graphite smeared on the walls of the vacuum chamber (a layer of "aquadag" or carbon black serves the same purpose in oscillographs). These secondary electrons close the "circuit" of the primary electrons, since the collector electrode is connected to the positive pole of the source of high tension; this electrode receives a current equal to the primary current, as any superfluous secondary electrons return to the screen, the potential of which adjusts itself automatically to the value for which the primary and secondary currents exactly balance.

The secondary emission phenomenon is highly stable for potentials of some thousands of volts, but at very low or very high accelerating potentials, it becomes erratic and the multiplication factor falls below unity. With very high potentials, however, we can dispense with the assistance which secondary emission provides at lower potentials, by designing screens which are both transparent and conducting. Such screens consist of the usual thin film of fluorescent material, one surface of which is attached to a glass support, but whose other surface, on which the beam impinges, is covered with a film of aluminium, evaporated onto the fluorescent film *in vacuo*. The aluminium is sufficiently thin for high energy electrons to pass through it with no appreciable retardation. This film of high conductivity metal closes the primary current circuit, and in addition, provides a mirror which returns towards the observer that part of the emitted light which would otherwise have been wasted. Further, this layer of aluminium is a most effective shield for the fluorescent material against the destructive impacts of the ions which are inevitable products of the residual gas in the tube. Screens of this kind, brilliant, stable, and robust, are most convenient for work on an electron optical bench.

It is far more difficult to conduct away the superfluous charge at the screen in the ion microscope. The fluorescent film, resting on a metal support, must be extremely thin, but very rich in fluorescent centres, and must be prepared with as little binding material as possible, perhaps with no cellulose binder at all (Gauzit, 1953). There is no question of lining the film with a layer of conductor as this would be completely opaque to an ion beam.

1.5 PHOTOGRAPHY

The photographic plate is even more sensitive than the fluorescent screen as a detector of electrons, provided that the energy of the electrons exceeds a few thousand volts. Intruments using electrons are, however, considerably less stable than their optical equivalents, and the faculty of the photographic plate of "integrating" automatically the incident energy is rarely exploited. In the electron microscope and in electron diffraction, the exposure time usually lies between 0·1 and 10 sec, and only in a few special situations has it been possible to produce lenses and sources of sufficient stability for long exposures—some hours for ions, a day for electrons—to be usefully employed. Examples of these exceptional cases are provided by the photographs taken with β-rays from gadolinium (Marton and Abelson, 1947) or from Th B (Barker, Richardson and Feather, 1950); by astronomy using the electron "telescope" (Lallemand and Duchesne, 1951); and by the first attempts at ion microscopy (Boersch, 1942c, 1948). Generally speaking, any advance which leads to a reduction in exposure time produces higher quality images.

The response characteristics of photographic emulsions remain qualitatively unchanged, but the classification of the various emulsions according to their optical sensitivity has to be completely re-evaluated for electron "illumination". Emulsions which are "slow" for light photographs often prove to be "fast" for electrons, and the film which is used commercially for recording sound is extensively used among electron microscopists; this particular emulsion is extremely sensitive to electrons, and is also very fine-grained.

A great deal has been published on the subject of electron sensitometry; the reader is referred to the books and bibliographies of electron microscopy mentioned at the end of this book. Here, we shall do no more than mention the orders of magnitude which are habitually used in electron microscopy, following a very careful examination by von Borries (1942, 1944) of German emulsions.

For Agfa "Normal", the photographic density D, defined by

$$D = \log_{10} \frac{I_{\text{Incident}}}{I_{\text{Transmitted}}},$$

is 1·5 for a charge density of 10^{-10} C cm^{-2} and an accelerating potential V of 24 kV; for $V = 75$ kV, D increases to 2·6, but decreases again to 1·5 for $V = 200$ kV. This form of variation with potential is common to all emulsions; only the optimum potential varies. For $V = 200$ kV, $D = 1·85$ for Agfa "Mikro" and 1·6 for Agfa "Kontrast"; in addition D is dependent upon the charge density.

The problem of recording ion images photographically is very difficult and largely unsolved since ions have only a very feeble penetrating power, and a highly idiosyncratic sensitometry.

Precise data are scant among the literature. The most numerous studies have been made with slow ions and mass spectrographs (Aston).

Some information about the relative sensitivity of a few different emulsions can be found among works on emission ion microscopy (Couchet, Gauzit and Septier, 1951) or transmission ion microscopy (Magnan and Chanson, 1951; Gauzit, 1951). Emulsions which are suitable for electrons are too rich in gelatine for ions. An emulsion with large surface density of sensitive grains is required; Kodak B 10, Ilford Q, and nuclear emulsion plates give tolerable results.

Yet another problem appears in emission microscopy, where the ion intensity is stronger: the flow of charge across the insulating support. For currents of the order of microampères, large potential differences build up at certain points, with the consequence that surface currents flow, sparking over to the metal wall of the camera occurs, and it becomes quite impossible to focus intense beams. Only with low intensities can the images be recorded.

1.6 THE OPTICAL BENCH

Varied though the applications of electron microscopy are, they can be divided, from the technological point of view, into two major categories. On the one hand, there are those in which monomolecular layers provide the electron source either photoelectrically or by secondary emission. These layers are excessively fragile and can only be produced within glass envelopes, sealed off after very careful evacuation (down to a pressure of 10^{-7} mm Hg) and baked at a high temperature to outgas the objects inside the envelope.

It is not, therefore, possible to make a preliminary study with an arrangement which can easily be dismantled and prototypes have in fact to be very similar to the final model, sealed off in the same way. There is, nevertheless, the possibility of moving some of the internal electrodes from outside, either through flexible membranes sealed to the glass, or with magnets which act on small pieces of soft iron which are free to move inside the evacuated envelope. Demountable joints are quite out of the question, and experimental work is time-consuming, exacting and expensive.

Quite distinct from these systems are those in which only robust sources of electrons are used, refractory metals for example, or else sources which can easily be replaced, like the oxide-coated cathode, and which have in addition comparatively long lifetimes even at pressures of the order of 10^{-5} mm Hg. Demountable metal devices can now be built, with neoprene joints and a number of interchangeable elements which can be varied at will, and a whole range of measurements and experiments is feasible with a single piece of apparatus, which is evacuated and is known as an electron optical bench by analogy with its light optical counterpart.

The arrangement of these benches is usually based on the layout of a standard electron microscope, which is the most highly-developed of present day electron devices; often, the bench is nothing more than a microscope which has been modified to be very flexible, but which at the same time remains quite unrefined. This kind of bench is well described in the literature, by Boersch (1951), by Marton et al. (1951) and by Fert (1954), for example.

The most satisfying form of electron optical bench consists of an assembly of separate interchangeable metal components which are assembled one above the other on the end-plate of a pump with a high pumping rate. Each element plays an individual role; there might well be, for example, an electron or ion gun, with a control for each of the variable parameters, electrostatic or magnetic lenses, magnetic screens, observation windows, a photographic chamber, and connecting cylinders. Each unit, of the most robust construction, is centred automatically so that the axes of all the units always coincide; to ensure that the junction between each pair of units is leak-proof, each joint consists of an annular ring of neoprene which is tightly clamped into an annular recess (which also provides accurate centring) cut into the end-plate of the adjacent unit.

Another version, which has been described by Marton et al. (1951), is particularly suitable for magnetic lenses, and requires only a single potential lead for the gun. The whole assembly is ranged within a metal cylinder coupled to a high output vacuum pump. The gun is fixed permanently to the cylinder, from which it is insulated to withstand up to 100 kV. The optical system is mounted on a trolley which can either be within the evacuated chamber, or outside it; the lenses and diaphragms can be shifted from the outside in three mutually perpendicular directions, by means of arms, set in Wilson joints distributed around the fluorescent

Usually, however, the lens potential varies in all three directions, and we are restricted to electrodes whose shapes are those of the coordinate surfaces of a triply orthogonal system, as it is then possible to find a solution by separating the variables, that is, by reducing the partial differential equation to three ordinary differential equations.

Alternatively, it is possible to make an approximate calculation of the potential distribution along the axis of electron optical systems with a symmetry axis, by making simplifying assumptions about the boundary conditions (Bertram, 1940, 1942). Nevertheless, in the majority of cases, there is no option but to obtain a numerical solution of $\nabla^2\Phi = 0$, not at all an easy task despite recent efforts (Shortley *et al.*, 1947, 1948; Motz and Klanfer, 1946; and Hesse, 1949). However complex be the system, with axial or translation symmetry, a knowledge of the potential can always be obtained by the so-called "relaxation method"; in this method, the partial derivatives in the Laplace equation are replaced by finite differences, and we estimate the value of the potential at the intersections of a square lattice. A series of successive approximations produces an ever closer estimate of the true potential at each lattice-point. A detailed analysis and extensive bibliography are to be found in the books by Durand, by Southwell (1946) and by Shaw (1953).

We now give a brief account of each of these methods of solving the Laplace equation, after which we consider in greater detail analogue methods of solution involving the electrolytic tank or the resistance network.

2.1.1 The Method of Conformal Transformation

Principle of the method. This method can be used only in the case of systems in two variables, that is, systems with translation symmetry; it depends upon the properties of analytic functions of a complex variable.

If we set $j = \sqrt{-1}$, the variable $Z = x + jy$ (in which x and y are two real variables) is known as a complex variable and may be represented as a point having coordinates (x, y) in the complex plane xOy (the Argand diagram, Fig. 7a). In polar coordinates, we write

$$r = \sqrt{x^2 + y^2} \quad \text{and} \quad \theta = \tan^{-1}(y/x)$$

so that

$$Z = x + jy = r(\cos\theta + j\sin\theta) = r\exp j\theta.$$

We now consider another complex variable,

$$W = u(x, y) + jv(x, y).$$

To every value of Z will correspond a value of W, and we can write

$$W = f(Z).$$

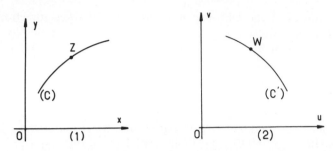

FIG. 7a. Complex variables. When Z traces out (C), W traces out (C').

If Z is continuously varied, the corresponding point will describe a curve (C) in the plane xOy and the point corresponding to W will trace out a curve (C') in the complex plane uOv (Fig. 7a). The function W is said to be "analytic" if the derivative

$$\frac{dW}{dZ} = \lim_{\Delta Z \to 0} \frac{f(Z + \Delta Z) - f(Z)}{\Delta Z}$$

exists and is *unique* (irrespective of the direction of ΔZ in the plane xOy). It is easy to show that this is the case when the real functions $u(x, y)$ and $v(x, y)$ satisfy the Cauchy–Riemann relations

$$\frac{\partial u}{\partial x} = \frac{\partial v}{\partial y} \qquad \frac{\partial v}{\partial x} = -\frac{\partial u}{\partial y}. \tag{2.3}$$

All functions that can be expressed as a series expansion in the neighbourhood of the origin ($W = \sum_{n} c_n Z^n$) belong to this family (examples are e^Z, $\sin Z$, ...).

On differentiating relations (2.3), we obtain the two equations

$$\frac{\partial^2 u}{\partial x^2} + \frac{\partial^2 u}{\partial y^2} = 0 \qquad \frac{\partial^2 v}{\partial x^2} + \frac{\partial^2 v}{\partial y^2} = 0$$

or
$$\tag{2.4}$$

$$\nabla^2 u(x, y) = 0 \qquad \nabla^2 v(x, y) = 0.$$

Thus each of the two component functions of W satisfies the Laplace equation and either may be regarded as the potential function in a two-dimensional system.

Since the function W is analytic, the derivative dW/dZ at a point is independent of the direction of ΔZ at this point. We can therefore write

$$\frac{dW}{dZ} = A \, e^{j\phi} \quad \text{or} \quad dW = A \, e^{j\phi} \, dZ.$$

The segment dW in the $W(u, v)$-plane is obtained from dZ by a change of scale A and a rotation ϕ. Hence, if two curves in the Z-plane intersect at an angle α, the transformed curves will intersect at the same angle in the W-plane; such a transformation is known as a conformal transformation. In particular, the curves $u = \text{const}$ and $v = \text{const}$ in the W-plane intersect at right angles; their transformations $u(x, y)$ and $v(x, y)$ in the Z-plane will therefore be orthogonal (Fig. 7b). If, then, $u(x, y)$ is selected to be the potential function, the family of curves $u(x, y) = \text{const}$ in the Z-plane will form a map of the equipotentials and if we plot the curves $v(x, y) = \text{const}$, we obtain the electric field lines immediately (and *vice versa*).

A conformal transformation will therefore relate a region of uniform field in the W-plane to a more complex situation in the Z-plane, in which certain curves $u(x, y) = \text{const}$ (or $v(x, y) = \text{const}$) will coincide with the real electrodes considered.

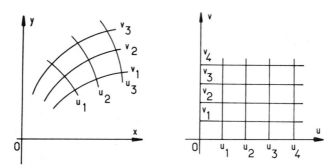

FIG. 7b. The curves $u = \text{const}$ and $v = \text{const}$ in the Z- and W-planes.

Two simple examples

(a) **Let $W = Z^n$.** We can then write W in the form

$$W = r^n e^{jn\theta}$$

so that

$$u = r^n \cos n\theta,$$

$$v = r^n \sin n\theta.$$

We select v as the potential function. Then $v = 0$ at $\theta = 0$ and $\theta = \pi/n$. For $n = 2$, the corresponding map is shown in Fig. 7c; we can therefore obtain a potential

$$\Phi = Ar^2 \sin 2\theta$$

by means of two plane electrodes coinciding with the axes Ox and Oy and held at zero potential together with a hyperbolic conducting electrode at a distance a from O and held at potential $+\Phi_0$. We then have

$$\Phi(r, \theta) = \Phi_0 \frac{r^2}{a^2} \sin 2\theta$$

or   (2.5)

$$\Phi(x, y) = \frac{2\Phi_0}{a^2} xy.$$

This distribution can also be obtained with the aid of four identical electrodes held at potentials $\pm\Phi_0$ (Fig. 7c), placed symmetrically with respect to the axes Ox and Oy: this transformation has thus provided us with a quadrupole lens (see Chapter 10). With $n = 3$, we should obtain a sextupole lens and with $n = 4$, an octopole, and so on; the case $n = 1$ corresponds to the parallel plate condenser.

The general solution of Laplace's equation will therefore be of the form

$$\Phi(r, \theta) = \sum_n A_n r^n \sin n\theta.$$

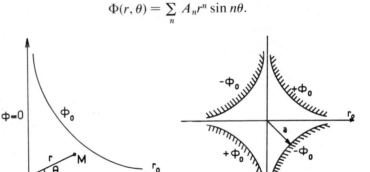

FIG. 7c. Electrodes creating the distribution $\Phi(r, \theta) = Ar^2 \sin 2\theta \; (W = Z^2)$.

(b) Let us now consider the function $W = A \log Z$. We have

$$u(r, \theta) + jv(r, \theta) = A \log r + Aj(\theta - \theta_0).$$

Taking $\Phi(r, \theta) = v(r, \theta)$, the curves $v = \text{const}$ will be straight lines emerging from the origin O, and the curves $u = \text{const}$ will be circles centred on O.

For the special case of the electrodes illustrated in Fig. 7d, consisting of two planes separated at O by an infinitesimally small gap and held at potentials V_1 and V_2, we find

$$v(r, \theta) = \frac{V_2 - V_1}{\theta_2 - \theta_1} \theta + \frac{V_1\theta_2 - V_2\theta_1}{\theta_2 - \theta_1},$$

$$\tag{2.6}$$

$$u(r, \theta) = \frac{V_2 - V_1}{\theta_2 - \theta_1} \log r.$$

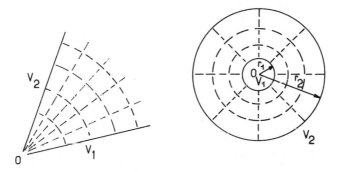

FIG. 7d. Plane wedge and coaxial cylinders ($W = A \log Z$).

When $\theta_2 - \theta_1 = \pi$, $V_1 = 0$ and $V_2 = V_0$ so that by choosing $\theta_1 = 0$, we obtain

$$v(\theta) = \frac{V_0}{\pi}\,\theta,$$

$$u(r) = \frac{V_0}{\pi} \log r.$$

If, on the contrary, we take $\Phi(r, \theta) = u(r, \theta)$ we shall obtain the solution of the Laplace equation between two coaxial metal cylinders of radii r_1 and r_2 held at potentials V_1 and V_2 respectively (Fig. 7d):

$$u(r) = \frac{V_2 - V_1}{\log (r_2/r_1)} \log \frac{r}{r_1} + V_1,$$

$$v(\theta) = \frac{V_2 - V_1}{\log (r_2/r_1)} \theta. \tag{2.7}$$

Other examples are to be found in the books by Ramo and Whinnery and Durand. An electrode system can be associated with every analytic function W in this way, and a list of these solutions may thus be prepared.

The Schwarz transformation. It would be convenient if one could (by means of a conformal transformation) transform a given set of electrodes into a simpler set, in which the equipotentials were expressible in terms of a known function, belonging to the preceding category. A few solutions of this difficult problem are available.

One of the most useful of these is the Schwarz transformation; this is applied when the electrodes are planes, intersecting around a polygonal boundary (Fig. 8a), which may indeed have some of its apices at infinity.

This transformation is written

$$\frac{dZ}{d\zeta} = A \prod_n (\zeta - \zeta_n)^{(\alpha_n/\pi - 1)}, \tag{2.8}$$

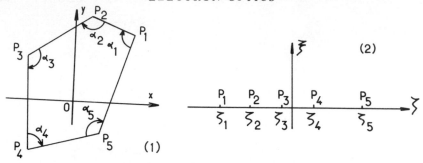

FIG. 8a. The Schwarz transformation. (1) The electrode system; (2) the same system after transformation.

in which

$$Z = A \int_0^\zeta (\zeta - \zeta_1)^{(\alpha_1/\pi - 1)} (\zeta - \zeta_2)^{(\alpha_2/\pi - 1)} \ldots (\zeta - \zeta_n)^{(\alpha_n/\pi - 1)} \, d\zeta + B.$$

The points $\zeta_i = \pm \infty$ do not enter into this formula.

Consider a simple example: the outline of the parallel plate condenser shown in Fig. 8b. From symmetry considerations, we need only calculate the function $u(x, y)$ in a simpler system consisting of two electrodes: one a plane, held at potential $\Phi = V_0$, and the other a half-plane, distant $y = h$ from the first for $x > 0$, and held at $\Phi = 0$. In this special case, we have $\alpha_1 = 0$, $\alpha_2 = 2\pi$ and $\alpha_3 = \pi$. In the complex ζ-plane, the apices A, B, C, D, E of the polygon (only D is not at infinity) all lie on the real axis Ox'. We may seek a particular transformation such that the first apex (CB) is situated at $\zeta = 0$, the second (D) at $\zeta = 1$ and the third (EA) at $\zeta = \pm \infty$ (Fig. 8c). We then have

$$Z = A \int_0^\zeta (\zeta - 0)^{(0/\pi - 1)} (\zeta - 1)^{(2\pi/\pi - 1)} d\zeta + B$$

or

$$Z = A \int_0^\zeta (1 - 1/\zeta) \, d\zeta + B = A(\zeta - \log \zeta) + B.$$

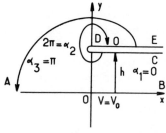

FIG. 8b. The edge of a plane condenser.

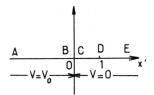

FIG. 8c. The system shown in Fig. 8b after transformation: we recover the plane wedge of angle π.

For $Z = 0 + jh$, $\zeta = 1$ so that $jh = A + B$:

$$Z = A(\zeta - 1 - \log \zeta) + jh.$$

In the vicinity of $\zeta = 0$, we have

$$\frac{dZ}{d\zeta} = -\frac{A}{\zeta}.$$

Integrating over a circle of radius r' about $\zeta = 0$, from $\alpha' = \pi$ to $\alpha' = 0$ (that is, from C to B), and using the relation

$$d\zeta = d(r' \, e^{j\alpha'}) = jr' \, e^{j\alpha'} \, d\alpha' = j\zeta \, d\alpha',$$

we have

$$\int_B^C dZ = jh = \int_\pi^0 \left(-\frac{A}{\zeta}\right) j\zeta \, d\alpha' = jA\pi,$$

so that

$$A = h/\pi.$$

The required transformation is therefore

$$Z = \frac{h}{\pi}(\zeta - 1 - \log \zeta + j\pi).$$

The potential function of the system shown in Fig. 8c, consisting of two half planes held at $\Phi = 0$ and $\Phi = V_0$, is given (see above) by the v term of the analytic function

$$W = \frac{V_0}{\pi} \log \zeta = u + jv$$

or

$$\zeta = \exp (\pi W/V_0) = \exp \frac{\pi(u + jv)}{V_0}.$$

Hence

$$Z = \frac{h}{\pi}\left(e^{\pi W/V_0} - 1 - \frac{\pi W}{V_0} + j\pi\right). \tag{2.9}$$

After extracting u and v from this formula, the curves $V = \text{const}$ can be plotted in the Z-plane; this provides us with the potential distribution in the fringe-field region of the parallel plate condenser. This distribution is of importance in particle deflectors; the fringe-field at the edges of the pole-pieces of an electromagnet may be treated in a similar way.

2.1.2 General Solutions. The Separation of Variables Method

This is a general method, and can be used to solve all partial differential equations; it is, however, only applicable for certain simple types of boundary conditions, and hence only for certain simple electrode shapes.

In essence, the solution is written as a product of functions, each of which contains only one of the variables of the coordinate system employed.

Cartesian coordinates

(a) *Two-dimensional systems*

$$\nabla^2\Phi(x, y) = \frac{\partial^2\Phi}{\partial x^2} + \frac{\partial^2\Phi}{\partial y^2} = 0.$$

We write

$$\Phi(x, y) = X(x) \cdot Y(y).$$

Substituting into Laplace's equation and writing $\partial^2 X/\partial x^2 = X''$ and $\partial^2 Y/\partial y^2 = Y''$, we find

$$\frac{X''}{X} = -\frac{Y''}{Y}.$$

The term X''/X is a function of x only and the term Y''/Y of y only. The equation will be satisfied everywhere only if these two terms are constant:

$$\frac{X''}{X} = k^2, \qquad \frac{Y''}{Y} = -k^2,$$

giving general solutions

$$X(x) = A_1 \cosh kx + B_1 \sinh kx,$$
$$Y(y) = A_2 \cos ky + B_2 \sin ky,$$

so that $\Phi(x, y)$ can be written

$$\Phi(x, y) = \begin{Bmatrix} A_1 \cosh kx \\ B_1 \sinh kx \end{Bmatrix} \begin{Bmatrix} A_2 \cos ky \\ B_2 \sin ky \end{Bmatrix}. \qquad (2.10)$$

The constants A_1, A_2, B_1, B_2 and k are determined for a specific system by the boundary conditions at the electrodes. If k is imaginary, the circular functions become hyperbolic functions, and conversely.

(b) Three-dimensional systems. We write

$$\Phi(x,\ y,\ z) = X(x) \cdot Y(y) \cdot Z(z)$$

and following the same procedure, we find

$$\frac{X''}{X} = \alpha^2, \quad \frac{Y''}{Y} = \beta^2, \quad \frac{Z''}{Z} = -\gamma^2,$$

with

$$\alpha^2 + \beta^2 = \gamma^2$$

and hence

$$\Phi(x, y, z) = \begin{Bmatrix} A_1 \cosh \alpha x \\ B_1 \sinh \alpha x \end{Bmatrix} \begin{Bmatrix} A_2 \cosh \beta y \\ B_2 \sinh \beta y \end{Bmatrix} \begin{Bmatrix} A_3 \cos \gamma z \\ B_3 \sin \gamma z \end{Bmatrix}. \qquad (2.11)$$

Other solutions are found by giving α, β, γ imaginary values, but the relation $\alpha^2 + \beta^2 = \gamma^2$ must always be satisfied; this means that at least one of the partial solutions X, Y or Z will contain hyperbolic functions.

(c) The idea of spatial harmonics. The term $\Phi(x, y) = A \sinh kx \sin ky$ is an elementary solution of (2.10). We now show that it is impossible to create such a potential distribution exactly, with electrodes of finite size; with finite electrodes, therefore, only an approximation to the theoretical field can be created.

For $x = 0$, $y = 0$ and $y = \pi/k = b$, we have $\Phi = 0$. We assume that the line $y = b/2$ intersects an electrode held at potential Φ_0 at $x = l$.

We then have

$$\Phi_0 = A \sinh kl, \quad \text{so that} \quad A = \Phi_0 \operatorname{cosech}(\pi l/b)$$

and

$$\Phi(x, y) = \frac{\Phi_0}{\sinh \pi l/b} \sinh \frac{\pi x}{b} \sin \frac{\pi y}{b}. \qquad (2.12)$$

The cross-section of the electrode at potential $\Phi = \Phi_0$ will be the curve defined by $\Phi(x, y) = \Phi_0$, or

$$\sinh \frac{\pi x}{b} \sin \frac{\pi y}{b} = \sinh \frac{\pi l}{b}.$$

This curve tends asymptotically to the straight lines $y = 0$ and $y = b$, and extends from $x = l$ to $x = +\infty$ (Fig. 8d); such an electrode cannot be built in practice. (The same can be shown to be true of all the elementary solutions of the Laplace equation.) If we discontinue the electrode beyond a finite distance, it is only possible to satisfy the boundary conditions by

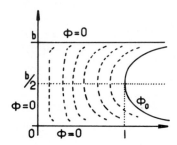

FIG. 8d. Electrodes corresponding to the simple solution $\Phi(x, y) = A \sinh kx \sin ky$.

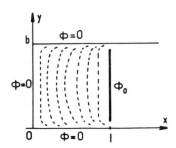

FIG. 8e. A rectangular box.

superimposing an infinite number of these elementary solutions. Consider, for example, the system shown in Fig. 8e, which can be made in practice, and which consists of planes $x = 0$, $y = 0$ and $y = b$ held at potential $\Phi = 0$ together with a planar region at $x = l$ and held at $\Phi = \Phi_0$. The gap between the latter and the other electrodes is assumed to be infinitesimally narrow.

At $x = l$, the elementary solution

$$\Phi(x, y) = A \sinh kx \sin ky \quad \text{(with } k = \pi/b)$$

cannot on its own satisfy the condition

$$\Phi = \Phi_0 \quad \text{for} \quad 0 < y < b.$$

If, however, we take the higher order solutions, with $k_n = n\pi/b$, we obtain a general solution of $\nabla^2\Phi = 0$ in the form

$$\Phi(x, y) = \sum_n A_n \sinh k_n x \sin k_n y.$$

This sum of elementary solutions can now be made to satisfy the condition:

$$\Phi_0 = \sum_n A_n \sinh \frac{n\pi l}{b} \sin \frac{n\pi y}{b} \quad 0 < y < b. \tag{2.13}$$

This expression is in fact the Fourier series expansion of the function

$f(y) = \Phi_0$. Here, this series contains only sine terms:

$$f(y) = \Phi_0 = \sum_{n=1}^{\infty} a_n \sin \frac{n\pi y}{b} \tag{2.14}$$

and it is easy to show that if n is odd, $a_n = 4\Phi_0/n\pi$, and if n is even, $a_n = 0$. Comparing the physical solution (2.13) and the series (2.14), we see that

$$a_n = A_n \sinh \frac{\pi l}{b},$$

so that finally,

$$\Phi(x, y) = \sum_{n \, \text{odd}} \frac{4\Phi_0}{n\pi} \frac{\sinh\,(n\pi x/b)}{\sinh\,(n\pi l/b)} \sin \frac{n\pi y}{b} . \tag{2.15}$$

This function, which converges rapidly (except in the immediate vicinity of $\pi = l$), can be used to determine the potential everywhere within the electrode system.

The term $n = 1$ is the fundamental, and the terms $n > 1$ are known as "spatial harmonics". They are necessary in order that the boundary conditions at the electrodes determining the field can be satisfied.

In practice, the preceding system, with no gaps between the electrodes, could not be built. If the electrode at $x = l$ is in the form of a band of width $d < b$, and is symmetrical about $y = b/2$, we need only know the function $\Phi(l, y)$ for $0 < y < b$ and calculate the coefficients a_n of its Fourier series in order to find the coefficients A_n and hence the expression for the general solution, $\Phi(x, y)$; in this special case, it is a very good approximation to replace $\Phi(l, y)$, as a function of y, by an isosceles trapezium.

Cylindrical polar coordinates

(a) **The general solution.** We now have

$$\nabla^2 \Phi(r, \theta, z) = \frac{1}{r} \frac{\partial}{\partial r} \left(r \frac{\partial \Phi}{\partial r} \right) + \frac{1}{r^2} \frac{\partial^2 \Phi}{\partial \theta^2} + \frac{\partial^2 \Phi}{\partial z^2} = 0,$$

into which we substitute

$$\Phi(r, \theta, z) = R(r)\Theta(\theta)Z(z).$$

Differentiating, we obtain

$$\frac{1}{rR} \frac{d}{dr}(rR') + \frac{1}{r^2} \frac{\Theta''}{\Theta} + \frac{Z'}{Z} = 0.$$

We can then write $\Theta''/\Theta = -n^2$ and $Z''/Z = k^2$; n is necessarily a real integer, since the physical solution can only contain $\cos n\theta$ or $\sin n\theta$, because the potential must take the same value when we advance through $\theta = 2\pi$ around a circle $r = $ constant. k may be real or imaginary.

We thus obtain

$$\frac{1}{r}\frac{d}{dr}(rR') + \left(k^2 - \frac{n^2}{r^2}\right)R = 0.$$

Introducing a new variable $v = kr$, this R-equation is transformed into a Bessel equation:

$$\frac{1}{v}\frac{d}{dv}\left(v\frac{dR}{dv}\right) + \left(1 - \frac{n^2}{v^2}\right)R = 0,$$

of which the general solution for real k is expressed as

$$R_n(kr) = AJ_n(kr) + BN_n(kr).$$

($J_n(kr)$ is a Bessel function of the first kind of order n in the variable kr and $N_n(kr)$ is a Bessel function of the second kind.)

If k is imaginary, the solution is of the form

$$R_n(kr) = AI_n(kr) + BK_n(kr),$$

in which I_n and K_n are modified Bessel functions of the first and second kinds. Since the functions $N_n(kr)$ and $K_n(kr)$ become infinite on the axis Oz $(r = 0)$, B is necessarily zero when there are no electrodes on the axis; this is the case for all electrostatic lenses used in practice.

The general solution of the Laplace equation is therefore of the following form:

$$\text{real } k: \quad \Phi(r, \theta, z) = A \begin{Bmatrix} J_n(kr) \\ N_n(kr) \end{Bmatrix} \begin{Bmatrix} \cos n\theta \\ \sin n\theta \end{Bmatrix} \begin{Bmatrix} \cosh kz \\ \sinh kz \end{Bmatrix},$$

$$\text{imaginary } k: \quad \Phi(r, \theta, z) = A \begin{Bmatrix} I_n(kr) \\ K_n(kr) \end{Bmatrix} \begin{Bmatrix} \cos n\theta \\ \sin n\theta \end{Bmatrix} \begin{Bmatrix} \cos kz \\ \sin kz \end{Bmatrix}. \tag{2.16}$$

There are two important cases to be considered.

SYSTEMS WITH TRANSLATION SYMMETRY $(k = 0)$: The differential equation for R is of the form

$$r\frac{d}{dr}\left(r\frac{dR}{dr}\right) - n^2R = 0,$$

which has the general solution

$$R(r) = a_n r^n + b_n r^{-n}$$

so that

$$\Phi(r, \theta) = \sum_n (a_n r^n + b_n r^{-n}) \begin{Bmatrix} \cos n\theta \\ \sin n\theta \end{Bmatrix}.$$

Notice that if $k = 0$, we may have $Z(z) = Az + B$, but the system no longer has translation symmetry. In a lens with no electrode at the axis, we have $b_n = 0$ (so that the potential remains finite at the axis). We recover the general solution

$$\Phi(r, \theta) = \sum_n A_n r^n \begin{Bmatrix} \cos n\theta \\ \sin n\theta \end{Bmatrix}$$

encountered in the preceding section.

SYSTEMS WITH ROTATIONAL SYMMETRY $\left(\dfrac{\partial}{\partial \theta} = 0, \ n = 0 \right)$: The Bessel functions are of zero order. If the axis is free of electrodes, we have

$$\Phi(r, z) = A I_0(kr) \begin{Bmatrix} \cos kz \\ \sin kz \end{Bmatrix}$$

or

$$\Phi(r, z) = A J_0(kr) \begin{Bmatrix} \cosh kz \\ \sinh kz \end{Bmatrix} = B J_0(kr) \begin{Bmatrix} \exp kz \\ \exp(-kz) \end{Bmatrix}.$$

(b) Spatial harmonics. As before, these expressions are only elementary solutions of the Laplace equation, and cannot be exactly reproduced in practice. The boundary conditions can be satisfied in a real system only by introducing an infinite number of spatial harmonics.

Consider, for example, the electrodes shown in Fig. 8f, consisting of

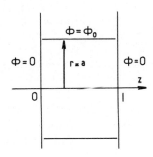

FIG. 8f. A cylindrical box (two planes held at $\Phi = 0$ and a cylinder at $\Phi = \Phi_0$).

two planes $z = 0$ and $z = l$, held at potential $\Phi = 0$, together with a tubular portion of radius $r = a$; the function $Z(z)$ can be represented by a series of terms of the form

$$Z(z) = A_m \sin k_m z \quad \text{with} \quad k_m = m\pi/l$$

so that

$$\Phi(r, z) = \sum_m A_m I_0(k_m r) \sin k_m z.$$

The coefficients A_m are determined by expanding the function $\Phi(a, z)$ for $0 < z < l$ as a Fourier series:

$$\Phi(a, z) = \sum_{m} a_m \sin\frac{m\pi z}{l}.$$

In the simple case where $\Phi(a, z) = \Phi_0 = const$, for $0 < z < l$, it is easy to show that

$$\Phi(r, z) = \sum_{m\,\text{odd}} \frac{4\Phi_0}{m\pi} \frac{I_0(m\pi r/l)}{I_0(m\pi a/l)} \sin\frac{m\pi z}{l}. \tag{2.17}$$

In this way, the majority of practical problems involving simple electrode shapes may be solved. With the aid of an electronic computer, the potential distribution can be plotted rapidly from solutions in series form. This method requires a knowledge of the potential distribution over a cylinder $r = a$. In the case of the system of Fig. 8g, we assume that, for

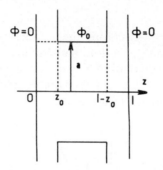

Fig. 8g. The three-electrode lens (the planes at $Z = 0$ and $Z = l$ are assumed to be transparent to particles).

$0 < z < z_0$ and $l - z_0 < z < l$, the potential varies linearly with z so that $\Phi(a, z)$ is represented by an isosceles trapezium; we accept a good approximation (often perfectly adequate), for want of a rigorously exact expression.

(c) *The use of Fourier integrals.* In a system in which $\Phi_0(z)$ is not bounded in the direction Oz (as is the case in lenses formed from cylindrical tubes or diaphragms with openings in them), the function $\Phi(a, z)$ extends in theory from $z = -\infty$ to $z = +\infty$. It can then be represented not by a Fourier series but by a Fourier integral.

If $\Phi(r, z)$ can be put into the form of an infinite series, thus:

$$\Phi(r, z) = \sum_{k} \{A(k)I_0(kr) \sin kz + B(k)I_0(kr) \cos kz\}, \tag{2.18}$$

we may write

$$\Phi(r, z) = \int_0^\infty I_0(kr)\{A(k) \sin kz + B(k) \cos kz\}\,dk$$

and

$$\Phi(a, z) = \int_0^\infty I_0(ka)\{A(k) \sin kz + B(k) \cos kz\}\,dk.$$

The coefficients $A_k I_0(ka)$ and $B_k I_0(ka)$ can be calculated from the relations

$$A(k)I_0(ka) = \frac{2}{\pi}\int_0^\infty \Phi(a, z) \sin kz\, dz,$$

$$B(k)I_0(ka) = \frac{2}{\pi}\int_0^\infty \Phi(a, z) \cos kz\, dz.$$

(2.19)

We now consider a few examples.

A lens, consisting of two cylinders of equal radius, and separated by an infinitesimally narrow gap (Fig. 8h). At $z = 0$, we take as origin of poten-

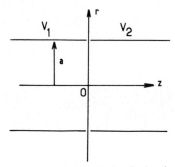

FIG. 8h. Lens consisting of two identical cylinders in contact.

tial the value $\Phi = (V_1 + V_2)/2$. The function $\Phi(a, z)$ becomes $-(V_2 - V_1)/2$ for $z < 0$ and $(V_2 - V_1)/2$ for $z > 0$; this odd function can be expressed in terms of $\sin kz$ alone:

$$\Phi(r, z) - \frac{V_1 + V_2}{2} = \int_0^\infty A(k)I_0(kr) \sin kz\, dk.$$

For $\Phi(a, z) = \dfrac{V_2 - V_1}{2}$ $(z > 0)$

$$A(k) = \frac{2}{\pi I_0(ka)}\int_0^\infty \frac{V_2 - V_1}{2} \sin kz\, dz = \frac{V_2 - V_1}{\pi k I_0(ka)}$$

and

$$\Phi(r, z) = \frac{V_1 + V_2}{2} + \frac{V_2 - V_1}{\pi} \int_0^\infty \frac{I_0(kr)}{kI_0(ka)} \sin kz \, dk \, ;$$

on the axis,

$$\Phi(0, z) = \frac{V_1 + V_2}{2} + \frac{V_2 - V_1}{\pi} \int_0^\infty \frac{\sin kz}{kI_0(ka)} dk = \frac{V_1 + V_2}{2} + \frac{V_2 - V_1}{\pi} F(k, z).$$

Bertram (1940) has pointed out that the integral $F(k, z)$ can be represented to a very good approximation by the function

$$F(k, z) \simeq \tanh (wz/a) \quad w = 1 \cdot 318,$$

so that

$$\Phi(0, z) \simeq \frac{V_1 + V_2}{2} + \frac{V_2 - V_1}{\pi} \tanh \left(\frac{wz}{a}\right). \tag{2.20}$$

If the two cylinders are separated by a distance 2d (Fig. 8i), *we have*

$$\Phi(a, z) \begin{cases} = V_1 & (z < d), \\ = \dfrac{V_1 + V_2}{2} + \dfrac{V_2 - V_1}{2d} z & (-d < z < +d), \\ = V_2 & (z > d). \end{cases}$$

$$A(k) = \frac{2}{\pi I_0(ka)} \left(\int_0^d \frac{V_2 - V_1}{2d} z \sin kz \, dz + \int_d^\infty \frac{V_2 - V_1}{2} \sin kz \, dz \right)$$

so that

$$A(k) = \frac{\sin kd}{\pi dk^2 I_0(ka)}$$

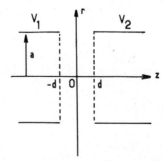

FIG. 8i. Lens consisting of two cylinders of the same radius.

and

$$\Phi(r, z) = \frac{V_1 + V_2}{2} + \frac{V_2 - V_1}{\pi d} \int_0^\infty \frac{I_0(kr)}{k^2 I_0(ka)} \sin kd \sin kz \, dk. \qquad (2.21)$$

On the axis,

$$\Phi(0, z) = \frac{V_1 + V_2}{2} + \frac{V_2 - V_1}{\pi d} \int_0^\infty \frac{\sin kd \sin kz}{k^2 I_0(ka)} dk.$$

The integral on the right-hand side can, however, be written in the form (Bertram, 1951a)

$$\int_0^\infty \frac{\sin kd \sin kz}{k^2 I_0(ka)} dk = \frac{1}{2} \int_{(z-d)/a}^{(z+d)/a} dz \int_0^\infty \frac{\sin kz}{k I_0(ka)} dk,$$

$$= \frac{1}{2} \int_{(z-d)/a}^{(z+d)/a} F(k, z) \, dz.$$

Replacing $F(k, z)$ by its approximate expression, we have

$$\Phi_0(z) = \Phi(0, z)$$

$$= \frac{V_1 + V_2}{2} + \frac{(V_2 - V_1)a}{4wd} \log\left(\cosh \frac{w(z+d)}{a} \Big/ \cosh \frac{w(z-d)}{a}\right). \qquad (2.22)$$

Using this method, the reader can recover the potential distribution on the axis of the lens of Fig. 87, given by formula (8.39), and hence derive the function $\Phi_0(z)$ for a three-diaphragm lens.

REMARK. Instead of the solution (2.18), we can equally well use a series of the form

$$\Phi(r, z) = \sum_k A(k) J_0(kr) \exp(-kz).$$

(The solution in $\exp(+kz)$ must be discarded, as $\Phi(r, z)$ must not become infinite at $z = \infty$.)

Taking $J_0(ka) = 0$, we can write the function $\Phi(r, z)$ as

$$\Psi(r, z) = V_2 + \sum_{n=1}^\infty A_n J_0(\mu_n r/a) \exp(-\mu_n z/a),$$

in which μ_n denotes the n-th root of $J_0(x)$.

At $z = 0$, we have $\Phi(r, 0) = (V_1 + V_2)/2$; we can regard this series as an expansion in Bessel functions in the variable $(\mu_n r/a)$, and following a procedure similar to that used for circular functions, the coefficients A_n may be calculated in terms of the known function $\Phi(a, z) = V_2$:

$$A_n = \left(\frac{V_1 + V_2}{2} - V_2\right)\int_0^a rJ_0\left(\mu_n\frac{r}{a}\right)dr$$

and

$$\Phi(r, z) = V_2 + (V_1 - V_2)\sum_{n=1}^{\infty} \frac{J_0(\mu_n r/a)}{\mu_n J_1(\mu_n)} \exp\left(-\frac{\mu_n z}{a}\right). \tag{2.23}$$

This expression is equivalent to (2.21) but is better adapted to calculations with a computer.

Superposition methods. By using elliptical coordinates (see Durand), it is possible to calculate rigorously the potential distribution, $\Phi(r, z)$, created by an infinitesimally thin conducting plane P held at a potential Φ_0, containing a circular opening of radius R:

$$\Phi(r, z) = \Phi_0 - \frac{1}{2}(E_1 - E_2)z + \frac{E_1 - E_2}{\pi}|z|\left(\frac{R}{\alpha} + \tan^{-1}\frac{\alpha}{R}\right), \tag{2.24}$$

where

$$\alpha = \left[\tfrac{1}{2}(r^2 + z^2 - R^2) + \tfrac{1}{2}\{(r^2 + z^2 - R^2)^2 + 4z^2 R^2\}^{1/2}\right]^{1/2}.$$

On the axis of the hole ($r = 0$), $\alpha = |z|$, and

$$\Phi_0(z) = \Phi_0 - \frac{1}{2}(E_1 + E_2)z - \frac{E_1 - E_2}{\pi}R\left(1 + \frac{z}{R}\tan^{-1}\frac{z}{R}\right)$$

or

$$\Phi_0(z) = a + bz + cz\tan^{-1}\frac{z}{R}. \tag{2.25}$$

For a family of parallel conducting planes, P_i, with abscissae z_i, held at potentials Φ_i, and containing holes of radii R_i all having the same axis Oz, Regenstreif (1951a) and Bertein (1952a) have shown that $\Phi_0(z)$ can be obtained by superimposing solutions of the foregoing type (see § 8.2). The constants are found from the boundary conditions: far from the axis, the electric fields between the planes are uniform and given by

$$E_i = -\frac{\Phi_{i+1} - \Phi_i}{z_{i+1} - z_i}.$$

2.1.3 Numerical Solution. Relaxation

The preceding methods yield rigorous or approximate solutions in analytic form; they make it possible to obtain the potential function or the flux function at every point $M(x, y, z)$. There are other methods, however, which yield the values of these functions only at a finite number of points situated at the intersections of a mesh of step-length h, which may be as small as required. These methods, which are called "relaxation methods", are particularly well-suited to calculations with a computer and can be applied to systems of any shape. It is only necessary to know the boundary conditions at the electrodes, and also between the electrodes a reasonable distance from the region of interest.

These methods are described in detail in the books by Durand and Southwell, who also discuss their accuracy and the methods of obtaining rapid convergence towards the final solution. By using finite difference equations (instead of partial differential equations), we can determine the value of the function at some given point in terms of its values at neighbouring points. Calculation of a potential "map" now becomes a process of successive approximations (or relaxation), each stage of which is more accurate than its predecessor. When two successive sequences of calculations have given results differing by less than some fraction $\triangle\Phi/\Phi$ decided beforehand, the calculation is halted.

We shall give only the principle of these methods.

Systems with translation symmetry. The function $V(x, y)$ can be expanded as a Taylor series in the neighbourhood of a point $M(x_0, y_0)$:

$$V = V_0 + \frac{1}{1!}\left\{(x-x_0)\left(\frac{\partial V}{\partial x}\right)_0 + (y-y_0)\left(\frac{\partial V}{\partial y}\right)_0\right\}$$

$$+ \frac{1}{2!}\left\{(x-x_0)^2\left(\frac{\partial^2 V}{\partial x^2}\right)_0 + 2(x-x_0)(y-y_0)\left(\frac{\partial^2 V}{\partial x \partial y}\right)_0 + (y-y_0)^2\left(\frac{\partial^2 V}{\partial y^2}\right)_0\right\} + \cdots.$$

$$(2.26)$$

If we draw a rectangular grid of mesh-length h, we need only know the values of the function V at five neighbouring points in order to calculate the five partial derivatives at M that appear in this formula.

Let us apply this relation to the points 1, 2, 3 and 4 of Fig. 8j $(x-x_0 = \pm h; y-y_0 = \pm h)$. We obtain:

$$\left(\frac{\partial V}{\partial x}\right)_0 = \frac{1}{2h}(V_1 - V_3), \qquad \left(\frac{\partial V}{\partial y}\right)_0 = \frac{1}{2h}(V_4 - V_2),$$

$$\left(\frac{\partial^2 V}{\partial x^2}\right)_0 = \frac{1}{h^2}\{(V_1 - V_0) - (V_0 - V_3)\}, \left(\frac{\partial^2 V}{\partial y^2}\right)_0 = \frac{1}{h^2}\{(V_4 - V_0) - (V_0 - V_2)\}.$$

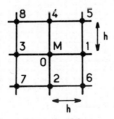

FIG. 8j. Plane network with square openings.

Hence,

$$(\nabla^2 V)_0 = \left(\frac{\partial^2 V}{\partial x^2}\right)_0 + \left(\frac{\partial^2 V}{\partial y^2}\right)_0 = \frac{1}{h^2}\,(V_1 + V_2 + V_3 + V_4 - 4V_0), \qquad (2.27)$$

and the condition that

$$(\nabla^2 V)_0 = 0$$

leads to

$$V_1 + V_2 + V_3 + V_4 = 4V_0. \qquad (2.28)$$

Using the points 5, 6, 7 and 8, we should find

$$\left(\frac{\partial^2 V}{\partial x \partial y}\right)_0 = \frac{1}{4h^2}\,(V_5 - V_6 + V_7 - V_8).$$

By this method, therefore, we can solve a partial differential equation containing first and second derivatives. If the first derivatives are discontinuous at the point $M(x_0, y_0)$, we calculate them using points all on the same side of the point $M: x - x_0 = h$ and $2h$, for example.

The calculation can be made more accurate by retaining higher degree terms in $(x - x_0)$ and $(y - y_0)$ in the expansion of $V(x, y)$. It is then possible to show, taking advantage of the fact that $\nabla(\nabla^2 V) = 0$ and $\nabla^2(\nabla^2 V) = 0$, that the relation

$$4(V_1 + V_2 + V_3 + V_4) + (V_5 + V_6 + V_7 + V_8) = 20\,V_0$$

is correct up to eighth order.

Relations of a similar kind involving the dielectric constant can be established for the surface of an insulator.

To solve a given problem in practice, we proceed as follows. The electrodes in question are drawn to a large scale (Fig. 8k), and a rectangular grid is ruled in such a way that the electrodes coincide with the

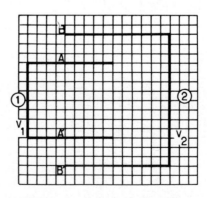

FIG. 8k. Solution of a planar problem.

lines of the mesh. (Sloping lines or curves will be represented by "steps" and the smaller the mesh-length, the greater the accuracy.)

The values of the potential at the outer points are inserted: over the electrodes, the potential is known and between the electrodes, it obeys a known law (along the segments AB and $A'B'$ of Fig. 8k, for example). To all the other points of the grid arbitrary values of the potential are then ascribed, but these should be as reasonable and self-consistent as possible. (Thus on Fig. 8k, the values should fall between V_1 and V_2, and from electrode 1 to electrode 2 they should vary from V_1 to V_2.)

Using this first set of values, $\Phi_{ij}^{(1)}$, a second set, $\Phi_{ij}^{(2)}$, is calculated, using the foregoing formulae, from which a third set $\Phi_{ij}^{(3)}$ is obtained, and so on until $\Phi_{ij}^{(n)}$ and $\Phi_{ij}^{(n-1)}$ scarcely differ. The length of the calculation obviously depends upon the difference between the arbitrary distribution $\Phi_{ij}^{(1)}$ and the real solution; with a lot of practice, this difference can be reduced to as little as a few percent, and the solution is then obtained after a few "relaxations" to the degree of accuracy, $\Delta\Phi/\Phi$, selected in advance. After interpolation between the values found at the mesh-points, the equipotentials can be inserted.

Greater accuracy can also be obtained, over a restricted region, by repeating the calculation with a much smaller step length and using the values $\Phi_{ij}^{(n)}$ as boundary conditions.

These methods were originally employed without the assistance of electronic computers, and a great deal of mathematical work went into determining rules for increasing the speed of "convergence" towards the final solution.

Rotationally symmetric systems. Here, only the r and z coordinates are involved. The expansion of $\Phi(r, z)$ in the neighbourhood of $M(r_0, z_0)$ is given by (2.26), when y and x are replaced by r and z respectively.

We first consider a point M such that $r_0 = ph$ (Fig. 8l). Laplace's

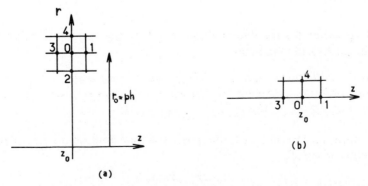

FIG. 8l. Rotationally symmetric network. (a) A point 0 not on the axis; (b) a point 0 on the axis.

equation takes the form

$$\frac{\partial^2\Phi}{\partial z^2}+\frac{\partial^2\Phi}{\partial r^2}+\frac{1}{r}\frac{\partial\Phi}{\partial r}=0,$$

from which it is easy to derive the relation

$$(\nabla^2\Phi)_0=\frac{1}{2ph^2}\{2p(\Phi_1-\Phi_0)+2p(\Phi_3-\Phi_0)+(2p-1)(\Phi_2-\Phi_0)$$
$$+(2p+1)(\Phi_4-\Phi_0)\} \tag{2.29}$$

so that at M,

$$8p\Phi_0=2p(\Phi_1+\Phi_3)+(2p-1)\Phi_2+(2p+1)\Phi_4. \tag{2.30}$$

When M lies on the axis, $r=0$, this formula no longer applies. On symmetry grounds, however, $(\partial\Phi/\partial r)_0=0$, so that

$$\frac{\partial\Phi}{\partial r}=0+r\left(\frac{\partial^2\Phi}{\partial r^2}\right)_0$$

and

$$\Phi_4=\Phi_0+0+\frac{h^2}{2}\left(\frac{\partial^2\Phi}{\partial r^2}\right)_0.$$

Hence

$$\left(\frac{1}{r}\frac{\partial\Phi}{\partial r}\right)_0=\left(\frac{\partial^2\Phi}{\partial r^2}\right)_0=\frac{2(\Phi_4-\Phi_0)}{h^2}.$$

The relation

$$(\nabla^2\Phi)_0=0$$

leads to

$$6\Phi_0=\Phi_1+\Phi_3+4\Phi_4. \tag{2.31}$$

Just as in the preceding case in the two variables x, y, improved formulae may be derived.

2.1.4 Expression for the Potential and Fields in the Vicinity of a Symmetry Plane or an Axis of Symmetry

Systems with translation symmetry, possessing a plane of symmetry Ox. The potential in the neighbourhood of Ox can be written in the form

$$\Phi(x, y)=\sum_n A_n(x)y^{2n}.$$

The term $A_0(x)$ denotes the potential distribution in the plane of symmetry, along Ox:

$$A_0(x)=\Phi_0(x).$$

Substituting this expansion into the Laplace equation and writing

$\nabla^2\Phi(x, y) = 0$, we obtain the recurrence relation

$$A''_{n-1} + 2n(2n-1)A_n = 0$$

or

$$A_n = -\frac{A''_{n-1}}{2n(2n-1)}.$$

This gives

$$A_1 = -\frac{1}{2!}A''_0,$$

$$A_2 = \frac{1}{4!}A_0^{(4)},$$

. . . .

Hence

$$\Phi(x, y) = \sum_n \frac{(-1)^n}{(2n)!} y^{2n}\Phi_0^{(2n)}(x)$$

so that near Ox, we have

$$\Phi(x, y) = \Phi_0(x) - \frac{y^2}{2}\Phi''_0(x) + \frac{y^4}{24}\Phi_0^{(4)}(x) - \cdots. \qquad (2.32)$$

With the aid of this expression, we can show that two equipotentials that intersect at Ox are at right angles to one another (Fig. 8m).

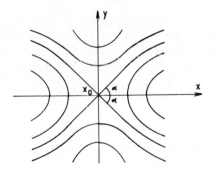

FIG. 8m. The intersection of two equipotentials (a two-dimensional system): $\alpha - 45°$.

If two equipotentials intersect at $x = x_0$, then necessarily,

$$\Phi'_0(x_0) = 0,$$

and hence

$$\Phi(x, 0) = \Phi(x_0, 0) + \frac{1}{2}(x-x_0)^2\Phi''(x_0, 0), \qquad (2.33)$$

$$\Phi''(x, 0) = \Phi''(x_0, 0).$$

Substituting into equation (2.32) and recalling that at all points along one of the equipotentials $\Phi(x, y) = \Phi(x_0, 0)$ we obtain

$$y = \pm x - x_0.$$

The equipotentials intersect the axis again at an angle of $\pm 45°$.

Rotationally symmetric systems. For symmetry reasons, the function $\Phi(r, z)$ takes the form

$$\Phi(r, z) = \sum_n A_n(z) r^{2n}$$

and Laplace's equation imposes the condition

$$A_n = -\frac{A''_{n-1}}{4n^2},$$

so that

$$\Phi(r, z) = \sum_n \frac{(-1)^n}{(n!)^2} \left(\frac{r}{2}\right)^{2n} \Phi_0^{(2n)}(z)$$

or

$$\Phi(r, z) = \Phi_0(z) - \frac{r^2}{4}\Phi_0''(z) + \frac{r^4}{64}\Phi_0^{(4)}(z) - \cdots. \tag{2.34}$$

Reasoning similar to that of the preceding sections shows that two equipotentials that intersect on the axis at $z = z_0$ are given by the equation

$$r = \pm \sqrt{2}(z - z_0)$$

(to first order in r). They intersect the axis at an angle of $54°44'$.

The curvature of an equipotential in the neighbourhood of the axis (Fig. 8n) can be determined simply with the aid of equation (2.34). We have

$$\Phi(r, z) = \Phi_0(z) - \frac{r^2}{4}\Phi_0''(z)$$

and a Taylor expansion gives

$$\Phi_0''(z) = \Phi_0''(z_0),$$
$$\Phi_0(z) = \Phi_0(z_0) + (z - z_0)\Phi_0'(z_0),$$

so that

$$\Phi(r, z) = \Phi_0(z_0) + (z - z_0)\Phi_0'(z_0) - \frac{r^2}{4}\Phi_0''(z_0).$$

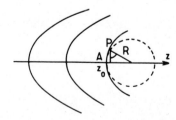

FIG. 8n. The curvature of an equipotential near the axis.

Since the points $P(r, z)$ and $A(0, z_0)$ lie on the same equipotential, we have

$$\Phi(r, z) = \Phi_0(z_0)$$

and so

$$r^2 = 4(z - z_0) \frac{\Phi_0'(z_0)}{\Phi_0''(z_0)}.$$

The radius of curvature R is thus given by

$$r^2 \simeq 2(z - z_0)R \quad \text{or} \quad R = \frac{2\Phi_0'(z_0)}{\Phi_0''(z_0)}. \tag{2.35}$$

Finally, relation (2.34) can be used to determine the shape of the electrode system needed to produce a given distribution $\Phi_0(z)$. Plass (1942) has calculated lenses such that, on the axis, we have

$$\Phi_0(z) = \Phi_0(1 - \tfrac{1}{2} e^{-z^2/2a^2})$$

and

$$\Phi_0(z) = \Phi_0\{1 - \tfrac{1}{2} \tanh (z/a)\}$$

(a denotes the radius of the aperture in one of the electrodes).

2.2 THE ELECTROLYTIC TANK

2.2.1 The Underlying Principle

Consider a homogeneous isotropic conductor, conductivity σ, carrying a current density i; let the potential distribution be V. The continuity equation gives

$$\text{div}\, i = 0. \tag{2.36}$$

Combining Ohm's law

$$i = \sigma E$$

with

$$E = -\,\text{grad}\, V,$$

we find

$$i = -\,\sigma\,\text{grad}\, V.$$

On writing equation (2.36) in Cartesian coordinates:

$$\frac{\partial}{\partial x}\left(\sigma \frac{\partial V}{\partial x}\right) + \frac{\partial}{\partial y}\left(\sigma \frac{\partial V}{\partial y}\right) + \frac{\partial}{\partial z}\left(\sigma \frac{\partial V}{\partial z}\right) = 0, \tag{2.37}$$

and substituting $\sigma = $ constant, we obtain

$$\frac{\partial^2 V}{\partial x^2} + \frac{\partial^2 V}{\partial y^2} + \frac{\partial^2 V}{\partial z^2} = 0. \tag{2.38}$$

V therefore satisfies the Laplace equation: there is an exact analogy between the distribution of electrostatic potential in a lens and the distribution of electrical potential in a homogeneous, isotropic conductor, provided that the boundary conditions are identical.

In electrostatics as in dynamic electricity, the boundary conditions between two media 1 and 2 relate the tangential components of the field E and the normal components of the current density (either displacement or conduction current). Each pair is continuous across the boundary, thus:

$$\left(\frac{\partial V}{\partial t}\right)_1 = \left(\frac{\partial V}{\partial t}\right)_2, \tag{2.39}$$

$$a_1 \left(\frac{\partial V}{\partial n}\right)_1 = a_2 \left(\frac{\partial V}{\partial n}\right)_2 \tag{2.40}$$

(in (2.40) a represents dielectric constant and conductivity in electrostatics and current electricity respectively). The electrolytic tank is in fact a very general tool, since any phenomenon which can be described by an equation of the form (2.37) or (2.38), or by an equation which can be transformed into one of these forms, is amenable to analysis with its aid. The metal electrodes, the conductivity of which is supposed infinite, are immersed in an electrolyte with a very low conductivity, σ_1, and are then raised to potentials proportional to those at which they are to be held in the real experiment. A dielectric body is represented in the tank by a substance with conductivity σ_2, such that σ_2/σ_1 is equal to the relative dielectric constant of the body.

In most cases, the electrolyte is simply tap-water with a resistivity of about $2500 \, \Omega$ cm; the impedances in the model are then of the order of thousands of ohms, and are thus comparable with those in the supply. The conductivity can be increased by adding a small amount of hydrochloric acid; it is also advantageous to add a few drops of some wetting agent (0·05 per cent Teepol, for example) to avoid the formation of a troublesome meniscus in the zones where the water is in contact with the electrodes or the walls of the tank.

For quick measurements, the electrodes are made of sheet-copper or brass, which can easily be bent into the required shape. For more precise results, however, it is necessary to use massive electrodes, as thin sheets are too easily deformed; the electrode surfaces must be cleaned with extreme care as surface tension would otherwise be liable to produce local distortion of the surface of the electrolyte. The cleaning may be performed electrolytically—a steady current of the order of 10 A cm^{-2} is passed, with the electrode acting as the anode. The electrodes must be positioned with the utmost precision.

2.2.2 The Various Types of Tank

Practical electron systems fall into one of two categories, those possessing axial symmetry, and those with planar or translational symmetry.

Axially symmetric systems

A priori it would seem necessary to build a model of the whole system which is to be studied, but in fact a much simpler model suffices, as we can demonstrate very simply—if an insulated plate is placed in the tank, the equipotential surfaces are all normal to it because the normal component of the electric field is zero; in the presence of such a plate, therefore, the potential distribution is identical to the distribution which would be produced by a combination of the original electrodes and their mirror images with respect to the plate. The surface of the liquid is a particular case of this effect, and by arranging that the axis of the system lies in the surface, only one half of the system, bisected along a meridian plane, needs to be constructed.

It is, however, possible to simplify matters still further, since the liquid in a tank with an insulated sloping bottom forms a wedge whose edge represents the axis of symmetry of the system. The model electrodes are reduced to a small fraction of their originals, and provided the wedge angle is small (5 to 15°), cylinders, with the meridian of the system under consideration as base, can be used (see Fig. 9a).

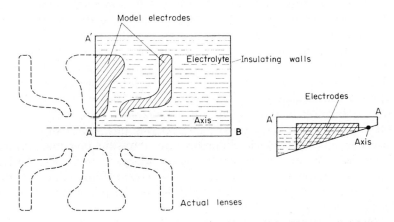

FIG. 9a. An example of the actual form of the model electrodes in a tank with a sloping bottom.

Unfortunately, owing to the shallowness of the electrolyte and the locally disturbing effects of the meniscus, measurements are least easily made in the most interesting region, close to the axis. To obtain the potential actually on the axis, however, we can use an extrapolation technique. This method is particularly tempting, in any case, as the model electrodes are of so simple a shape.

Introducing insulated walls into the tank, we can make use of such further symmetry as the actual system may have, to make one further simplification of the model—the symmetrical unipotential lens is such a case; only the quadrant BAA' need be built, the half of the central electrode being attached to an insulating sheet which is normal to the symmetry axis (see Fig. 9a).

Systems with translation symmetry

To represent the electrodes of the slit lenses which are used in mass-spectrometry, we have only to place portions of right cylinders on the bottom of a tank with a flat horizontal bottom; the corresponding electrodes would be infinitely long in the vertical direction.

2.2.3 The Measurement of Potential and Field

In order to obviate polarization phenomena, it is better in fact to supply the electrodes with a low frequency sinusoidal current, between 50 c/s and one kc/s, from an oscillator of some tens of watts, which provides a potential of a few tens of volts. The potentials between the electrodes should not exceed a few volts so that strong currents which would decompose the electrolyte are avoided.

The potential Φ_S at any point of the liquid surface is measured by inserting a fine metal probe (a platinum wire of 0·05 mm or 0·1 mm diameter) at the point, and using a null method to compare its potential with that provided by an accurate potentiometer across the terminals of the oscillator. By annulling the difference between Φ_S and Φ_A (the potentiometer potential), there is no risk of the introduction of a probe disturbing the potential in the electrolyte. The most sensitive way of detecting the null point is by applying the oscillator output to the horizontal deflector plates of a cathode ray oscillograph and a signal proportional to $(\Phi_S - \Phi_A)$ to the vertical plates; this signal is obtained by amplifying $(\Phi_S - \Phi_A)$ with the aid of a selective amplifier which eliminates any parasitic potentials which have frequencies different from that of the useful signal. In theory, the Lissajous figure which appears on the screen is an inclined straight line, and to obtain $\Phi_S = \Phi_A$, either the probe or the slider of the potentiometer is adjusted until the line is horizontal.

In practice, stray interelectrode capacities and the stray capacities which are produced by the meniscus wherever there is a liquid–solid contact

transform the straight line into an ellipse. In equilibrium, we should obtain
an ellipse with its major axis horizontal; the minor axis must be reduced
to the minimum to increase the accuracy with which the zero can be
detected, and in particular, the precautions mentioned above must be taken
in the cleaning. Alternatively, a variable capacitance can be introduced
into one of the other arms of the measuring bridge.

The electrodes are supplied with the aid of two potentiometers which are
placed in parallel across the main generator. Their potential is adjusted
while they are connected to the input of the amplifier attached to the
probe; as a result of the currents which are drawn, however, the potentials

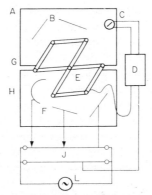

Fig. 9b. An electrolytic tank with its accessories.
A: drawing board for recording. *B*: pen. *C*: meter to detect the null point *D*: ampli-
fier. *E*: probe. *F*: electrodes. *G*: pantograph. *H*: electrolytic tank. *J*: potentiometer.
L: A.C. source.

Fig. 9c. Photograph of the tank and its accessories.

of the electrodes are not independent of one another—we are forced to proceed by a series of successive adjustments. In addition to an amplitude control, it is useful to provide a phase control, which entails inserting either a variable resistance or a phase-change circuit in series into the electrode supply lead.

If we select values of Φ_A between 0 and 100, say (in steps of 10, for example), we can follow equipotentials $\Phi_S = \Phi_A$ with the probe. With the aid of a pantograph coupled to the system which moves the probe, a direct point by point plot of these equipotentials can be made on a drawing board. Great care is necessary in the construction of the mechanical parts if recording errors are to be avoided.

The tank provides, in this way, a map of the potential distribution Φ, in the form of equipotentials numbered from 0 to 100, numbers which represent the two extremes of potential which are applied to the system. In Fig. 9b, the arrangement is shown schematically, while Fig. 9c is a photograph of a tank actually in use.

Since the early work of Fortescue and Farnsworth (1913) and Peres and Malavard (1938), the apparatus has undergone numerous improvements. Ingenious devices for ensuring that the recording-pen or marker follows the probe faithfully have been introduced: for example, the pen may be fixed directly above the probe, where it traces out the equipotentials on a sheet of paper which rests against a glass plate parallel to the surface of the electrolyte.

In the best tanks, Φ can be determined to within a few parts in 10^4. Devices have been built to trace the equipotentials automatically, by mounting the probe on a trolley which is driven parallel to one side of the tank by means of a motor M_1, and perpendicular to this direction by a second motor M_2; this latter stops, however, immediately $(\Phi_S - \Phi_A)$ becomes zero. Given Φ_A, therefore, the probe is forced to follow the equipotential $\Phi = \Phi_A$. An installation of this type is fully described in an article by Picquendar and Cahen (1960).

Direct measurement of the field

One could calculate the field at any point, given the distribution of potential Φ, from the relation $E = -\,\mathrm{grad}\,\Phi$. Alternatively, we can conceive of a method of measuring the field electrically, by using two separate probes maintained a known distance apart (ΔS), and measuring the potential difference $\Delta \Phi$ between their tips, after amplification. The field would then be given by

$$E = \frac{\Delta \Phi}{\Delta S}.$$

The presence of the meniscus in the liquid between the two probes makes such a measurement extremely delicate, however. If accurate and reproducible results are to be obtained, special probes must be employed, consisting of rigid wires embedded in glass, only the tips of which are exposed; the meniscus is then formed around the insulated parts. The electronic equipment is here somewhat more complicated than that which is required for the measurement of potential, as the two extremities of the signal ($\triangle \Phi$) which is to be amplified are "floating". A great deal of information about devices which have been designed specially for field measurement is to be found in the articles by Hollway (1955) and Sanders and Yates (1953, 1956). The two latter authors have made a thorough examination of the influence of the meniscus, and have been able to eliminate it completely by constructing a tank to a very special design; the pair of probes is fixed in the bottom of the tank, and the model is displaced mechanically with respect to the double probe. In this way, an accuracy of a few parts in 10^3 can be reached.

2.2.4 Space Charge

Several different methods are at present available for determining the potential distribution of a beam in the presence of space charge. These consist in solving an equation of the form

$$\nabla^2 \Phi = \frac{\partial^2 \Phi}{\partial x^2} + \frac{\partial^2 \Phi}{\partial y^2} = -K\frac{i}{\sqrt{\Phi(x, y)}}$$

in the region through which the beam passes (the system considered is two-dimensional). The principle of the earliest solution, suggested by Musson-Genon (1947), is as follows.

The local conductivity (σ) in the tank is proportional to the depth $h(x, y)$: $\sigma = a\,h(x, y)$. The relation $\mathrm{div}\,\sigma E = 0$ thus takes the form

$$-\nabla^2 \Phi = \mathrm{div}\,E = -E \cdot \frac{\mathrm{grad}\,h}{h} = -E \cdot \mathrm{grad}\,(\log h).$$

We obtain an analogy between the two equations, and hence between $\Phi(x, y)$ in the beam and $V(x, y)$ in the tank, if we write

$$E \cdot \mathrm{grad}\,(\log h) = -K\frac{i}{\sqrt{\Phi}}.$$

In a tank the bottom of which is shaped in such a way that h varies and this relation is satisfied (see Fig. 10a), the potential distribution across the surface of the electrolyte ($h = 0$) will provide a chart of the potential $\Phi(x, y)$.

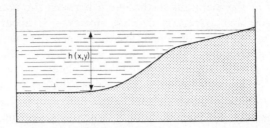

FIG. 10a. Cross-section of the tank with a shaped bottom, as used by Musson-Genon.

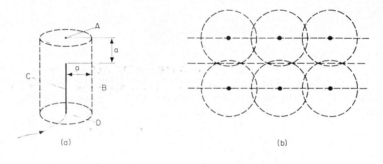

(a)   (b)

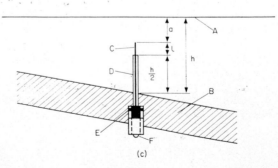

(c)

FIG. 10b. The simulation of space charge in the electrolytic tank (after van Duzen and Brewer, 1959): the probes for injecting currents. (a) A single probe, and the corresponding cylinder of charge. A: water surface; B: cylinder of charge; C: probe; D: tank floor. (b) The arrangement of the probes, seen in plan. (c) Wire probe current source—details. A: surface of the electrolyte; B: tank floor; C: conductor; D: insulating varnish; E: O-ring water-seal; F: electrical connection.

The equation for *h*, however, contains functions which are *a priori* unknown, namely Φ and $E = -\operatorname{grad}\Phi$. We have, therefore, to proceed by successive approximations and a tank of constant depth is first of all used to solve the equation without a space charge term, $\nabla^2 \Phi = 0$. This gives the distributions $\Phi_0(x, y)$ and $E_0(x, y)$ with which a first approximation to the expression for the variation of *h* is obtained:

$$h_1(x, y) = h_0 \exp\left[-K\int_{s_0}^{s} \frac{i}{E_0\sqrt{\Phi_0}}\, ds\right].$$

A bottom of this shape is then built, Φ and E again measured, and second approximations $\Phi_1(x, y)$ and $E_1(x, y)$ obtained. A more precise expression for the depth function, $h_2(x, y)$, can then be calculated; this procedure is repeated until a self-consistent solution is found, which requires no further modification to be made to the bottom of the tank.

In practice, however, this is a long process, and a solution which is simpler is nowadays employed (Hollway, 1955). The space charge in the beam is simulated in the tank by introducing a current density with the same volume distribution. This current is, in fact, led along a finite number of very fine electrodes, each of which simulates a cylinder of charge which is homogeneous in density. The juxtaposition of a number of such cylinders must create the same field distribution at the surface of the electrolyte in the tank as the distribution which prevails across a plane cross-section of the real beam.

Hollway, and later Brewer (1957) and van Duzen and Brewer (1959) have studied the problem of choosing the shape and size of electrodes which best simulate reality, both for flat-bottomed tanks and tanks with sloping bottoms. Each electrode should consist of a thin insulated rod which passes through the insulating bottom of the tank in a direction normal to the liquid surface; the end of the rod is stripped of insulation over a length *l*, and lies at a depth *a* beneath the surface of the liquid. If the depth of the liquid is *h* at this point (see Fig. 10b), and if we obey the rule

$$a + l = \frac{h}{2},$$

experiment shows that this localized charge is practically equivalent to a cylinder of current, its axis along the wire, its height *h* and its radius *a*. We have, thus, simply to provide the bottom of the tank with a regular array of electrodes, distance *a* apart, and to supply them with the intensities which are appropriate to the zone through which the beam passes and which will be proportional to the charge density in the real beam. Here again, a series of successive approximations is necessary, but we pass from one to the next simply by adjusting the intensities of the various currents.

2.3 THE RESISTANCE NETWORK

In many laboratories, the elaborate installation represented by a tank is replaced with advantage by a resistance network, comparable in accuracy and simpler to use.

The possibility of using such networks to solve partial differential equations had been suggested in 1943 by Hogan, but it was not until considerably later (de Packh, 1947; Hennequin, 1948; Charles 1949; Liebmann, 1949, 1950a) that the idea of actually using them to solve the Laplace equation—or rather, the corresponding finite difference equation—was exploited. Provided the mesh-length of the network is short enough, the solution which is obtained can be regarded as very close to the theoretical solution.

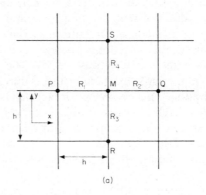

(a)

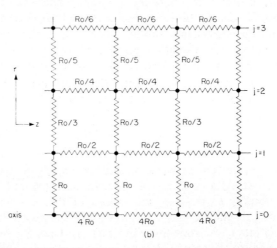

(b)

FIG. 11. Resistance networks. (a) Calculation of Φ at a particular point M. (b) The values of the resistances in a network designed for axially symmetrical systems (after Francken, 1959/60).

We shall consider first the case in which the system possesses translational symmetry. At any point M, the electrostatic potential satisfies the equation

$$(\nabla^2 \Phi)_M = \left(\frac{\partial^2 \Phi}{\partial x^2}\right)_M + \left(\frac{\partial^2 \Phi}{\partial y^2}\right)_M = 0. \tag{2.41}$$

In the plane Oxy of the system, we draw an imaginary network of mesh-length h, which consists of two sets of straight lines, parallel to Ox and Oy respectively; M is situated at a point of intersection or node of the network. We denote by P, Q, R, S the four nodes which surround M (see Fig. 11 a). Given the potentials at these points, we can calculate the second derivatives which appear in equation (2.41), see §2.2.3, and we obtain

$$(\nabla^2\Phi)_M = \frac{1}{h^2}\left[(\Phi_P - \Phi_M) + (\Phi_Q - \Phi_M) + (\Phi_R - \Phi_M) + (\Phi_S - \Phi_M)\right] = 0. \tag{2.42}$$

We construct a resistance network with the same interval h as the fictitious network above. Each point M is linked to the neighbouring nodes by four resistances R_1, R_2, R_3 and R_4. Applying Kirchhoff's law to the node M, we have

$$\Sigma i_M = 0$$

or

$$\frac{V_P - V_M}{R_1} + \frac{V_Q - V_M}{R_2} + \frac{V_R - V_M}{R_3} + \frac{V_S - V_M}{R_4} = 0. \tag{2.43}$$

Equation (2.43)—as regards the value of the potential $V(x, y)$ at the point M in the resistance network—is analogous to (2.42)—as regards the electrostatic potential $\Phi(x, y)$—if we set

$$R_1 = R_2 = R_3 = R_4.$$

All the resistances in the network are, for this case, equal. Electrodes of infinite conductivity are represented by short-circuiting the nodes of the network which lie along the curve which corresponds to a magnified version of the electrostatic system, and applying potentials proportional to those which are actually present in this system.

This same technique can equally well be used to study systems with axial symmetry, provided a network in which the resistances vary between the axis and the periphery is used; the network corresponds to one of the half-planes of radial symmetry of the system.

The Laplace equation now has the form:

$$\frac{\partial^2 \Phi}{\partial r^2} + \frac{1}{r}\frac{\partial \Phi}{\partial r} + \frac{\partial^2 \Phi}{\partial z^2} = 0, \tag{2.44}$$

and the corresponding finite difference equation for the point M is (see §
2.2.3)

$$(\nabla^2\Phi)_M = \frac{1}{2ph^2}\left\{2p(\Phi_P - \Phi_M) + 2p(\Phi_Q - \Phi_M)\right.$$

$$\left.+ (2p-1)(\Phi_R - \Phi_M) + (2p+1)(\Phi_S - \Phi_M)\right\} = 0 \quad (2.45)$$

when M is at a distance $r_M = ph$ from the axis and

$$(\nabla^2\Phi)_M = \frac{1}{h^2}\left\{(\Phi_P - \Phi_M) + (\Phi_Q - \Phi_M) + 4(\Phi_S - \Phi_M)\right\} = 0 \quad (2.46)$$

if M is located on the axis.

In the corresponding resistance network, we have the relation

$$\frac{V_P - V_M}{R_1} + \frac{V_Q - V_M}{R_2} + \frac{V_R - V_M}{R_3} + \frac{V_S - V_M}{R_4} = 0 \quad (2.47)$$

when M is not on the axis. This relation will be analogous to (2.45) if the
resistances are given the following values:

$$\begin{aligned}
R_1 = R_2 &= R_0/2p, \\
R_3 &= R_0/(2p-1), \\
R_4 &= R_0/(2p+1).
\end{aligned} \quad (2.48)$$

For $p = 1$ (so that M is a distance $r_M = h$ from the axis), we shall have
$R_3 = R_0$. The resistances that are connected at one end to the axis are
all equal to R_0, the value of which is chosen arbitrarily.

When M is on the axis, Kirchhoff's law gives us

$$\frac{V_P - V_M}{R_1} + \frac{V_Q - V_M}{R_2} + \frac{V_S - V_M}{R_4} = 0, \quad (2.49)$$

and this is the analogue of (2.46) provided we set

$$R_1 = R_2 = 4R_4 = R_0. \quad (2.50)$$

In conclusion, therefore, we have

$$R(Oz) = 4R_0$$

on the axis. Along rows parallel to Oz,

$$R_{p=1} = R_0/2, \quad R_{p=2} = R_0/4, \ldots$$

and along columns parallel to Or (see Fig. 11(b))

$$R_{0,1} = R_0, \quad R_{1,2} = R_0/3, \quad R_{2,3} = R_0/5, \quad \ldots.$$

The network can be constructed from a sheet of insulating bakelite, punched at regular intervals of length h. Each node consists of a conducting rod attached to the board, and the resistances are soldered to these rods on one side of the board; the connexions which represent electrodes, and the measurements are made on the other side.

An alternating supply may be used, although if the resistances are of high precision, a direct current supply to the network is preferable to avoid self-induction effects. With resistances of ordinary quality which have been chosen to be equal to the calculated values to within 1 per cent, the accuracy which can be attained in the measurements is considerably better than 1 per cent (Liebmann, 1950a). If manganin resistances are used, chosen to be accurate to within two parts in 10^3, an accuracy of one part in 10^4 is to be hoped for, or even better with a larger network. Any errors will then originate in the fact that electrodes which have in fact a curved meridional cross-section must be represented by a series of straight lines in the network; the interval h must in consequence be small in comparison with the dimensions of the system. The accuracy of the measurements can also be increased by "enlarging" a region of special interest. The complete system is first represented on a small scale on the network, and subsequently the interesting region is produced on a larger scale; the first experiment suggests the appropriate potentials which should be applied at the boundaries.

CHAPTER 3

THE FIELD DISTRIBUTION
IN MAGNETIC LENSES

3.1 DIRECT CALCULATION OF THE MAGNETIC FIELD

In the various systems that are used in electron optics, the magnetic fields can be produced in a number of different ways.

We may use current-carrying coils, with or without a high permeability magnetic circuit; the latter enables us to concentrate the magnetic flux within a limited region, and to obtain higher fields with a given coil. Unfortunately, these circuits, made of some ferromagnetic material, saturate when the magnetic induction B reaches values from 20 to 22 kG (2 to 2·2 tesla) for soft iron, 22 to 23 kG for pure iron and 23 to 25 kG for certain cobalt steels. Unshielded coils can be used to obtain fields considerably higher than these values but the electrical power dissipated in the windings then becomes prohibitive and in practice only pulsed operation is feasible. The recent discovery of superconducting materials with high critical fields has, however, brought iron-free coils back into favour, for with them steady fields greater than 100 kG can be achieved (see Chapter 9).

Far rarer is the use of permanent magnets, in conjunction with circuits of some ferromagnetic material. The limitations are the same as those of ordinary electromagnets. Furthermore, the field variations in the useful zone can only be of limited amplitude; they are produced by moving a soft iron shunt in the neighbourhood of the magnetized blocks (see Chapter 9).

3.1.1 General Equations

It is extremely difficult to calculate the magnetic induction everywhere in a system containing magnetic materials. In electron optics, this problem need only be solved *in vacuo* (or in air), that is to say, in a medium of magnetic permeability μ equal to $\mu_0 = 4\pi \times 10^{-7}$ (R.M.K.S. units), in which the induction B and the magnetic field H are connected by the simple relation

$$B = \mu_0 H. \tag{3.A}$$

At a point in a region in which no currents are flowing, B and H satisfy the relations

$$\operatorname{div} B = 0, \tag{3.B}$$
$$\operatorname{curl} H = 0. \tag{3.C}$$

The latter can also be written

$$\operatorname{curl} B = 0. \tag{3.C'}$$

Within a conductor carrying current density i, we should have

$$\operatorname{curl} H = i, \tag{3.D}$$
$$\operatorname{curl} B = \mu_0 i. \tag{3.D'}$$

By analogy with electrostatics, we may assume that, if $i = 0$ in the region in question, B (or H) is derived from a function Φ_m called the "scalar magnetic potential".

$$B = -\operatorname{grad} \Phi_m, \tag{3.E}$$

and this potential Φ_m satisfies Laplace's equation

$$\nabla^2 \Phi_m = 0. \tag{3.F}$$

All the methods described in the preceding chapter of solving equation (3.F) will therefore be applicable, but now we must always take the boundary conditions into account. In particular, the surfaces of ferromagnetic substances of very high permeability can be considered, with the approximation $\mu_r = \infty$, as equipotential surfaces. The usual shapes of the pole-pieces are different from those of the electrodes encountered in electrostatics, however, and new problems arise.

In addition, a magnetic "vector potential" function is defined at every point by the relations

$$\begin{aligned} B &= \operatorname{curl} A, \\ \operatorname{div} A &= 0. \end{aligned} \tag{3.G}$$

At points where $i = 0$, we shall have

$$\nabla^2 A = 0.$$

Over an arbitrary closed curve (C), the circulation of A, given by $\oint A \cdot \mathrm{d}s$, is equal to the flux Ψ of the vector B through any surface Σ bounded by this curve. Thus

$$\oint_C A \cdot \mathrm{d}s = \iint_\Sigma \operatorname{curl} A \cdot \mathrm{d}s = \iint_\Sigma B \cdot \mathrm{d}s = \Psi_{(\Sigma)}. \tag{3.H}$$

The vector potential is particularly useful in calculating the fields of iron-free coils.

3.1.2 The Special Case of a System Containing no Magnetic Material (Iron-free Coils)

If B is produced by a number of conductors of volume V, through which currents of density i flow, A is given at a point $M(x, y, z)$ by

$$A = \frac{\mu_0}{4\pi} \iiint_V \frac{i\,dV}{r},\tag{3.I}$$

in which r denotes the distance of the point M from the volume element dV. For a *filamental conductor* of cross-section S, we have (Fig. 12a)

$$i\,dV = |i|S\,ds = J\,ds$$

so that

$$A = \frac{\mu_0 J}{4\pi} \int \frac{ds}{r}.\tag{3.J}$$

This gives

$$B = \frac{\mu_0 J}{4\pi} \operatorname{curl}\left(\int \frac{ds}{r} \right).$$

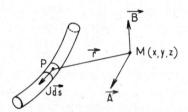

FIG. 12a. Field of a conducting wire.

For every point M, the integration must be performed along the whole length of the conductor. From the relation

$$\operatorname{curl}\left(\frac{ds}{r}\right) = \frac{1}{r}\operatorname{curl} s + \left(\operatorname{grad}\frac{1}{r}\right) \times ds$$

and using the fact that s does not depend upon the position of M, we obtain

$$\operatorname{curl}\left(\frac{ds}{r}\right) = \left(\operatorname{grad}\frac{1}{r}\right) \times ds = -\frac{r \times ds}{r^3},$$

with $r = \overrightarrow{MP}$. Using instead the convention $r = \overrightarrow{PM}$, we have

$$B = \frac{\mu_0 J}{4\pi} \int \frac{r \times ds}{r^3}.\tag{3.K}$$

This expression is the law of Biot and Savart. To calculate the components of B, we can use (3.K) directly or, alternatively, we may use (3.J) together with (3.G).

In Cartesian coordinates, we have

$$B_x = \frac{\partial A_z}{\partial y} - \frac{\partial A_y}{\partial z},$$

$$B_y = \frac{\partial A_x}{\partial z} - \frac{\partial A_z}{\partial x}, \qquad (3.L)$$

$$B_z = \frac{\partial A_y}{\partial x} - \frac{\partial A_x}{\partial y},$$

and in cylindrical polars,

$$B_r = \frac{1}{r}\frac{\partial A_z}{\partial \theta} - \frac{\partial A_\theta}{\partial z},$$

$$B_\theta = \frac{\partial A_r}{\partial z} - \frac{\partial A_z}{\partial r}, \qquad (3.L')$$

$$B_z = \frac{1}{r}\left\{\frac{\partial (rA_\theta)}{\partial r} - \frac{\partial A_r}{\partial \theta}\right\}.$$

In the two simple cases which follow, these relations simplify considerably.

(a) *Straight coils, infinitely long in the Oz direction (translation symmetry).* The only non-zero component of **i** is i_z, and we have

$$A_x = A_y = 0, \quad A = A_z,$$

$$B_z = 0, \quad B_x = \frac{\partial A_z}{\partial y}, \quad B_y = -\frac{\partial A_z}{\partial x}. \qquad (3.M)$$

(b) *Circular coils having a common axis Oz (rotational symmetry).* We now have $i = i_0$, so that

$$A_r = A_z = 0, \quad A = A_\theta,$$

and

$$B_\theta = 0, \quad B_r = \frac{\partial A_\theta}{\partial z}, \quad B_z = \frac{1}{r}\frac{\partial (rA_\theta)}{\partial r}. \qquad (3.N)$$

In the case of a closed current loop (C) carrying current J, we can show that the scalar magnetic potential Φ_m at a point M is given (Fig. 12b) by

$$\Phi_m = \frac{\mu_0 J}{4\pi}\Omega \qquad (3.O)$$

in which Ω denotes the solid angle subtended at M by the curve (C). We then find

$$B = -\frac{\mu_0 J}{4\pi}\text{ grad }\Omega. \qquad (3.P)$$

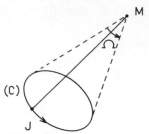

FIG. 12b. A circular turn.

EXAMPLE 1. The field produced by a single circular turn of negligible cross-section and radius a (Fig. 12c).

On symmetry grounds, $B_r = 0$. Let z be the abscissa of M and z_0 that of the turn (S); $x = a \cos \theta$ and $y = a \sin \theta$ are the coordinates of

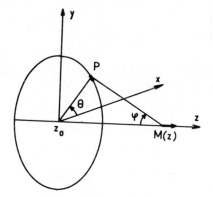

FIG. 12c. Calculation of the axial field of a circular turn.

a point P on the turn. The projections of $\overrightarrow{PM}$ on the axes are equal to x, y and $(z - z_0)$.

Thus

$$dx = -a \sin \theta \, d\theta, \quad dy = a \cos \theta \, d\theta,$$

and (3.K) gives

$$B_z = \frac{\mu_0 J}{4\pi} \int_{(S)} \frac{x dy - y dx}{r^3} = \frac{\mu_0 J}{4\pi} \int_{(S)} \frac{a^2 d\theta}{r^3} = \frac{\mu_0 J a^2}{2 r^3}$$

and

$$B_z = \frac{\mu_0 J a^2}{2\{(z - z_0)^2 + a^2\}^{3/2}}. \tag{3.Q}$$

With the aid of (3.Q), we can calculate the magnetic induction B_z on the axis of a thin solenoid of radius a and length $2l$ centred on $z = 0$; by considering the induction dB_z produced by an element of the solenoid of length $d\zeta$ situated at $z = \zeta$ and containing n identical turns per unit length, and integrating from $-l$ to $+l$, we find

$$B_z = \frac{\mu_0 n J a^2}{2} \int_{-l}^{l} \frac{d\zeta}{\{(z-\zeta)^2 + a^2\}^{3/2}}.$$

Setting $(z - \zeta)/a = \cot \phi$, we obtain

$$B_z = \frac{\mu_0 n J}{2} (\cos \phi_1 - \cos \phi_2) \tag{3.R}$$

or

$$B_z = \frac{\mu_0 I}{4l} \left[\frac{z+l}{\{a^2 + (z+l)^2\}^{1/2}} - \frac{z-l}{\{a^2 + (z-l)^2\}^{1/2}} \right], \tag{3.R'}$$

in which

$$I = 2lnJ.$$

If l tends to infinity, we have $\phi_1 = 0$ and $\phi_2 = \pi$, and so

$$B_z = \mu_0 n J.$$

For a coil of appreciable thickness, we must integrate in the two directions Oz and Or. From (3.R) we know the contribution from a layer of thickness dr, radius r and length $2l$, carrying a total current

$$I = 2lnJ = 2l \, dr \cdot i.$$

(i is the current density and is independent of z.)
Writing (see Fig. 12d)

$$\phi_1 = \alpha(r), \qquad \phi_2 = \beta(r),$$

we integrate (3.R) from r_1 to r_2,

$$B_z = \frac{\mu_0 i}{2} \int_{r_1}^{r_2} \{\cos \alpha(r) - \cos \beta(r)\} \, dr.$$

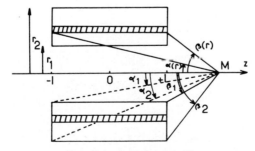

FIG. 12d. A thick solenoid.

We have

$$\tan \alpha = \frac{r}{z+l}, \qquad \tan \beta = \frac{r}{z-l},$$

$$\frac{d\alpha}{\cos^2 \alpha} = \frac{dr}{z+l}, \qquad \frac{d\beta}{\cos^2 \beta} = \frac{dr}{z-l},$$

so that

$$B_z = \frac{\mu_0 i}{2} \left\{ (z+l) \int_{\alpha_1}^{\alpha_2} \frac{d\alpha}{\cos \alpha} - (z-l) \int_{\alpha_1}^{\alpha_2} \frac{d\beta}{\cos \beta} \right\}$$

or

$$B_z = \frac{\mu_0 i}{2} \left\{ (z+l) \log \frac{\tan \left(\frac{\alpha_2}{2} + \frac{\pi}{4} \right)}{\tan \left(\frac{\alpha_1}{2} + \frac{\pi}{4} \right)} - (z-l) \log \frac{\tan \left(\frac{\beta_2}{2} + \frac{\pi}{4} \right)}{\tan \left(\frac{\beta_1}{2} + \frac{\pi}{4} \right)} \right\}. \qquad (3.S)$$

If I denotes the total current in the thick solenoid, we have

$$i = \frac{I}{2l(r_2 - r_1)}. \qquad (3.S')$$

Knowing the field distributions along the axis Oz, we shall see later that it is possible to calculate B_r and B_z in the neighbourhood of the axis.

REMARKS. (i) We may calculate B_z on the axis Oz, starting from equations (3.O) and (3.P). For a single turn, we should have

$$\Omega = 2\pi(1 - \cos \phi)$$

(see Fig. 12c), and so

$$\Phi_m = \frac{\mu_0 J}{2} (1 - \cos \phi) = \frac{\mu_0 J}{2} \left\{ 1 - \frac{z - z_0}{\sqrt{a^2 + (z - z_0)^2}} \right\}.$$

Hence

$$B_z = -\frac{\partial \Phi_m}{\partial z} = -\frac{\mu_0 J}{2} \left\{ -\frac{1}{r} + \frac{(z - z_0)^2}{r^3} \right\} = \frac{\mu_0 J a^2}{r^3}$$

with

$$r = \sqrt{a^2 + (z - z_0)^2}.$$

Alternatively,

$$B_z = \frac{\mu_0 J}{2a} \sin^2 \phi.$$

For a thin solenoid,

$$B_z = \frac{\mu_0 n J}{2a} \int_{\phi_1}^{\phi_2} \sin^3 \phi \, d\phi$$

and we recover the formula

$$B_z = \frac{\mu_0 nJ}{2}(\cos\phi_1 - \cos\phi_2).$$

(ii) If the field created throughout the whole space is required, it is preferable first to calculate $A_\theta(r, z)$ (cf. Durand, 1952) and to deduce B from it. The expression for A_θ (and hence, those for B_z and B_r) contains the elliptic integrals of Legendre; these are well tabulated.

EXAMPLE 2. The expression for the field produced at a point $M(x, y)$ by a straight wire of cross-section $dS = d\xi\, d\eta$, carrying current $I = jd\xi\, d\eta$ (Fig. 12e(a)), is of the form

$$dB_x = -\frac{\mu_0 j}{2\pi}\frac{y-\eta}{(x-\xi)^2+(y-\eta)^2}\, d\xi\, d\eta,$$

$$dB_y = +\frac{\mu_0 j}{2\pi}\frac{x-\xi}{(x-\xi)^2+(y-\eta)^2}\, d\xi\, d\eta.$$

From this we can calculate the magnetic induction produced by a narrow band of thickness $d\xi$ and width $2b$ (Fig. 12e(b)):

$$\triangle B_x = \frac{\mu_0 j}{2\pi}d\xi \log\frac{(x-\xi)^2+(y-b)^2}{(x-\xi)^2+(y+b)^2},$$

$$\triangle B_y = -\frac{\mu_0 j}{2\pi}d\xi\left(\tan^{-1}\frac{x-\xi}{y+b} - \tan^{-1}\frac{x-\xi}{y-b}\right),$$

or

$$\triangle B_x = -\frac{\mu_0 I}{4\pi b}\log\frac{R_2}{R_1},$$

$$\triangle B_y = \frac{\mu_0 I}{4\pi b}(\alpha_2 - \alpha_1),$$

in which $I = 2bj\, d\xi$; R_1, R_2, α_1 and α_2 are defined in Fig. 12e(b).

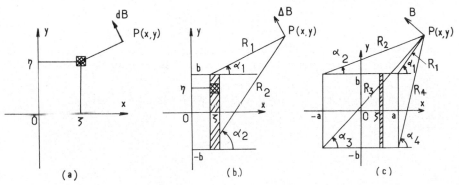

FIG. 12e. A bar of rectangular cross-section. (a) A wire of square cross-section; (b) a thin ribbon; (c) a thick bar.

Finally, integrating with respect to x, we obtain the induction produced by a bar of rectangular cross-section $2b \times 2a$ (Fig. 12e(c)), carrying a current $I = 4abj$:

$$B_y = \frac{\mu_0 I}{8\pi ab}\left\{(x+a)\log\frac{R_2}{R_3} + (x-a)\log\frac{R_4}{R_1}\right.$$
$$\left. + (y+b)(\alpha_3 - \alpha_4) + (y-b)(\alpha_1 - \alpha_2)\right\},$$

$$B_y = \frac{\mu_0 I}{8\pi ab}\left\{(y+b)\log\frac{R_3}{R_4} + (y-b)\log\frac{R_1}{R_2}\right.$$
$$\left. + (x+a)(\alpha_3 - \alpha_2) + (x-a)(\alpha_1 - \alpha_4)\right\}.$$

With four thin sheets or four rectangular bars (Septier, 1960), we can produce field distributions with quadrupolar symmetry, having a transverse field gradient $G = \partial B/\partial r$ that is virtually constant in all directions; iron-free quadrupole lenses can thus be designed in this simple way (see Chapter 10).

EXAMPLE 3. *The use of complex functions, and the creation of a multipole field in a system with translation symmetry* (Beth, 1966). From the relations

$$B_x = -\frac{\partial\Phi_m}{\partial x} = \frac{\partial A_z}{\partial y},$$
$$B_y = -\frac{\partial\Phi_m}{\partial y} = -\frac{\partial A_z}{\partial x}$$

we have

$$\nabla^2\Phi_m = 0 \quad \text{and} \quad \nabla^2 A_z = 0,$$

so that the function

$$W(z) = -(A + j\Phi_m)$$

is an analytic function in the complex variable $z = x + jy$.

In consequence, the "complex potential" $W(z)$ may be written in the form

$$W(z) = -(A + j\Phi_m) = W_0 + W_1 z + W_2 z^2 + \cdots W_n z^n.$$

We then have

$$B(z) = \frac{dW}{dz}$$

or

$$B(z) = W_1 + 2W_2 z + 3W_3 z^2 + \cdots nW_n z^{n-1}.$$

Successive terms in this expression correspond to constant fields, dipole fields, quadrupole fields and so on. Thus for $n = 2$, for example,

$$W(z) = z^2 = r^2 \exp 2j\theta = u(x, y) + jv(x, y),$$

where

$$u = x^2 - y^2 = r^2 \cos 2\theta,$$
$$v = 2xy - r^2 \sin 2\theta.$$

The function $u(x, y)$ represents $A_z(x, y)$ and $v(x, y)$ gives $B(x, y)$.

For an infinitely long wire carrying current I situated at $z_0 = x_0 + jy_0$ we should have

$$W(z) = -(A + j\Phi_m) = \frac{\mu_0 I}{2\pi} \log(z - z_0),$$

$$B(z) = B_y + jB_x = \frac{\mu_0 I}{2\pi(z - z_0)}.$$

Along a curve (C) surrounding the wire,

$$\oint_{(C)} B(z)\,dz = \oint_{(C)} \frac{\mu_0 I}{2\pi(z - z_0)}\,dz = j\mu_0 I.$$

Consider now an infinitesimally thin ribbon of width $\triangle z$ carrying a total current $\triangle I$. The values of $B(z)$ at two infinitesimally close points A_1 and A_2 on either side of the ribbon are not the same, and letting $\triangle z$ tend to zero, we have

$$B_1(z) - B_2(z) = j\mu_0 \frac{dI}{dz},$$

in which dI/dz represents the current density per unit length in the ribbon. Integrating, we find

$$W_1(z) - W_2(z) = j\mu_0 I + \text{const.}$$

Let us apply this relation to the following specific problem: we consider a circular cylinder of radius r_0 centred on $z = 0$. The equation of this cylinder is $z = r_0 e^{j\theta}$. The current at each point is denoted by $I(\theta)$. Let us assume that the total current is zero and that the dependence of I upon θ is such that $I(\theta) = I(\theta + 2\pi)$. $I(\theta)$ can be written

$$I(\theta) = \sum_{n=1}^{\infty} C_n \sin(n\theta + \theta_n)$$

and

$$\sin(n\theta + \theta_n) = \frac{1}{2j}\{e^{j(n\theta + \theta_n)} - e^{-j(n\theta + \theta_n)}\}.$$

Replacing $e^{jn\theta}$ by $(z/r_0)^n$, we have

$$\sin(n\theta + \theta_n) = \frac{1}{2j}\left\{e^{j\theta_n}\left(\frac{z}{r_0}\right)^n - e^{-j\theta_n}\left(\frac{r_0}{z}\right)^n\right\}.$$

At every point z on the (outer and inner) surface of the curve,

$$W_{ext} - W_{int} = \frac{\mu_0}{2}\sum_{n}^{\infty} C_n\left\{e^{j\theta_n}\left(\frac{z}{r_0}\right)^n - e^{-j\theta_n}\left(\frac{r_0}{z}\right)^n\right\} + \text{const.}$$

At a point $M(z)$, $W(z)$ must be finite when $|z| \to 0$ and when $|z| \to \infty$. We must therefore have

$$W_{ext}(z) = -\frac{\mu_0}{2}\sum_{n} C_n e^{-j\theta_n}\left(\frac{r_0}{z}\right)^n + \text{const.}$$

$$W_{int}(z) = \frac{\mu_0}{2}\sum_{n} C_n e^{j\theta_n}\left(\frac{z}{r_0}\right)^n + \text{const.}$$

Inside the cylinder, the field will therefore be of the form

$$B_{int}(z) = -\frac{\mu_0}{2}\sum_{n} nC_n r^n e^{j\theta_n} z^{n-1}$$

and outside,

$$B_{ext}(z) = \frac{\mu_0}{2}\sum_{n} nC_n r_0^n e^{j\theta_n} z^{-(n+1)}.$$

EXAMPLE. We wish to create inside the cylinder a field with constant gradient:

$$B_x = Gy, \qquad B_y = Gx.$$

Thus

$$B(z) = B_y + jB_x = G(x + jy) = Gz$$

$$-\tfrac{1}{2}n\mu_0 C_n r_0^{-n} e^{j\theta_n} = 0 \qquad \text{for } n = 1,$$
$$= G \qquad \text{for } n = 2,$$
$$= 0 \qquad \text{for } n > 2.$$

Hence

$$C_2 = -\frac{G}{\mu_0} r_0^2 \sin 2\theta,$$

and therefore

$$I_\theta = -\frac{G}{\mu_0} r_0^2 \sin 2\theta.$$

The current density per unit length (in A/m) on the cylinder is

$$j(\theta) = \frac{dI}{r_0 d\theta} = -\frac{2G}{\mu_0} r_0 \cos 2\theta = -j(0)r_0 \cos 2\theta.$$

The field gradient is given by

$$G = \frac{\mu_0 j(0)}{2},$$

and the maximum induction at the layer is

$$B_0 = \frac{\mu_0 j(0) r_0}{2}.$$

If, for example, we wish to obtain $G = 1$ tesla/m, then over a circle of radius $r = 0 \cdot 10$ m we must satisfy the relation

$$j(\theta) - -\frac{0 \cdot 2}{\mu_0} \cos 2\theta = -j(0) \cos 2\theta$$

or

$$j(0) \simeq 160,000 \text{ A/m}.$$

Current densities as high as this can be attained by using superconducting windings (see Chapter 7).

3.1.3 Magnets with Magnetic Material

The general relations given above are still valid. To calculate B in the gap of an unsaturated electromagnet (or a permanent magnet), we assume that the permeability μ_r is infinite. The surfaces of the pole-pieces are then equipotentials, and the Laplace equation

$$\nabla^2 \Phi_m = 0$$

is solved in whichever system is employed (Hesse, 1949; Lenz, 1950). Thus the field B_z of a rotationally symmetric lens similar to the one shown schematically in Fig. 13b will effectively be given by the expression for E_z in an electrostatic lens consisting of two equal tubes. It is far more difficult to determine the scalar potentials at the poles, however: we merely determine the potential difference between the two poles. The distribution of B_z on the axis is such that

$$\triangle\Phi_m = \int\limits_{-\infty}^{\infty} B_z \mathrm{d}z = \mu_0 nI,$$

where nI denotes the total current through the magnetizing coils; when $\mu = \infty$, this potential difference is wholly between the poles.

It is, in fact, frequently better to measure this potential difference; for this, a small long coil is wound onto a flexible support, each end is placed in contact with one of the pole-pieces, and then withdrawn. The two ends are then brought into contact with each other, and it is easy to show that the difference of flux across the coil, (which is measured with a fluxmeter) is proportional to $\triangle \Psi$, the magnetic potential difference between those points on the two pole-pieces to which the ends of the winding were applied. If the magnetic circuit saturates, μ becomes small (μ/μ_0 tends to unity); as soon as $\mu < 100$, the "potential drop" in the magnetic material can no longer be neglected, and this drop will depend upon the path followed in determining the loop-integral of H in the iron: the pole-pieces will no longer be "equipotentials".

If, nevertheless, the distribution of B is the same as that in iron, μ can be deduced from the characteristic curve of the material employed. We can then proceed as follows. Consider an electromagnet (see Fig. 12f) consisting of a large yoke and two poles surrounded by coils each carrying total current $nI/2$. Over the path $ABCDEA$ we have

$$\oint H dl = nI$$

and for $\mu \simeq \infty$ we have $\int_{C} H dl = 0$ except between B and C; hence $nI = \int_{B} H dl$ and we can regard the potential difference $\Phi_B - \Phi_C$ as equal to $\mu_0 nI$, and the pole-pieces as equipotentials.

If, on the contrary, μ is small, we have $\int_{A}^{B} H dl \neq 0$ with $H = B/\mu$. We write

$$\int_{A}^{B} H dl = \alpha_1 nI, \qquad \int_{C}^{D} H dl = \alpha_2 nI.$$

We can always arrange that $\int_{D}^{A} H dl \simeq 0$ by making the cross-section of the

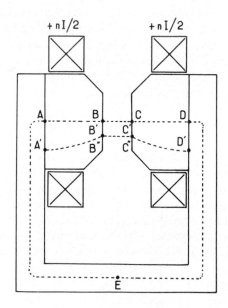

FIG. 12f. Cross-section of an electromagnet with a soft iron yoke.

yoke big enough. We now have

$$\Phi_B - \Phi_C = \mu_0\left\{1 - (\alpha_1 + \alpha_2)nI\right\}.$$

If the value of B is known everywhere in the cross-section, the potential difference between various pairs of points $(B'C', B''C'')$ can be calculated and thus we can find the potential distribution over the surfaces of the pole-pieces. We then proceed by successive approximations; we start from the distribution corresponding to the unsaturated case. For a given value of B_0 in the gap, we can calculate the flux in the poles, and hence obtain a first approximation to the distribution of B in the iron. We then carry out the calculation above and again solve $\nabla^2\Phi_m = 0$ with the new boundary conditions; this gives a second, more exact approximation to B. After two or three "relaxations", B is known very accurately.

This procedure can only be used when a computer or a resistance network (see below) is available.

3.1.4 The Expressions for the Field Components and the Vector Potential in the Neighbourhood of the Axis (or of a Symmetry Plane)

Just as in electrostatics, we can expand in series near the axis of a rotationally symmetric lens, since $\Phi_m(r, z)$ is a solution of the Laplace equation.

We obtain the expression

$$\Phi_m(r, z) = \sum_{n=0}^{\infty} \frac{(-1)^n}{(n!)^2} \cdot \left(\frac{r}{2}\right)^{2n} \Phi_0^{(2n)}(z),$$

in which

$$\Phi_0(z) = \Phi(0, z).$$

Hence

$$B_z(r, z) = -\frac{\partial \Phi(r, z)}{\partial z},$$

$$B_r(r, z) = -\frac{\partial \Phi(r, z)}{\partial r}.$$

We know, however, that

$$B_r(r, z) = -\frac{\partial A(r, z)}{\partial z},$$

so that

$$\frac{\partial A}{\partial z} = \sum_{n=1}^{\infty} \frac{(-1)^n}{n!(n-1)!} \left(\frac{r}{2}\right)^{2n-1} \Phi_0^{(2n)}(z),$$

and integrating with respect to z,

$$A(r, z) = \sum_{n=1}^{\infty} \frac{(-1)^n}{n!(n-1)!} \cdot \left(\frac{r}{2}\right)^{2n-1} \Phi_0^{(2n-1)}(z).$$

In the neighbourhood of the axis, therefore, we have

$$B_z(r, z) = -\frac{\partial \Phi(r, z)}{\partial z} = \sum_{n=0}^{\infty} \frac{(-1)^n}{(n!)^2} \cdot \left(\frac{r}{2}\right)^{2n} B_0^{(2n)}(z),$$

$$B_z(r, z) = B_0(z) - \frac{r^2}{4} B_0''(z) + \frac{r^4}{64} B_0^{(4)}(z) - \cdots,$$

$$B_r(r, z) = -\frac{r}{2} B_0'(z) + \frac{r^3}{16} B_0^{(3)}(z) - \cdots,$$

$$A_\theta(r, z) = \frac{r}{2} B_0(z) - \frac{r^3}{16} B_0''(z) + \cdots.$$

If we consider a circle (C) of radius r, centred on the axis Oz, we have

$$\oint_{(C)} A_\theta r \, \mathrm{d}\theta = \iint B_z \, \mathrm{d}S = \Psi(r, z) = 2\pi r A_\theta,$$

where Ψ denotes the flux of B_z through this circle.

Hence

$$\Psi(r, z) = \pi r^2 \left\{ B_0(z) - \frac{r^2}{8} B_0''(z) + \cdots \right\}.$$

In a similar way we could obtain B_x, B_y and A_z in the vicinity of a symmetry plane in a system possessing translation symmetry.

3.2 DIRECT METHODS OF MEASURING THE INDUCTION B AND ITS GRADIENT

A knowledge of the distribution of magnetic induction along the axis of a lens, $B(z)$, is essential, and the magnitude of B may vary from zero to several kilogauss over a distance of the order of millimetres; another difficulty is that we are very often obliged to use extremely tiny measuring probes, as the dimensions of the lenses are themselves so small. For both of these reasons, high precision measuring techniques which depend upon nuclear resonance are not applicable here. From among the numerous methods available for measuring strong magnetic fields (see Symonds, 1955, for example), we shall therefore describe only those which can usefully be employed to measure the fields in lenses.

3.2.1 Methods Based upon Induction Phenomena

When the lenses are free of iron, the simplest method is to use a small coil, very carefully positioned, and to supply the lens with alternating current. The potential difference which is induced by the alternating current across the ends of the coil is measured after amplification in a linear amplifier or in an amplifier which has been calibrated beforehand to give values of the field directly.

If the lenses are shielded, however, alternating current can no longer be used, because of the hysteresis effects which would be produced; the lenses are supplied with a *highly stabilized* direct current, and the distribution of **B** is then measured by one of the various methods described below.

(i) *Direct measurement of the flux through a small coil.* The coil is moved slowly in the zone to be explored—along the Oz axis of a lens, for example— and the potential difference induced across its terminals is transmitted to a highly sensitive fluxmeter (Sauzade, 1958); the signal, V, at the output may reach one volt for $\psi = 100$ maxwells. By recording V directly onto a sheet of paper which unrolls in synchronism with the motion of the search coil, we obtain automatically a plot of the distribution of flux $\psi(z)$. We can obtain the absolute value of $B(z)$ by calibrating the coil beforehand in a known homogeneous field, always provided the dimensions of the coil are considerably smaller than those of the lens, in which the field is inhomogeneous. Laudet (1957) has, however, developed a method of calculation with the aid of which we can derive the values of $B(z)$, $\dfrac{dB}{dz}$ and $\dfrac{d^2B}{dz^2}$ quite accurately from the curve produced by a relatively large search coil.

We can equally well plot the value of B, point by point, with a calibrated search coil connected to a fluxmeter or to a ballistic galvanometer. The coil is placed in position in the field which is to be charted, with its turns normal to B; it is then suddenly withdrawn from the field, or alternatively, the coil is left untouched but the sense of the current in the coil is reversed (Dosse, 1941. See Fig. 12g).

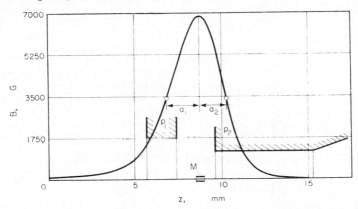

FIG. 12g. The principle behind Dosse's measurements (1941). The coil M is shifted along the axis, and in each position, the change of flux which is produced by reversing the magnetizing current is measured.

(ii) *An alternating e.m.f. is created across the terminals of a small rotating coil which is placed in the field; this e.m.f., which is proportional to **B** and to the frequency of rotation, is then measured.* The coil is supported on the non-magnetic axle of a synchronous motor; for preference, the motor should be a synchronous three-phase motor, virtually free of mechanical vibration (see Fig. 13a). For the measurements to be accurate, the coil must be extremely small and must be placed at its successive positions within the gap between the pole-pieces with great care (see Langer and Scott, 1950, for example). This method can be made more accurate by using two coils driven by the same motor; one coil, B_1, rotates within the lens being studied, and gives a signal v_1, whilst the other, B_2, rotates within a homogeneous magnetic field which has been very accurately standardized (with a nuclear resonance probe, for example). The signal v_1 is then counteracted by a known fraction k of the signal v_2 given by B_2; by adjusting the relative orientations of the symmetry axes of B_1 and B_2, the phase difference between v_1 and v_2 can be exactly annulled. Finally, we suppress as far as possible any vibration of the coils by using special vibration-proof bearings, and thus avoid parasitic signals being superimposed onto the fundamental (Germain, 1955a). We can thus arrange that the signals v_1 and kv_2 are equal, and hence deduce the value of the unknown induction from the quantity k which is given by an accurate potentiometric method. In these conditions, we can easily reach an accuracy of 0·1 per cent.

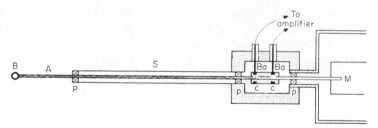

FIG. 13a. Sketch of a rotating coil.
A: rotating axle. *B*: coil. *C*: contacts (silver). *P*: bearing. *S*: rigid fixed support.
Ba: brushes (carbon–silver). *M*: synchronous motor.

FIG. 13b. Fert and Gautier's arrangement (1954), with an oscillating coil.
L: lens. *S*: solenoid for measuring B at the centre of the first turn. B_2: coil for measuring dB/dz. B_1: coil for measuring d^2B/dz^2. P: vibrating plate.

(iii) *A third possibility is to set a long coil in vibration along the direction of the axis; as before, the induced e.m.f. is proportional to B at the end face of the solenoid.* However, if we have a short fixed coil of small diameter, dB/dz can be obtained, and with two identical coils mounted in opposition, $\dfrac{d^2B}{dz^2}$ (Fert and Gautier, 1951; Gautier, 1952. See Fig. 13b).

A description of a large number of devices of this type for measuring the induction and its gradient in magnetic lenses with large dimensions is to be found in an article by Grivet and Septier (1960). In particular, the gradient of B in a direction, OX, normal to B can be measured directly by using a short coil which has its axis parallel to B and is attached to a Perspex rod which vibrates transversely along OX; the oscillation frequency of the rod is selected to be equal to one of its resonant frequencies, and a large amplitude, $\triangle x$, can thus be obtained with only a small expenditure of power.

3.2.2 Method Using the Force which Is Exerted by the Induction B on a Current-carrying Conductor

In this method (Ments and Le Poole, 1947) a long slender solenoid, S, wound on a non-magnetic arm and carrying a current i, is suspended horizontally along the lens axis. The end-face which lies within the field experiences a force proportional to the value of the induction B at the point. The

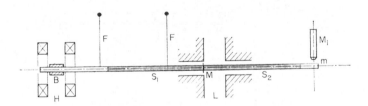

Fig. 13c. Durandeau's design (1953) in which the field is measured by the force which it exerts on a coil.
L: lens. F: pendular suspension. S_1, S_2: identical solenoids. B: compensating coil. H: Helmholtz coil. M_i: microscope with graduated objective m. M: the axial point at which the measurement is effectively made.

displacement of the whole mobile assembly is measured in a microscope. An improvement on this design has been made by Durandeau (1953), using two solenoids, S_1 and S_2, identical save in that they are wound in opposite senses (Fig. 13c), each carrying a current i. Either the force exerted by the induction on the common face of the two coils is measured, or else the displacement of the mobile part is annulled with the aid of an auxiliary coil B_1, which also carries a current i, and which is placed in the known field of a system of Helmholtz coils; the Helmholtz coil current is then proportional to the magnitude of B.

3.2.3 The Hall Effect

The probe is now a small thin rectangular plate $ABCD$, thickness a (see Fig. 14). AB and CD are the short sides which are covered with a conductive layer, and between these a steady current I flows. If we label the midpoints of the longer sides, BC and DA, M and N respectively, and if we place the plate in a region where the magnetic induction is B and is normal to the plane $ABCD$, we shall find a potential difference between electrodes placed at M and N which is proportional to the magnitude of B and to the current I:

$$V_M - V_N = K_H \frac{IB}{a},$$

in which K_H is the "Hall constant" of the substance employed, normally a semi-conductor such as indium arsenide or indium antimonide. When $I \sim 0.1$ amps and $B \sim 10$ kG, a plate of a few square millimetres can produce a potential difference of a few tens of millivolts.

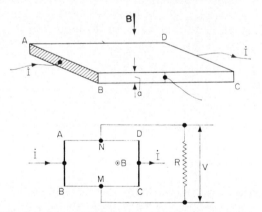

Fig. 14. A Hall plate for measuring an induction B perpendicular to the face $ABCD$; AB and CD are metallized.

In fact, the terminals at M and N are connected through an external resistance, R, and the potential which is measured, $(V_M - V_N)$, is a "load" potential; we must, therefore, make allowance for the internal resistance of the plate between M and N, R_{MN}. Due to the magneto-resistance effect however, the latter is a function of B, and hence $(V_M - V_N)$ is never perfectly proportional to B, especially when the induction reaches high values. For accurate measurements (between one part in 10^3 and 10^4), I has to be stabilized to better than one part in 10^4, and the temperature of the plate has to be held constant to within less than 0·1°C. (K_H varies with temperature—from this point of view, indium arsenide is superior to every other semi-conductor, as its temperature coefficient is very slight at room temperature).

The potential difference between the edges of the plate is measured by a standard method in which the potential to be measured is opposed by a potential which has been calibrated to within about one part in 10^4. The plate itself is accurately calibrated by using nuclear resonance, and a known electrical circuit which provides a curve $B = f(I)$ to which we subsequently refer.

By juxtaposing two identical plates, we can measure the gradient of B directly. This technique has as yet only been used in electromagnets and lenses of large dimensions even though we now know how to construct plates on a miniature scale with which point by point measurements become possible.

3.3 THE USE OF ELECTRICAL ANALOGUES

3.3.1 The Standard Transposition

The results of §2.3 suggested that both the electrolytic tank and the resistance network are convenient tools for studying fields; provided that the hypothesis that the pole-piece surfaces are equipotentials proves to be valid, there is a formal identity between electrostatic and scalar magnetic potentials.

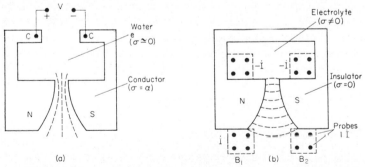

FIG. 15a. The direct analogy. The pole-pieces and the magnetic circuit are represented by metal electrodes. No allowance is made for the windings of the electromagnet and the lens. A chart of the "scalar equipotential surfaces", $\varphi =$ constant, is traced out.

FIG. 15 b. The conjugate analogy. The iron is represented by a block of paraffin; the coils B_1, B_2 are represented schematically by injecting currents. We obtain "vector equipotentials", $A =$ constant.

The pole-piece profiles are equivalent to "perfect" conductors, held at electrostatic potentials proportional to their magnetic counterparts. Coils are represented, very roughly, by a break in the magnetic circuit across whose ends an overall magnetic potential difference NI is applied, to represent the effect of the magnetizing current (see Fig. 15a). In the tank, such a break is represented by an electrical gap across whose ends an electric potential proportional to NI is applied.

Even when certain portions of the magnetic circuit are saturated, or in the process of becoming saturated, it is still possible to obtain the distribution of magnetic induction, B (Liebmann, 1953). Once the characteristics of the magnetic material in question are available, we are in a position to know the value of the permeability μ for each local value of B. If the value of the resistance at a point at which $\mu = \mu_0$ (air) is R_n, then a permeability μ_1 at the same point will be represented by a resistance R'_n such that

$$R'_n = \left(\frac{\mu_0}{\mu_1} \right) R_n.$$

If $\mu_1 = \infty$ then clearly $R'_n = 0$ (the nodes are short-circuited). In practice, we proceed by successive approximation; a chart of the scalar equipotentials is drawn up for the infinite permeability case, and the induction B at the surface of the steel is calculated starting from a value B_0 at some point on the axis, $B_0 = 20 \text{ kG}$ for example. The resistances which represent the boundaries of the pole-pieces and the interior of the pole-pieces themselves can then be modified, a new chart of the potential can be drawn and fresh values of B extracted. Continuing in this way, we tend gradually towards a self-consistent solution, for which the network has no further need of modification.

The method depending upon a direct analogy can, however, only be used to determine the distribution of B within the narrow gaps in magnetic lenses—the windings lie far enough away from the useful zone to produce no appreciable perturbation. This is by no means true of an electromagnet with a wide opening and for this case, the "conjugate" analogy which we are about to describe must be invoked; *this analogue is also capable of handling the case of systems free of iron.*

3.3.2 The Conjugate Tank

We now pass on to a more recent method, again using an electrical analogy, in which it is possible to make allowance for the magnetizing coils. The method is due to Peierls and Skyrme (1947, 1949), and has been tried in practice by Bracher (1950)—it is known as "the conjugate tank".

Consider the equations which govern the ordinary tank, valid for the whole system of electrodes, current sources, and insulators which is immersed in the electrolyte. If i be current density at some point, and J the intensity *emitted* per unit volume, then

$$\mathrm{div}\, i = J.$$

Elsewhere than at point sources, div $i = 0$.

Further,

$$\mathrm{curl}\, E = 0,$$

and

$$i = \sigma\, E.$$

The electrical potential V satisfies the equations

$$E = -\,\mathrm{grad}\, V, \tag{3.1}$$

$$\nabla^2 V = -\,J/\sigma, \tag{3.2}$$

in which σ is the conductivity of the medium.

On the boundary between two media, free of surface sources, div $i = 0$ provides the boundary conditions for the normal components of current density:

$$i_{2n} = i_{1n},$$

while curl $E = 0$ implies

$$E_{2t} = E_{1t}$$

(that is, that the tangential component of the electric field is continuous).

As indicated in § 2.2.1, these two relations can be put in the form

$$\sigma_1 \left(\frac{\partial V}{\partial n} \right)_1 = \sigma_2 \left(\frac{\partial V}{\partial n} \right)_2 \tag{3.3}$$

and

$$\left(\frac{\partial V}{\partial t} \right)_1 = \left(\frac{\partial V}{\partial t} \right)_2. \tag{3.4}$$

Consider now the magnetic field of an electromagnet. The magnetic flux B is conservative, that is.

$$\mathrm{div}\, B = 0.$$

If j is the current density in the coils, Ampère's theorem relates the magnetic field H and j by

$$\mathrm{curl}\, H = j.$$

Defining vector potential A by

$$B = \mathrm{curl}\, A,$$

$$\mathrm{div}\, A = 0,$$

substituting $B = \mu\, \mu_0\, H$, and using the standard identity

$$\mathrm{curl}\,\mathrm{curl}\, u = \mathrm{grad}\,\mathrm{div}\, u - \nabla^2 u,$$

we find

$$\nabla^2 A = -\,\mu\, \mu_0\, j. \tag{3.5}$$

If we suppose that the magnetizing current and the boundary between the two media are parallel to the axis Oz of a system of Cartesian coordinates, this implies

$$j_x = j_y = 0, \quad \boldsymbol{j} = (0, \ 0, \ j_z).$$

Since $\boldsymbol{B}$ lies in a plane normal to $\boldsymbol{j}$, we have

$$A_x = A_y = 0, \quad \boldsymbol{A} = (0, \ 0, \ A_z),$$

so that

$$\nabla^2 A_z = -\mu \mu_0 j_z. \tag{3.6}$$

Let $\boldsymbol{n}$ be the unit vector, normal to some separating surface, and $\boldsymbol{b}$ the unit vector perpendicular both to $\boldsymbol{n}$ and to Oz; the boundary conditions at the surface become:

$$\left(\frac{\partial A_z}{\partial t}\right)_1 = \left(\frac{\partial A_z}{\partial t}\right)_2, \tag{3.7}$$

and

$$\frac{1}{\mu_1 \mu_0}\left(\frac{\partial A_z}{\partial n}\right)_1 = \frac{1}{\mu_2 \mu_0}\left(\frac{\partial A_z}{\partial n}\right)_2. \tag{3.8}$$

Comparing equations (3.2) and (3.6), (3.3) and (3.8), and (3.4) and (3.7), we find the correspondence:

$$\text{Electrolytic tank} \left\{ \begin{array}{l} V \text{———} A_z \\ J \text{———} j_z \\ \sigma \text{———} \dfrac{1}{\mu\mu_0} \end{array} \right\} \text{Electromagnet.}$$

In practice, this correspondence enables us to interpret the equipotential curves which the electrolytic tank has provided as vector equipotentials of A_z, provided that pole-pieces (assumed in this case to be of infinite permeability) are represented by insulators ($\sigma = 0$), and coils by a line of current-sources of current density proportional to j_z lying parallel to Oz (see Fig. 15b).

From the A_z curves, it is simple to deduce $\boldsymbol{B}$, which is tangential to the equipotentials at every point, and whose magnitude is proportional to grad A_z. The triad: grad A_z, $\boldsymbol{B}$ and Oz is right-handed.

One further result is of interest—if the values of the equipotentials are equally spaced on the potential scale, the same flux of magnetic induction lies in each of the tubes defined by each pair of equipotential surfaces.

This method is not suitable for direct application to systems with axial symmetry. Nevertheless, it has been applied successfully to the study of very large magnets whose curvature can be neglected (those of synchrotrons, for example). The results provided by the tank accord well with those obtained by direct measurement (Germain, 1955a).

Attempts have been made to apply this technique to systems with axial symmetry, always assuming infinite permeability (Germain, 1955b). The only components of the current density $\boldsymbol{j}$ and the vector potential $\boldsymbol{A}$ of the induction $\boldsymbol{B}$ in this case are of the form j_θ and A_θ . The relation

$$\operatorname{curl} \boldsymbol{B} = \operatorname{curl} \operatorname{curl} \boldsymbol{A} = \mu_0 \mu_1 \boldsymbol{j}$$

reduces to a single scalar expression

$$\frac{\partial}{\partial z}\left[\frac{1}{r}\frac{\partial}{\partial z}(r A_\theta)\right] + \frac{\partial}{\partial r}\left[\frac{1}{r}\frac{\partial}{\partial r}(r A_\theta)\right] = -\mu_0 \mu_1 j_\theta \qquad (3.9)$$

in which μ_1 represents the relative permeability of the medium.

Apart from a factor of 2π, the quantity $(r A_\theta)$ represents the magnetic flux ψ which threads a turn of radius r:

$$\psi = 2\pi\, r A_\theta, \qquad (3.10)$$

so that (3.9) is equivalent to

$$\frac{\partial^2 \psi}{dz^2} - \frac{1}{r}\frac{\partial \psi}{\partial r} + \frac{d^2 \psi}{\partial r^2} = -2\pi\, r\, \mu_0 \mu_1 j_\theta .$$

At the frontier between two media of relative permeabilities μ_1 and μ_2, Ampère's theorem states that

$$\operatorname{curl} \boldsymbol{H} = 0,$$

and Gauss's theorem that

$$\operatorname{div} \boldsymbol{B} = 0,$$

which in this case leads to the relations:

$$\left[\frac{1}{\mu_0 \mu_1}\frac{\partial}{\partial n}(r\, A_\theta)\right]_1 = \left[\frac{1}{\mu_0 \mu_1}\frac{\partial}{\partial n}(r\, A_\theta)\right]_2 \qquad (3.11\,\text{a})$$

between the normal components, and

$$\left[\frac{\partial}{\partial t}(r\, A_\theta)\right]_1 = \left[\frac{\partial}{\partial t}(r\, A_\theta)\right]_2 \qquad (3.11\,\text{b})$$

between the tangential components.

We now consider a two-dimensional conducting medium, M, $(O\,x\,y)$, the conductivity, σ, of which is a function solely of y:

$$\sigma = \frac{k}{y}. \qquad (3.12)$$

Current sources giving a current density d are distributed across the conducting surface. The relations

$$\operatorname{div} \boldsymbol{i} = d,$$

$$\boldsymbol{i} = -\sigma \operatorname{grad} V$$

give

$$\frac{\partial}{\partial x}\left(\frac{1}{y}\frac{\partial V}{\partial x}\right) + \frac{\partial}{\partial y}\left(\frac{1}{y}\frac{\partial V}{\partial y}\right) = -\frac{d}{k}. \qquad (3.13)$$

The boundary conditions between two homogeneous media, of conducti-
vities $\sigma_1 = \dfrac{K_1}{y}$ and $\sigma_2 = \dfrac{K_2}{y}$ respectively, take the form:

$$\left(k_1 \frac{\partial V}{\partial n} \right)_1 = \left(k_2 \frac{\partial V}{\partial n} \right)_2 , \tag{3.14 a}$$

$$\left(\frac{\partial V}{\partial t} \right)_1 = \left(\frac{\partial V}{\partial t} \right)_2 \tag{3.14b}$$

(using equations (3.3) and (3.4)).

An examination first of equations (3.10) and (3.11) and then of (3.13) and (3.14) reveals that they are identical if we make the following corre-spondences:

$$
\begin{aligned}
r\,A_\theta &\rightarrow V, \\[4pt]
\mu\,\mu_0 &\rightarrow \frac{1}{k}, \\[4pt]
j_\theta &\rightarrow d, \\[4pt]
z &\rightarrow x, \\[4pt]
n &\rightarrow y.
\end{aligned}
\tag{3.15}
$$

Once we possess the conducting medium (M) we have simply to measure V by a standard method; the equipotential curves will represent the curves

$$r\,A_\theta = \text{constant}$$

or, what is equivalent, curves of equal flux, $\psi = \text{constant}$. At a pair of corresponding points of the medium (M) and the actual magnetic field, the electric field E and the induction B will be related in the following ways:

(i) $V = k_r\,A_\theta$ with $k > 0$,
(ii) E is normal to B (since $E \cdot B = 0$),
(iii) $|E|$ is proportional to $r\,|B| : |E| = k\,r\,|B|$,
(iv) The three vectors E, B and k form a right-handed triad (k is a unit vector normal to the plane $x0y$).

The medium (M) can be constructed in an approximate fashion by filling a tank of variable depth y with an electrolyte (more precisely, the bottom should be hyperbolic in form) provided that the depth does not change too rapidly; measurements which have been made in the region near the axis (where the liquid has to be very deep) cannot be regarded with any confidence (Germain, 1955b). Infinitely permeable pole-pieces are made of paraffin wax ($k = 0$).

In conclusion, we should mention that pole-pieces with a permeability which is not infinite can be represented in the conjugate tank analogy by replacing the electrodes which have zero conductivity by others made of some substance (paraffin wax mixed with powdered graphite, for example) which has a conductivity which is related to that of the electrolyte in the following way:

$$\frac{\sigma_{\text{electrode}}}{\sigma_{\text{electrolyte}}} = \frac{\mu_{\text{air}}}{\mu_{\text{iron}}}.$$

3.3.3 The Resistance Network

The resistance network is a far more flexible tool than the conjugate tank, and with its aid, the induction B can be obtained for systems with axial symmetry, with or without iron.

The simplest analogy is between a network and the medium (M) defined above; the various quantities are thus related in the following wav·

$$V \rightarrow r\, A_\theta = \frac{\psi}{2\pi}.$$

Networks of this kind have been described by Liebmann (1950b), and constructed by Wakefield (1958) and by Christensen (1959).

The resistances which are placed in rows parallel to the axis Ox are of the form $R_n = r R_0$ (in which R_0 is a constant). Along the axis, $R = 0$; the nodes are short-circuited. The resistances which are placed in columns parallel to the straight line Oy (perpendicular to Ox) on either side of each node P_n which lies a distance r_n from the axis Ox are given the values:

$$R_n = (r_n^2 - r_{n-1}^2)\frac{R_0}{2}, \qquad\qquad R_{n+1} = (r_{n+1}^2 - r_n^2)\frac{R_0}{2}.$$

The analogy (3.15) between magnet and network cannot be used directly, as we have both to change the geometrical scale, writing

$$\frac{y}{r} = \frac{x}{z} = s,$$

and to scale the currents which are fed into the network. We suppose that (NI) is the number of ampère-turns in one of the coils of the magnet, and i is the total current fed into the network in the zone which represents this coil, and we write

$$p = \frac{NI}{i}.$$

If we compare the current densities j in the real coil and d in the network, we find

$$p = \frac{1}{s^2}\frac{j}{d}.$$

If the analogy between equations (3.10) and (3.11) and equations (3.13) and (3.14) is to be complete, therefore, we must write:

(a) $\dfrac{\psi}{2\pi} = r\,A = \dfrac{k\,\mu\,p}{s}\,V,$

(b) $\qquad\qquad j = p\,s^2\,d,$

(c) $\qquad\qquad \mu\,\mu_0 = \dfrac{1}{k},$ $\qquad\qquad\qquad\qquad$ (3.16)

(d) $\qquad\qquad z = \dfrac{x}{s},$

(e) $\qquad\qquad r = \dfrac{y}{s}.$

$\underbrace{\qquad\qquad\qquad}_{\text{magnet}}\quad\underbrace{\qquad\qquad\qquad}_{\text{network}}$

Measurement of V, therefore, gives the value of ψ directly from equation (3.16a); it will be expressed in webers. The magnitude of the induction, $|B|$, can then be obtained as a function of the field E measured *on the network*; at a point associated with the value y,

$$|B| = \frac{k\,\mu\,p\,s}{y}\,|E| \quad \text{webers m}^{-2}.$$

The full calculation of the values which the resistances should be given both in the case when the scale is the same in the x and y directions and when an interesting region of the system is to be expanded are to be found in the article by Christensen.

Just as in the tank, coils are represented by currents which are fed into the network at the corresponding nodes. In complex systems, we take the sense of the currents in the coils with respect to the axis Ox into account by supplying the network from a current source with two polarities and earthing the axis of the network. An alternating current is usually employed; the detection device is commonly a valve voltmeter with a very high impedance, provided with a detector which indicates the polarity of the potential being measured with respect to earth (Wakefield, 1958).

Along the boundaries of the pole-pieces, of infinite permeability, $R = \infty$; the circuit is open, since the condition $\dfrac{\partial \psi}{\partial n} = 0$ must be satisfied.

Another type of network of a more complicated design exists with which the curves $A_\theta = $ constant can be obtained directly, just as they can in the conjugate tank when the system possesses a plane of symmetry (Liebmann, 1950b).

CHAPTER 4

THE OPTICAL PROPERTIES OF ELECTROSTATIC LENSES

4.1 IMAGE FORMATION

4.1.1 The Focusing Effect of the Radial Field

Having considered static electric and magnetic fields from a general point of view, we are now in a position to establish the properties which such fields must necessarily possess if they are to be capable of focusing into a point image all the electrons which emerge from a point object. In this chapter, only electrostatic lenses will be considered, and, as a simple case with which to begin, we select the three-electrode lens of Fig. 1. The

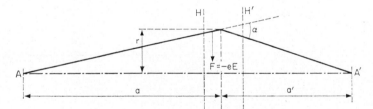

Fig. 16. Image-formation in a thin lens.

two outer grids are held at the accelerating potential of the electrons, while the central electrode is held at a quite different value. In this way, the electric field is localized between the two grids, and outside the lens the electrons move in straight lines. If all the electrons emitted by an (axial) point A are to pass through another (axial) point A', each ray must be deflected towards the axis through some angle α as it passes through the lens (Fig. 16). Since an electric field E produces a force $F = e\,E$ on each electron, the lens has to produce a radial field directed away from its centre. For a ray which reaches the lens at a distance r from the axis, the angular deviation α has to be proportional to r, thus

$$\alpha = r\left(\frac{1}{a} + \frac{1}{a'}\right). \tag{4.1}$$

85

As the deviation is proportional to the force, the intensity of the radial field must increase with the distance r from the axis according to a law of the form $E_r = K(z)\,r$, where $K(z)$ is either a constant or a function of z alone. In every lens, the field can be described by an expression of this kind, as we shall subsequently see.

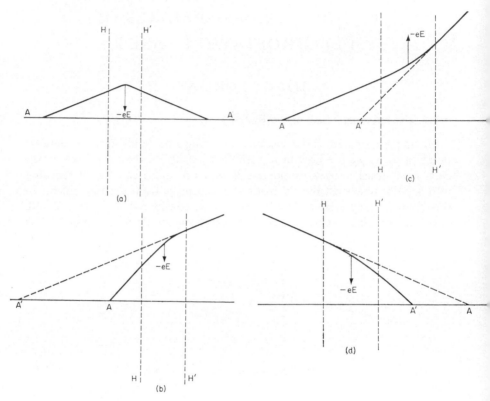

FIG. 17. (a) Convergent lens—real image. (b) Convergent lens—virtual image. (c) Divergent lens—virtual image. (d) Convergent lens—the real image of a virtual object.

The consequence is that for every ray, the ratio of α to r is a constant, $1/f$, such that the object and image distances satisfy the elementary lens relation

$$\frac{1}{a} + \frac{1}{a'} = \frac{1}{f}. \tag{4.1'}$$

If, therefore, a system of electrodes produces a *radial field, whose intensity at any point is proportional to the distance of the point from the axis*, it will focus charged particles just as an optical lens focuses light rays. The rays will converge onto a point such as A', if the radial force is directed towards the axis, and if the lens is powerful enough; if both of these conditions are satisfied simultaneously, we shall have a converging lens producing a real image, as in Fig. 17a. In such a lens, however, it would be possible to reduce the field to such an extent that it became no longer capable of turning the rays which have originated at A through an angle sufficiently large for them to intersect the axis again at A', however distant. Nevertheless, if we produce backwards the rays which diverge rectilinearly beyond H' (a dotted line represents the produced section of an actual ray), the "virtual rays" thus created will converge into a point A'. The actual rays seem to have originated in A', which is called the "virtual" image of A (Fig. 17b). The same convention makes it possible to speak of "imaging" when the radial force repels the particles from the axis; the lens is called "divergent", in this case, and a real object always has a virtual image (see Fig. 17c). Figure 17d shows how this concept of virtuality can be extended to the object; a lens would have a virtual object if, for example, it were placed between another lens and the (real) image which this lens has produced.

For each field intensity, therefore, and for each particle velocity, an image is produced; these are the two factors which describe the action of the lens, or, optically speaking, its convergence. The slower the electrons and the stronger the radial field, the more convergent will be the lens—this must not be carried too far, however, as there is a point at which the lens turns into a mirror, as we shall see later.

4.1.2 The Impossibility of Separating the Longitudinal and Radial Field Components

The electrodes of such lenses as the three-electrode lens produce a satisfactory field configuration—at least in the axial region—but with a structure less simple than we have supposed in the previous section. In fact, *each radial component is associated with a longitudinal component of electric field, parallel to the axis.*

The map of lines of equal field strength shows that the radial component is appreciable only where the lines of force have a large curvature, and where, consequently, the longitudinal component of the field is changing rapidly over short distances. The laws which govern the electric field provide a relation between the radial and longitudinal field components in the neighbourhood of the axis; the flux across the surface of a small cylinder, whose axis coincides with that of the lens, is zero, so that for a field of axial symmetry,

$$E_r = -\frac{1}{2} r \frac{\partial E_z}{\partial z} = \frac{1}{2} r \frac{d^2\varphi}{dz^2}, \qquad (4.2)$$

provided that second and higher order terms are negligible within the cylinder. We see that the radial component is a function of the rate at which the longitudinal component of the electric field is varying with z, or of the axial potential $\varphi(z)$. E_r depends upon r in the requisite fashion, and the factor $K(z)$ which was introduced earlier is proportional to the gradient of the longitudinal field component. This suggests that the extent of the lens should be defined as the region over which the derivative of E_z is still appreciable.

The effect of the longitudinal component is most important, as it retards or accelerates the electrons and thus has an indirect influence over the deviation of the beam which the radial component can produce; in short, it affects the convergence.

4.1.3 Ordinary Lenses without Grids

The longitudinal component has a far more pronounced effect in lenses without grids, which are much more common in practice, as lenses with grids are unsuitable for a number of reasons. Firstly, it is difficult

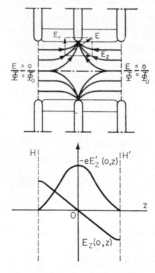

FIG. 18. The behaviour of the lines of force, the axial field, and the gradient of the axial field in a lens with two grids.

to construct a sufficiently transparent grid, combining extreme slimness of the meshes with mechanical rigidity; further, secondary electrons are produced at the wires, and their superfluous illumination at the final screen reduces the contrast of the image proper. In addition, the grid produces a slight haziness at the image.

So far, we have been assuming that the planes H and H' are effectively continuous conductors, at which lines of force will terminate normally (Fig. 18). In reality, when H and H' are grids, each line of force has a small kink (Fig. 19) just before it is terminated; an estimate of the influence of this deformation can be made by comparing the variation of potential along a line of force with the potential difference along the bend near the grid. Small though this effect is, its effect is almost invariably harmful, however much we try to diminish it by reducing the mesh-length of the grid.

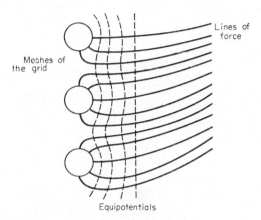

FIG. 19. The distortion of the field which the wires of the grids produce.

Subsequently, therefore, we shall consider only normal (grid-free) lenses, a typical example of which is the three-electrode lens, in which the grids are replaced by circular holes cut in plane electrodes ("Einzellinse" in German).

The longitudinal component now plays a considerably more important role. To make this more readily comprehensible, consider the symmetrical two-electrode lens of Fig. 20a.

In passing from the first half of the lens to the second, the sense of the radial field is reversed: if the field had a radial component only, it would have no overall effect. There is however, a longitudinal component which accelerates the electrons as they pass through the lens, which are as a result moving faster in the second half of the lens than in the first half. In the first half, therefore, the converging power is increased, and in the second, the diverging power diminished—the whole lens is convergent. In the three electrode lens, the action can be analysed similarly (see Fig. 20b) into an assembly consisting of a convergent unit between two divergent units.

Many of the features of ordinary electron lenses can be explained in terms of *the influence of this longitudinal component*. Lenses are, for example, *always convergent*. In order to explain this, we consider the shape of the lines of force of the electric field, which link the positive charges of one electrode with the negative charges of an adjacent electrode. The hole in the middle of each electrode make the lines of force curved, convex towards the axis, a consequence of their mutual repulsion. A region of convergence must always, therefore, be accompanied by a region in which the lens action is divergent; but electrons move more slowly in a convergent region, which slows them down, than in a divergent region, which accelerates them, so that the convergent effect always exceeds the divergent. A three-electrode lens remains convergent even though the central electrode be at a positive

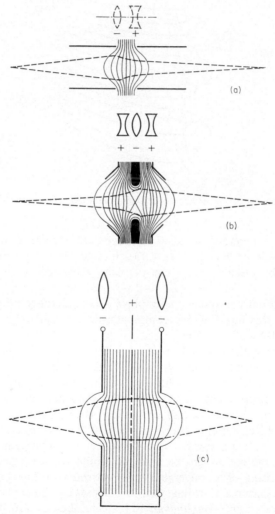

FIG. 20. Some electron lenses, with their glass optical counterparts. (a) A two-cylinder lens, which corresponds to a doublet. (b) A three-electrode lens, equivalent to a symmetrical triplet. (c) The practical example of a lens with a central mesh electrode, of which the analogue is a pair of convergent lenses.

potential with respect to the outer diaphragms. In the special case of a plane cathode near an accelerating diaphragm, however, the rule is no longer applicable, as the convergent region has disappeared—it is the only grid-free lens which is divergent, No application of these lenses, studied by Davisson and Calbick (1931, 1932) and MacNaughton (1952) has as yet been found.

Another feature of electron lenses which is explicable in terms of the effect of the longitudinal component is the fact that it is always found necessary to apply a potential difference across the two electrodes which is of the same order of magnitude as the beam accelerating potential, despite the differential retardation. This is quite different from the grid lens where the action of the field is everywhere uniform, and where only one tenth of the potential is necessary to produce the same convergence. Figure 20c shows a simple design for a lens with a single grid possessing the same optical properties as those of the version with two grids considered earlier. More details of these lenses, which are better than ordinary lenses in this respect, will be found in § 8.4.

4.1.4 The Electron Mirror; Minimum Focal Distance

We consider first a lens of the usual kind, with electrodes symmetrical about an axis $z'Oz$. The field in such a lens is then completely described by a set of equipotentials in a meridian plane, $\Phi(r, z) = $ constant. Our interest in Φ is caused by the fact that the velocity of the electrons (each of mass m and carrying charge $-e$) is given (in the non-relativistic case) by

$$v = \sqrt{\frac{2e\,\Phi}{m}} \qquad (4.3)$$

provided the value of the potential at the cathode is chosen as origin ($V_c = 0$) since the electrons are emitted with a negligible velocity. Then

$$e/m = 1 \cdot 759 \times 10^{11} \, \text{C kg}^{-1},$$

$$\sqrt{\frac{2e}{m}} = 0 \cdot 5932 \times 10^6 \, \text{MKS units.}$$

The basic relation (4.3) suggests immediately how an electron lens can be converted into an electron mirror—all that we have to do is to apply a potential to the central electrode of a three electrode lens which is not only negative with respect to the neighbouring electrodes, but also with respect to the cathode. The field which ensues will produce a lens action just so long as the minimum potential which the electrons encounter, and which is situated in the middle of the central electrode ($r = z = 0$), is positive (if $\Phi(0, 0) > 0$, that is). A complete examination is to be found in § 8.2.

When the negative potential applied to the central electrode is numerically sufficiently large, however, the electrons are decelerated and begin to return towards their source without having reached the centre of the lens. This situation is called an "electron mirror"; its optical properties are analogous to those of glass optical mirrors.

When $\Phi(0, 0)$ is negative and very large, the electrons are turned about before reaching the central region of the lens where the field is convergent, and the mirror is divergent; if, on the other hand, $\Phi(0, 0)$ is negative but only slightly so, the electrons penetrate deep into the lens, and before they are reflected, they will have experienced the converging action of the field—the mirror is convergent.

If $\Phi(0, 0)$ is negative but very close to zero (within a few tenths of a volt) oscillations appear in the electron trajectories. This phenomenon is intermediate between the lens and mirror regions. The potential zone which produces oscillating rays (which are known as "transgaussian" rays) is virtually useless for image formation, though it has a few rather specialized applications.

$\Phi(0, 0)$, therefore, cannot be reduced below a certain limit, and a very high convergence is not, as a result, attainable. In practice, the least focal length that one can obtain is about as large as the radius of the central electrode when the inter-electrode spacing is also of the order of this radius, which is frequently the case.

A similar phenomenon appears when $\Phi(0, 0)$ approaches zero from the positive side—the convergence of an electron lens too, therefore, is bounded.

Should the electron beam cross-sectional area not be small relative to the area of the holes, it is possible that peripheral rays will be reflected, while axial rays experience only lens action—the trajectories, in such a case, become most involved.

4.1.5 The Lens Considered as an Assembly of Prisms. Similitude Properties

Just as a glass lens can be decomposed into an assembly of prisms, so can an electron lens, and although the method is more artificial than those which we have so far used, the analogy provides suggestive results. The basic electron prism is the parallel plate condenser, the standard type of deflector in cathode ray oscillographs. This deviating property suggests a simple description of the action of the radial component E_r of the electric field. Each volume element of the field is to be thought of as a small condenser, with plates distance d apart, of length l parallel to the axis, supporting a potential difference $U = E_r d$, and with a mean potential of Φ (see Fig. 21). This elementary condenser deflects the ray through a small angle α given by

$$\tan \alpha = \frac{1}{2} \frac{Ul}{\Phi d} \simeq \alpha. \tag{4.4}$$

We can draw a number of important conclusions from this law, which hold for both thick and thin lenses. Firstly, since neither the charge nor the mass of the particles appears in the deviation law, the ray path is independent of e and m, and the lens would focus no differently if an ion source replaced the electron source. This property provides the possibility of focusing ions, which is a most important operation in particle accelerators, and was brilliantly introduced into mass spectrography in France by Cartan (1937). Ion sources are also beginning to be introduced into microscopy (Gauzit, 1951, 1953, 1954; Magnan and Chanson, 1951).

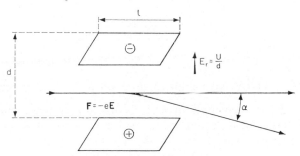

FIG. 21. The electrostatic prism.

It is worth mentioning in passing that electrostatic ion focusing can be hazardous in tubes which contain thermo- or photo-emissive cathodes, or luminous films, since an excessive bombardment by returning ions can deteriorate or destroy the delicate films (Broadway and Pierce, 1939; Schaeffer and Walcher, 1943). The positive ions which are produced all along the beam by the impacts of the electrons on the residual gas molecules retrace the electron paths and strike the cathode; negative ions, either produced by the cathode or originating in the residual gas, flow towards the fluorescent screen. In cathode ray tubes, these latter are caught in an ion trap, which consists of a pair of crossed deflecting fields, one magnetic, the other electrostatic, which act for a short distance on the beam. Their intensity is so adjusted that on electrons of mass m, their effects just cancel out; on ions with a mass M very different from m ($M \simeq 1840$ AM, where A is the atomic number of the ion), the two fields have a large residual deflecting effect, and the ions are carried away either to an electrode or towards the wall of the tube.

Another consequence of equation (4.4) is that the trajectory is independent of the absolute dimensions of the lens: only the ratio l/d matters. Similarly, only the ratio U/Φ, the ratio of electrode potential to accelerating potential, intervenes in (4.4) and not the absolute value of either potential. The action of an electrostatic lens will not be affected by small variations of U and Φ, which can even be oscillatory provided their ratio remains unaltered. It seems as though it should be possible to connect the electrodes to a potentiometer fed by an unstable or even oscillatory supply. In certain of the cruder applications of electron optics, high tension X-ray tubes, for example,

advantage is taken of this property by supplying the tube with the alternating
voltage from a transformer. In more delicate applications, on the other hand,
such as the electron microscope, only an extremely carefully stabilized high
tension can be used, not varying by more than one per cent; so fine are the
images which such an instrument produces that a defect due to relativity
would otherwise appear. This is explained by the fact that for very fast electrons, equation (4.4) requires a relativistic correction, and becomes:

$$\alpha \simeq \frac{Ul}{2\Phi d} \frac{1 + (e/m_0 c^2)\Phi}{1 + (e/2m_0 c^2)\Phi} \qquad (4.5)$$

(where c is the velocity of light and m_0 is the rest mass of the electron).

Equation (4.4) ceases to be linear as soon as $\dfrac{e\Phi}{2m_0 c^2}$, which is equal to

$0.977 \times 10^{-3}\,\Phi$ (Φ measured in kV), is no longer negligible (Ramberg,
1942). For the optical properties to remain constant it is no longer sufficient that such ratios as Φ_1/Φ_0, Φ_2/Φ_0 should be invariant, but the actual
potentials Φ_1, Φ_2 too must be narrowly stabilized; otherwise, the image
will be slightly hazy as a result of the slight variations of these potentials
about their mean values. In fact, however, this relativistic effect is most
difficult to observe, masked as it is by more ordinary defects caused by
imperfections in the source of high tension; for example, the capacities
at the high tension entry and between the electrodes may be appreciable,
and, by destroying the validity of the proportionality law in the presence
of a varying potential, may well dwarf the relativistic effect.

4.2 THE GAUSSIAN APPROXIMATION

4.2.1 Gaussian Conditions

We have seen that the necessary conditions for a good image to be
formed by an electron optical system obtain only within a tubular surface
of revolution enclosing the axis of the system rather closely. This is the
situation in which we can neglect terms in r and dr/dz of orders higher
than the first in the differential equations of the electron trajectories. This
first order approximation is known as "Gaussian optics", and the approximate trajectories which are obtained are called "Gaussian trajectories".
  Equation (4.2) gives the lower order terms in the series expansion for
E_r and E_z, which we shall write for convenience

$$E_z = -\varphi'(z), \qquad E_r = \frac{1}{2} r\, \varphi''(z); \qquad (4.6)$$

$\varphi(z)$, $\varphi'(z)$ and $\varphi''(z)$ represent the value, $\Phi(0, z)$, of the potential on the
axis, and its derivatives with respect to z.

4.2.2 The Equations of Motion

Let us consider now the detailed calculation of trajectories to the Gaussian approximation for non-relativistic electrons. Applying Newton's law in the radial direction we obtain

$$f_r = m \frac{\mathrm{d}^2 r}{\mathrm{d} t^2} = -\frac{1}{2} e\, r\, \varphi''(z) \tag{4.7}$$

(remembering that the charge on the electron is $-e$).

The radial force has a focusing action when $\varphi''(z) > 0$ and a defocusing action when $\varphi''(z) < 0$. If the curve which represents $\varphi(z)$ is plotted with the cathode potential as the origin of coordinates ($\varphi = 0$), the lens will be locally convergent at a point with abscissa z if the curve is concave upwards at this point, and divergent if the curve is concave downwards (Fig. 22).

For the motion in the z direction, we obtain a first order differential equation directly from the conservation of energy:

$$\frac{1}{2} m \left(\frac{\mathrm{d} z}{\mathrm{d} t} \right)^2 = e\varphi. \tag{4.8}$$

With the aid of equations (4.7) and (4.8), we can study such electron trajectories as remain within a meridian plane; this is the case for all rays originating from an axial object point, and for a particular class of rays originating in an off-axial object point. For this case, we find, eliminating the time variable between the equations,

$$\sqrt{\varphi} \cdot \frac{\mathrm{d}}{\mathrm{d} z} \left(\sqrt{\varphi} \cdot \frac{\mathrm{d} r}{\mathrm{d} z} \right) = -\frac{1}{4} r\, \varphi'', \tag{4.9}$$

or

$$\frac{\mathrm{d}^2 r}{\mathrm{d} z^2} + \frac{\varphi'}{2\varphi} \frac{\mathrm{d} r}{\mathrm{d} z} + \frac{\varphi''}{4\varphi} r = 0. \tag{4.10}$$

Skew trajectories are, in fact, no problem as it is simple to transform their differential equations into an equation of the form (4.9) or (4.10). If Ox and Oy are mutually perpendicular axes in a plane perpendicular to Oz, we can describe quantities which depend linearly upon x and y in a condensed fashion by introducing the complex variable $u = x + j y$:

$$\frac{\mathrm{d} u}{\mathrm{d} t} = \frac{\mathrm{d} x}{\mathrm{d} t} + j \frac{\mathrm{d} y}{\mathrm{d} t},$$

$$\frac{\mathrm{d}^2 u}{\mathrm{d} t^2} = \frac{\mathrm{d}^2 x}{\mathrm{d} t^2} + j \frac{\mathrm{d}^2 y}{\mathrm{d} t^2}, \tag{4.11}$$

$$E_u = E_x + j E_y = -\frac{1}{2} u \frac{\partial E_z}{\partial z}.$$

The equation which relates u and z in space relates r and z in a meridian plane. Writing u for r and interpreting the results with the aid of equations (4.11), oblique trajectories are found to have the same optical properties as the meridian trajectories to which we had previously restricted ourselves.

The similarity properties are immediately obvious if we notice that the equations are homogeneous and of zero degree with respect to φ, and that neither e nor m appears (we shall consider the effect of relativity upon the homogeneity later).

We now return to the Gaussian equation in the form (4.9), and integrate between object space (characterized by the index a) and image space (index b). We find

$$\left(\sqrt{\varphi}\,\frac{\mathrm{d}r}{\mathrm{d}z}\right)_b - \left(\sqrt{\varphi}\,\frac{\mathrm{d}r}{\mathrm{d}z}\right)_a = -\frac{1}{4}\int_a r\,\frac{\varphi''}{\sqrt{\varphi}}\,\mathrm{d}z.$$

For an incident trajectory parallel to the axis, $r_a' = 0$, and hence

$$r_b' = -\frac{1}{4\sqrt{\varphi_b}}\int_a^b r\,\frac{\varphi''}{\sqrt{\varphi}}\,\mathrm{d}z.$$

The image focal length, f_b, is given by

$$\frac{1}{f_b} = -\frac{r_b'}{r_a},$$

and hence

$$\frac{r_a}{f_b} = \frac{1}{4\sqrt{\varphi_b}}\int_a^b r\,\frac{\varphi''}{\sqrt{\varphi}}\,\mathrm{d}z. \tag{4.12a}$$

In a weak lens, r varies only slightly through the lens, and we can write approximately

$$r = r_a.$$

For a thin lens, therefore, we obtain the following expression for the image focal length, f_b:

$$\frac{1}{f_b} = \frac{1}{4\sqrt{\varphi_b}}\int_a^b \frac{\varphi''}{\sqrt{\varphi}}\,\mathrm{d}z. \tag{4.12b}$$

If φ' is zero on both sides of the lens, the second derivative can be eliminated on integrating by parts, and we obtain

$$\frac{1}{f_b} = \frac{1}{8\sqrt{\varphi_b}}\int_a^b \frac{\varphi'^2}{\varphi^{3/2}}\,\mathrm{d}z. \tag{4.12c}$$

4.2.3 The Reduced Equation

The Gaussian equation is difficult to use due to the presence of φ'', as for most lenses the only information available about the potential distribution has been obtained from an electrolytic tank or a resistance network and a second derivative determined graphically is far too inaccurate. Instead, therefore, it is often considered preferable to transform the Gaussian equation (Picht, 1939) by writing

$$R = r\,\varphi^{1/4}. \tag{4.13}$$

The curve $R(z)$ is called a "reduced" ray. Equation (4.10) then becomes

$$\frac{\mathrm{d}^2 R}{\mathrm{d}z^2} + \frac{3}{16}\left(\frac{\varphi'}{\varphi}\right)^2 R = 0, \tag{4.14}$$

which, apart from the absence of an R' term, displays the interesting feature of possessing a single expression characteristic of the lens. All the Gaussian optical properties of the lens are determined by the characteristic function

$$T(z) = \frac{\varphi'}{\varphi}.$$

For a weak lens, we can treat R in the same way as earlier we treated r. A ray incident parallel to the axis corresponds to $R_a =$ constant and $R'_a = 0$. The "reduced focal length", F, is defined in terms of the quotient R'_b/R_a thus:

$$\frac{1}{F} = -\frac{R'_b}{R_a}.$$

If we assume that $R \simeq$ constant $\simeq R_a$ throughout the lens, we find

$$\frac{1}{F} = \frac{3}{16}\int_a^b \left(\frac{\varphi'}{\varphi}\right)^2 \mathrm{d}z. \tag{4.15a}$$

If the potentials are constant on both sides of the lens, we have

$$R'_b = r'_b\,\varphi_b^{1/4},$$

$$f_b = \frac{r'_b}{r_a} = \left(\frac{\varphi_b}{\varphi_a}\right)^{1/4} F_b,$$

and hence

$$\frac{1}{f_b} = \frac{3}{16}\left(\frac{\varphi_a}{\varphi_b}\right)^{1/4} \int_a^b \left(\frac{\varphi'}{\varphi}\right)^2 \mathrm{d}z, \tag{4.15b}$$

which reduces, if $\varphi_a = \varphi_b$, to the following expression:

$$\frac{1}{f_b} = \frac{3}{16}\int_a^b \left(\frac{\varphi'}{\varphi}\right)^2 \mathrm{d}z. \tag{4.15c}$$

Formula (4.15c) differs from formula (4.12c) even though *a priori* equivalent approximations have been made; in practice, (4.12c) is found very often to lead to incorrect results, whereas (4.15b) yields values of f_b which are very close to those which are obtained by integrating the equations of motion numerically. This somewhat subtle point has been discussed by Sturrock at the London Conference on Electron Optics (1954) and more recently by Felici (1959), who gives the limits within which these formulae are in practice valid.

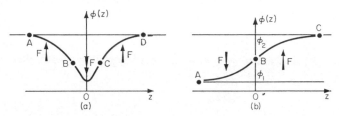

FIG. 22. The relation between the form of $\varphi(z)$ and the force F which acts on an electron. (a) "Einzel" lens. (b) Accelerating lens.

The approximation which leads to formula (4.12c) is valid only when $\varphi''(z)$ has the same sign throughout the region of integration, and this is never the case with ordinary lenses—only lenses with a central grid satisfy this requirement (§ 8.4). If, however, we break down the curve $\varphi''(z)$ into regions where $\varphi'' > 0$ and regions where $\varphi'' < 0$, we can find a thin lens equivalent to each region and hence calculate the focal length. The thin lens will lie at the centre of gravity of the curve (that is, the point such that the areas on either side are the same). By applying the rules obeyed by combinations of thin lenses, we can determine the elements of the real lens. This method has been used mostly for two-cylinder lenses which reduce to a convergent lens followed by a divergent lens when the complete lens is accelerating (Spangenberg and Field, 1942).

The same method of breaking down into thin lenses can be applied to the function $T^2(z) = \left(\dfrac{\varphi'}{\varphi}\right)^2$. We first obtain the centre of gravity, G, of this function and then the centres of gravity of the two regions of the curve which lie on either side of G; at the latter we place two thin lenses L_1 and L_2, with the focal lengths given by (4.15b). The real lens is equivalent to the combination of L_1 and L_2, to within a few per cent, even for very strong lenses (Felici, 1959). The various integrations can be performed graphically and the method is particularly rapid in the symmetrical einzel lens case.

4.2.4 The Immersion Objective

A new difficulty appears when the object itself is a source of thermal electrons, or of electrons produced either photo-electrically or by secondary emission, for these particles are emitted in all directions and the conditions under which the Gaussian approximation is valid seem no longer to be satisfied. The problem is only apparent, however, as the initial velocities are very low (corresponding to energies of the order of a few tenths of an electron volt) and the field is initially normal to the surface (which is equivalent to being parallel to the axis), very intense and almost uniform. In such conditions, the initial trajectory is parabolic in shape, and the velocity changes direction very rapidly along the curve until it becomes nearly parallel to the axis in a plane $z = z_1$, a few tenths of a millimetre from the

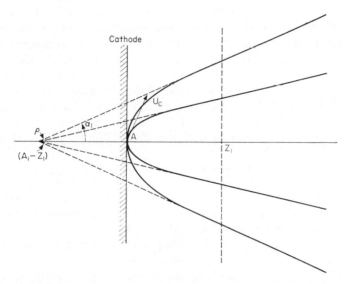

FIG. 23. The paths which the electrons follow as they leave the cathode. (U_c is one of the possible initial directions.)

cathode. At this plane, rays which have in fact originated in a point on the cathode such as A (Fig. 23) appear to be coming in a straight line from a small disc around A_1, situated at $z = -z_1$; the radius of this disc is $\varrho = \dfrac{2\varphi_c}{E_c}$, in which φ_c is the voltage equivalent to the most probable initial electron velocity ($\varphi_c = T/11600$ where T is the absolute cathode temperature) and E_c is the field on the cathode (Langmuir, 1937).

The radius of this disc is usually of negligible extent, lying as it does between $2\,\mu$m and $0.02\,\mu$m. The angular aperture of the beam corresponding to each point on the cathode is also small, of the order of some hundredths of a radian: $2\alpha = 2\sqrt{\dfrac{\Phi_c}{E_c z_1}}$. Beyond $z = z_1$, the conditions are totally Gaussian, and the lens needs no exceptional treatment. Experimental work has confirmed this theory, and electron or ion emission microscopes working at 30 kV give finer images than the best optical microscope (Mecklenburg,

1942; Kinder, 1944; Septier and Gauzit, 1950; Couchet, Gauzit and Septier, 1951, 1952, 1954; Septier, 1953; Möllenstedt and Hubig, 1954; Fert, 1956; see also Chapter 13).

4.2.5 Electron Mirrors

The usual reasoning seems to be inapplicable here as the region in which the electrons are reflected is a region in which the momentum is perpendicular to the axis, so that although dr/dz is initially finite, it becomes infinite at the point of reflexion. In any event, it certainly does not remain infinitesimally small while the electrons are within the lens. Since it was as a conseqence of this hypothesis that we were able to establish equation (4.9), it is to be feared that neither (4.9) nor the reduced equation which has been derived from it is capable of describing the behaviour of electrons in a mirror. However, only the reduced equation proves to be invalidated, while, as a careful examination of the higher order terms shows, equation (4.9) remains sound.

There are two alternative ways of studying mirrors. In the first, the system is decomposed into two regions, a narrow zone around the reflexion equipotential, where the field is regarded as uniform and the trajectories as parabolae, and a large zone consisting of the remainder of the system where the Gaussian conditions are satisfied; the usual equations can then be used with impunity, and by joining up the Gaussian and parabolic trajectories continuously, we can enumerate the optical properties.

The alternative way of studying electron mirrors is to choose another parameter to describe the trajectory. Up till now, we have used as abscissa the distance z, which increases monotonically with time within the lens. In mirrors, on the contrary, z is no longer a monotonic function of time, but it would be advantageous to choose one that is, and to express z and r in terms of this parameter. The time itself could be selected for this parameter, or, as Bernard has suggested (1952a), we could make the substitution $z = s^2$, with the origin of z at the point of reflexion so that the incident ray is characterized by $-\infty < s < +0$. The quantity dr/ds remains small, and never goes to infinity. An equation giving r as a function of s is easily derived, with solutions which are parametric representations of the trajectories.

Again, it is possible to obtain a reduced equation, analogous to the Picht equation, from which the optical properties of the mirror prove to be wholly defined by the single function:

$$\mathfrak{T} = \left(\frac{1}{\varphi} \frac{d\varphi}{ds} \right)^2 - \frac{4}{s^2},$$

where $\varphi(s)$ is the axial potential, with origin at the point of reflexion.

Electron mirrors are rarely used in practice, interesting though some of their properties are as we shall show later, since it is difficult to insert a receiver (fluorescent screen or photographic plate) into the path of the reflected beam which is in practice mingled with the incident beam. Mahl (1943), using electron mirrors in a microscope, obtained images comparable in quality with the images given by an ordinary lens.

The analysis when the point of reflexion is close to the centre of a symmetrical lens is more difficult, as the field falls to zero with the potential. This is a very rare situation in practice, however, as the images which result are unstable.

4.3 OPTICAL FORMALISM

4.3.1 Image Formation

For every thick lens, the ability to form an image is a general consequence of the linearity of equation (4.10).

If we know the trajectories $r_1 = X(z)$, $r_2 = Y(z)$ of two independent rays (rays, that is, between the equations of which no linear relation exists), we can deduce immediately from the theory of linear differential equations that the equation of every other ray is of the form

$$r = \lambda X(z) + \mu Y(z) \qquad (4.16)$$

where λ and μ are arbitrary constants.

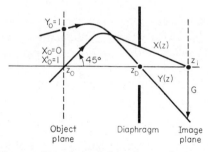

FIG. 24. The two characteristic rays of Gaussian optics (G is the magnification).

A priori, the trajectories r_1, r_2 are arbitrary, but they will reappear frequently, and it is of interest to select the two particular functions shown in Fig. 24: $X(z)$ intersects the axis at $z = z_0$ (where the object plane intersects the z-axis) with slope unity—i.e. $\left(\dfrac{dX}{dz}\right)_{z=z_0} = 1$; $Y(z)$ passes through the object plane at unit distance from the axis, which it subsequently intersects in the plane of the aperture diaphragm, the function of which is to restrict the lateral extent of the beam.

This choice of $Y(z)$ is standard in electron optical calculations, but is not, it must be noticed, the second fundamental trajectory habitually used in elementary optics for graphical constructions. There, a trajectory which leaves the object plane parallel to the axis is chosen.

We can make a number of straightforward deductions from equation (4.16). Every ray leaving a given point object A converges into a real point image A', after passing through the lens. If the rays do not converge, then we can produce back the rays which in fact emerge from the lens as straight lines (these lines are broken in the diagrams, to conform with the glass optical convention); mathematically, $X(z)$ and $Y(z)$ represent both the actual rays and their produced sections, and the "virtual" rays too, therefore, obey a linear relation similar to equation (4.16) (no problem arises from the double role which X and Y play), and converge into a virtual image point. Object point and image point are known as "conjugate points". The points which are conjugate to the points on the axis at plus or minus infinity are called the two "foci" of the lens.

From equation (4.16), we can also deduce information about the position and magnification of the image. The image of some object lying in a plane perpendicular to the axis is situated in the plane, perpendicular to the axis, containing the conjugate point (the frontal plane); and the image is similar to the object. For every pair of conjugate planes, the linear magnification of the image is a constant; the angular magnification, which is defined as the ratio of the angular aperture of an elementary pencil at the image to that of the corresponding pencil at the object, is also constant between conjugate planes.

We shall now establish an interesting relation between $X(z)$ and $Y(z)$, which will prove to be useful when we calculate the spherical aberration in § 7.3. $X(z)$ and $Y(z)$ are both solutions of equation (4.9), thus:

$$\frac{d}{dz}\left(\sqrt{\varphi}\frac{dX}{dz}\right) + \frac{1}{4}\frac{\varphi''}{\sqrt{\varphi}}X = 0,$$

$$\frac{d}{dz}\left(\sqrt{\varphi}\frac{dY}{dz}\right) + \frac{1}{4}\frac{\varphi''}{\sqrt{\varphi}}Y = 0.$$

We can eliminate the term which contains φ'' by multiplying the equations by $Y(z)$ and $X(z)$ respectively and subtracting:

$$Y\frac{d}{dz}\left(\sqrt{\varphi}\frac{dX}{dz}\right) - X\frac{d}{dz}\left(\sqrt{\varphi}\frac{dY}{dz}\right) = 0,$$

or

$$\frac{d}{dz}\left[\sqrt{\varphi}\left(Y\frac{dX}{dz} - X\frac{dY}{dz}\right)\right] = 0, \qquad (4.16')$$

or, integrating,

$$\sqrt{\varphi}\left(Y\frac{dX}{dz} - X\frac{dY}{dz}\right) = \sqrt{\varphi_0}\left[Y_0\left(\frac{dX}{dz}\right)_0 - X_0\left(\frac{dY}{dz}\right)_0\right], \qquad (4.16'')$$

in which the index *zero* refers to the object point in question.

For the two particular trajectories which we envisage here, we have

$$X_0 \left(\frac{dY}{dz} \right)_0 = 0$$

and

$$Y_0 \left(\frac{dX}{dz} \right)_0 = 1,$$

so that finally,

$$Y \frac{dX}{dz} - X \frac{dY}{dz} = \sqrt{\frac{\varphi_0}{\varphi}}. \tag{4.16'''}$$

4.3.2 The Standard Cardinal Points

In classical electron optics, we postulate that the object and the image shall lie in regions where the field is zero and the trajectories are straight lines. Physically, this is a most useful approximation as the field has effectively vanished beyond a few lens-diameters from the lens, although it still exists mathematically. In general, in electron optics, lenses have a finite and well-defined thickness, which is the region throughout which E—and later, B^2—has an appreciable effect upon the electrons. Mathematically speaking, electron optics treats not the trajectories themselves, but only their asymptotes, the straight lines to which the rays tend. The classical method leaves us with no information about the curved parts of the trajectories which lie within the lens; nor may we consider objects or images lying within the lens, for in this region we can only deal with virtual objects and images formed by producing real rays in straight (dotted) lines. Provided, therefore, we are willing to sacrifice *real immersed* objects and images, the standard cardinal points provide a most fruitful method of characterizing the optical properties of a lens system. Nevertheless, immersed virtual images and objects are intrinsically interesting—the objective of an electron microscope, for example, produces an intermediate image which is usually located within the projector lens, and plays, for this lens, the role of a virtual object.

Let us now consider a ray $y(z)$ which is parallel to the axis in object space; in image space, it intersects the axis at the image focus. Another ray, $U(z)$, also parallel to the axis in object space, will then be defined by $r = U(z) = \mu\, y(z)$. Whatever the value of μ or U, therefore, $U(z_0)/U'(z_i) = $ constant. In optical notation, this is equivalent to $h_0/\alpha_i = f_i$, which defines the image side focal length f_i (Fig. 25). These equations express the fact that if the beginning and end of each ray U is produced, the points of intersection lie in a fixed plane, known as the image (side) principal plane; an object (side) principal plane is defined in an exactly similar way by considering the family of rays which are concurrent at the object focus and parallel to the axis in image space [$V(z)$].

Once the two principal planes have been located and the two focal lengths are known, the optical properties of the system are completely determined; with the aid of simple geometrical constructions and the formulae to be found in every elementary book on optics, we can obtain the emergent ray corresponding to any given incident ray, and the position and magni-

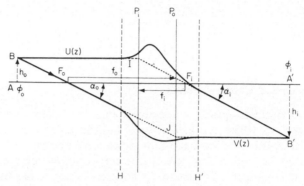

FIG. 25. The conventional principal planes and foci (H and H' are the approximate limits beyond which the lens field is negligible).

fication of the image of a given object. The relevant formulae are the following:

(i) With the origins at the principal planes,

$$\overrightarrow{P_0 F_0} = f_0, \qquad \overrightarrow{P_i F_i} = f_i, \qquad \overrightarrow{P_0 A} = p_1, \qquad \overrightarrow{P_i A'} = p',$$

$$\frac{f_0}{p} + \frac{f_i}{p'} = 1, \qquad \frac{h_i}{h_0} = -\frac{f_0}{f_i}\frac{p'}{p}, \qquad \frac{f_i}{f_0} = -\sqrt{\frac{\varphi_i}{\varphi_0}}. \tag{4.17}$$

(ii) With the origins at the foci,

$$\overrightarrow{F_0 P} = f_0, \qquad \overrightarrow{F_i P'} = f_i, \qquad \overrightarrow{F_0 A} = \pi, \qquad \overrightarrow{F_i A'} = \pi', \qquad \pi\pi' = f_i f_0,$$

$$\frac{h_i}{h_0} = \frac{f_0}{\pi} = \frac{\pi'}{f_i}, \qquad \frac{f_i}{f_0} = -\sqrt{\frac{\varphi_i}{\varphi_0}}. \tag{4.18}$$

A rather more subtle consequence of the differential effect which has already been described in the elementary explanation of lens convergence is that the principal planes are always crossed, as in Fig. 25. The reader will find a proof of this property, which follows straightforwardly from the structure of the reduced equation, in the book by Zworykin *et al.*

4.3.3 The Lagrange–Helmholtz Relation

This is the name given to the formula

$$\frac{f_i}{f_0} = -\sqrt{\frac{\varphi_i}{\varphi_0}}, \tag{4.19}$$

which relates the two focal distances. It is obtained by combining the differential equations (4.10) for the rays $U(z)$ and $V(z)$ symmetrically, thus eliminating the term in r, and integrating, which gives

$$U\frac{\mathrm{d}V}{\mathrm{d}z}\sqrt{\overline{\varphi}} - V\frac{\mathrm{d}U}{\mathrm{d}z}\sqrt{\overline{\varphi}} = \text{constant}. \tag{4.20}$$

Writing out this expression in the object and image planes in optical notation (Fig. 25), we find:

$$\sqrt{\overline{\varphi_0}} \cdot h_0\alpha_0 = \sqrt{\overline{\varphi_i}} \cdot h_i\alpha_i. \tag{4.21}$$

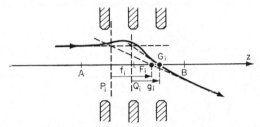

FIG. 26. Image cardinal elements (F and P are the normal cardinal elements, G and Q their immersion counterparts).

Equation (4.21) summarizes the fact that the product of angular magnification (α_i/α_0) and linear magnification (h_i/h_0) is equal to the constant $\sqrt{\overline{\varphi_0/\varphi_i}}$. Expressing the magnifications in terms of the cardinal elements, we obtain equation (4.19); two values of φ suffice to characterize object space and image space.

4.3.4 Objects and Images Lying within the Lens Field ("Immersed" Elements)

In glass optics, the curved parts of the rays are of no special interest, since they degenerate into sudden bends at the boundaries between each pair of media. The standard cardinal points are wholly adequate. In electron optics, on the other hand, it often happens that a lens is thick and sufficiently convergent to produce an image which possesses all the properties described in section 4.3.1 inside the lens. For example, a symmetrical lens may be perfectly capable of refracting a ray which is incident parallel to the axis so sharply that it intersects the axis inside the lens at some point G, which can be thought of as a real immersed focus (Fig. 26). This class of images and objects is of practical interest only for magnetic lenses; in electrostatic lenses, a specimen (the object), a fluorescent screen, or a photographic plate (as receiver) would perturb the electric field to such an extent that an image such as A_1B_1 (Fig. 27) could not be observed.

So long as we consider only the rectilinear asymptotes to the rays, the analysis remains simple, concerned as it is with linearly related straight lines which cut the axis in points which are defined mathematically by an elementary homographic relation. Generally speaking, however, no method of this kind can be devised which will allow us to consider effects produced by the curved part of the rays. No simple relation exists between the real immersed images (or objects) of the preceding paragraph and their corresponding objects (or images). If we wish to consider image formation actually inside the lens, we must resort to direct calculation of the two rays, starting from the differential equation. The Lagrange–Helmholtz relation (4.21) is still valid, but φ_0 and φ_i now represent φ at the particular pair of conjugate points in question.

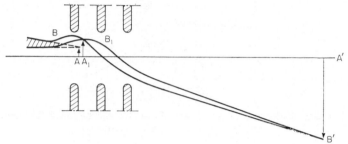

FIG. 27. The "classical" image $A'B'$ and the immersion image $A_1 B_1$ of the virtual object AB.

A class of fields exists, however, known as "Newtonian", for which relations of the classical form remain applicable, even for immersion objects and images (Glaser, 1941b; Glaser and Lammel, 1941; Hutter, 1945; Glaser, 1950b). The most interesting feature of this theory is that one of the members of the class provides rather a good approximation to the field within an ordinary magnetic lens. Whether the points lie within or without the lens, the relations remain valid, and the Newtonian expressions give the magnification and position of the image.

The relations can be represented geometrically by the usual constructions, provided that we define the foci as the two real points G_i and G_0 in which rays, initially parallel to the axis, effectively cross the axis for the first time, and that we define Q_i and Q_0, the principal planes, by $\overrightarrow{G_0 Q_0} = g_0$ and $\overrightarrow{G_i Q_i} = g_i$; the quantities g_0 and g_i are the focal lengths defined by

$$g_0 = \frac{r(+\infty)}{r'(G_0)}, \qquad g_i = \frac{r(-\infty)}{r'(G_i)}$$

(see Fig. 27). It is to be noticed, however, that these straight lines are not at all an indication of the path of the actual curved rays, and are not related in any direct way to the classical points F_i, F_0, P_i, and P_0. The Lagrange–Helmholtz relation (4.21) may no longer be put into the form (4.19).

4.3.5 Photometry; Brightness

Electron photometry can be developed from the Lagrange–Helmholtz relation in the form (4.21)—the angular magnification constant takes the same value for oblique rays as it does for meridian rays.

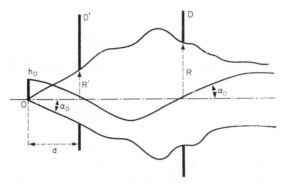

FIG. 28. The role of the diaphragms in electron optics.

A more suggestive form of the result is obtained by considering the area which the whole beam covers in a particular forward plane; this surface can be regarded as a diaphragm (of radius R) in a plane. If we introduce the image D' (of radius R') of D, which is defined by that part of the system which precedes D, the rays between D' and the object are straight lines; if d is the distance between the object and D', the value of α at the object is $\alpha_0 = R'/d$ (Fig. 28). We can consider the first term, $\sqrt{\varphi_0}\, h_0 \dfrac{R'}{d}$, of the Lagrange–Helmholtz relation as the product of $R' \sqrt{\varphi_0}$ and the aperture $\alpha' = h_0/d$ of an elementary beam in the D' plane:

$$h_0 \alpha_0 \sqrt{\varphi_0} = R' \alpha' \sqrt{\varphi_0}.$$

Applying the Lagrange–Helmholtz relation to the conjugate planes D and D', and recalling that D' lies in object space, we obtain

$$R' \alpha' \sqrt{\varphi_0} = R \alpha_D \sqrt{\varphi_D},$$

and thus

$$h_0 \alpha_0 \sqrt{\varphi_0} = R \alpha_D \sqrt{\varphi_D}. \tag{4.22}$$

The product $R \alpha_D \sqrt{\varphi_D}$ is constant irrespective of the choice of the plane D (R is the radius of the cross-section of the beam, and α_D is the angular aperture of an elementary beam in the plane D). The most interesting application of formula (4.22) is to electron sources (definitions are given in § 8.3). If we assume that the cathode is effectively plane, and of radius r_c, and take into account the features peculiar to immersion objectives already discussed ($h_0 = r_c$; $\alpha_0 = \sqrt{\varphi_c/E_c z_1}$; $\varphi_0 = E_c z_1$), we obtain an expression for the minimum possible radius R_s of the spot in a cathode ray tube working at final voltage φ_s and final angular aperture α_s, which is of the form

$$r_c \sqrt{\frac{T}{11,600}} = R_s \alpha_s \sqrt{\varphi_s} \qquad (4.23)$$

where we have substituted $\varphi_c = \dfrac{kT}{e} = \dfrac{T}{11,600}$.

In terms of specific emission, i_c, and current density at the screen, i_s (defined by $I_s = \pi r_c^2 i_c = \pi R_s^2 i_s$, I_s being the actual total current), we find

$$\frac{i_s}{i_c} = \left(\frac{11,600\,\varphi_s}{T}\right) \alpha_s^2, \qquad (4.24)$$

the celebrated formula of Langmuir (1937), perfected by Pierce (1939–45). Since α_s and r_c are related, however, (4.24) can be simplified. Let us consider, first of all, the image focal plane of an immersion objective, where the beam crosses the axis for the first time (at the "crossover"). Applying (4.23), we obtain

$$\alpha_f = \frac{r_c}{f}, \qquad r_f = f \sqrt{\frac{T}{11,600\,\varphi_f}} \qquad (4.25)$$

where f is the image focal distance, r_f the crossover radius, α_s the angular aperture of the beam at the focus, and φ_f the potential at this point.

If the second immersion lens divides the distance between the image focal plane and the screen into two parts, the ratio of whose lengths is m, then (see equation 4.17)

$$R_s = m\,r_f \sqrt{\frac{\varphi_f}{\varphi_s}} = m f \sqrt{\frac{T}{11,600\,\varphi_s}}. \qquad (4.26)$$

This expression is valid so long as r_c and α are small enough for spherical aberration to be negligible in the objective and in the second lens (otherwise, the Lagrange–Helmholtz relation loses its simple first order form). We can satisfy this requirement, at least partially, by choosing a suitable cathode design, or by stopping down the beam in the second lens.

Following Law (1937), a more detailed study would take into consideration the Maxwellian distribution of the emission velocities, and give us the current distribution in the spot. The result is that at a distance r from the centre, the density is given by

$$i(r) = \left(\frac{I_s}{\pi m^2 f^2}\right)\left(\frac{11,600\,\varphi_s}{T}\right) \exp\left\{\left(\frac{11,600\,\varphi_s}{T}\right)\left(-\frac{r^2}{m f^2}\right)\right\} \qquad (4.27)$$

or

$$i(r) = i_s\, e^{-(r/R_s)^2}.$$

The corresponding curve is bell-shaped, with a "natural width" of $2R_s$. In the crossover plane, we have a similar expression: $i(r) = i_f e^{-(r/r_f)^2}$. Experiment confirms this result perfectly, save for the absolute value of f which depends upon the space charge correction which is applied (Dosse, 1940; Jacob, 1939). The saturation current density i_c has been obtained in a number of guns with tungsten hairpin filaments; so strong is the influence of space charge on oxide-coated cathodes, however, that the emission is controlled by the quality of the vacuum within the sealed-off tube (Klemperer and Mayo, 1948). Finally, complex oscillatory phenomena may appear, which are only just beginning to be understood, which break up the beam into separate pencils (Klemperer, 1947; Ellis, 1947). Another example of the application of this geometrical photometry is the reduction of the size of the spot by "post-acceleration" (see equation 4.27); Pierce (1942) has furnished a complete treatment of this effect.

Just as in ordinary optics, we can define the brightness of an electron source as the electron intensity emitted per unit area of the surface into unit solid angle. In many electron optical instruments, the crossover can be regarded as the real source of the electrons. Its theoretical brightness in an infinitely small central zone of radius r_s is defined by

$$B_{th} = \frac{I}{\pi\, r_s^2\, \pi\alpha_s^2} = \frac{i_s}{\pi\,\alpha_s^2}, \qquad (4.27')$$

or using (4.24),

$$B_{th} = \frac{i_c}{\pi}\, \frac{\varphi_s}{\varphi_c},$$

or finally

$$B_{th} = i_c \frac{e\varphi_s}{\pi\, k\, T} = i_c \frac{11{,}600}{\pi\, T}\, \varphi_s. \qquad (4.27'')$$

As an example, suppose we have a tungsten filament, at a temperature of $2700°K$, and $i_c = 2\,A\,cm^{-2}$; if $\varphi_s = 30\,kV$, we find

$$B_{th} \sim 8 \times 10^4\, A\, cm^{-2}\, sterad^{-1}.$$

Similarly, we can define a mean brightness B_m for the whole area of the crossover; experiment, however, shows that B_m is considerably lower than B_{th}:

$$B_m \sim \frac{1}{10}\, B_{th}.$$

If we form successive images of the crossover with the aid of an aberration-free optical system in regions *in which the potential φ_s is the same, the theoretical brightness of the successive images remains constant* and equal to B_{th}; this is, in fact, a consequence of the Lagrange–Helmholtz relation.

To increase B_{th} for a given value φ_s, we can only increase i_c, the specific emission of the cathode, and reduce T. If an extremely bright source is required, we use only the central region of the crossover, rejecting the remainder with the aid of a diaphragm very small in diameter $\left(d \leq \frac{1}{10} \text{ mm}\right)$; the measured brightness is of the order of 80 or 90 per cent of B_{th}. Brightnesses of the order of 2×10^5 A cm^{-2} sterad^{-1} have recently been attained, with the aid of heated point cathodes (T.F. emission, see §17.1.2).

4.4 RELATIVISTIC LENSES

At high speeds, relativistic mechanics is necessary to describe the behaviour of the electrons. Although the equations are more complex, the general line of reasoning and the results are the same. Starting rather from the Einstein equations than from the Newtonian ones, the existence of first order focusing can be demonstrated, whatever the electron velocity. In effect, the differential equations remain linear.

If we write

$$\varepsilon(z) = \left(\frac{e}{2m_0 c^2}\right) \varphi(z)$$

with

$$\frac{e}{2m_0 c^2} = 0\cdot977 \times 10^{-3} \quad (\text{kV})^{-1},$$

the relativistic Gaussian equation has the form

$$\frac{1+\varepsilon}{1+2\varepsilon} \varphi \, r'' + \frac{1}{2} \varphi' \, r' + \frac{1}{4} \varphi'' \, r = 0. \tag{4.28}$$

The loss of homogeneity appears in the coefficient of the first term. If we substitute

$$R = r \, \varphi^{1/4}(1+\varepsilon)^{1/4}, \tag{4.29}$$

we obtain a reduced equation which resembles (4.14) save that the coefficient $\frac{3}{16}\left(\frac{\varphi'}{\varphi}\right)^2$ which appears is now multiplied by the factor $K(\varepsilon)$, defined by

$$K(\varepsilon) = \frac{\left(1 + \dfrac{4\varepsilon}{3} + \dfrac{4\varepsilon^2}{3}\right)}{(1+\varepsilon)^2}. \tag{4.30}$$

As a first approximation, we can neglect the variation of z in $\varepsilon(z)$, and replace the latter everywhere by a mean value $\bar{\varepsilon}$. Equation (4.30) shows that if this mean value rises, the correction factor K decreases from the value unity at low voltages to a minimum of $0\cdot88$ at 600 kV, then begins to rise, passing unity again at 2 MV and tending towards $4/3$ at still higher voltages. We ought, therefore, to be able to produce an initial increase of focal length followed by a slow decrease towards an asymptotic value by multiplying gun and lens potentials by the same (increasing) factor. Exact calculation confirms this qualitative prediction, as shown by Fig. 29 which represents the behaviour of a three-electrode unipotential lens (Laplume, 1947; Cotte, 1939).

The danger of breakdown prevents the use of very high potential differences. The systematic investigations of Arnal (1953) allow the optimum choice of potential to be made in a given electrostatic system. It seems

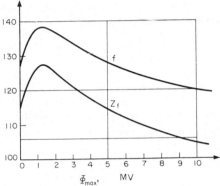

FIG. 29. The focal length and the position of the focus as a function of the maximum potential in the system; the other parameters are constant.

that adjacent electrodes will not support much more than 100 kV. In the megavolt region, therefore, grid lenses, with their lower potentials, become of interest (Gabor, 1947; Cartan, 1937; Knoll and Weichart, 1938; Bernard, 1951a, 1952a, 1953a).

4.5 PRACTICAL DETERMINATION OF ELECTRON TRAJECTORIES

4.5.1 Integration of the Ray Equation

The expression for the axial potential is rarely sufficiently simple for the Gaussian equation which results from it to be directly integrable. Worse, in most of the experimental arrangements of interest, it is not even possible to obtain an expression for the potential; only the experimental curves provided by an analogue method are available. On the other hand, it is often possible to find an expression which is both a very good approximation to the (calculated or measured) axial potential and sufficiently simple for the Gaussian equation to be soluble in terms of tabulated functions. This solves the problem, as we then have expressions for the trajectories.

In the chapter devoted to lens models, more details of these representations are to be found, and in particular, details of the procedures by which model and experiment may be matched. Here, we shall simply mention the works of: Rüdenberg (1948) and Regenstreif (1951 a), whose representation of the axial potential by a chain of parabolic segments allows the Gaussian equation (4.10) to be solved; of Hutter (1945) and Bernard and Grivet (1952a), who have expressed the trajectories in a two-electrode lens in terms of sinusoidal functions; and of Grivet (1952a) and Lenz (1951) who have obtained the trajectories in the same lens in terms of Legendre polynomials.

The latter methods effect an integration of the reduced equation (4.14).

4.5.2 Methods of Step-by-step Integration

In this section, the necessary practical information, with the aid of which the electron trajectories in a known potential distribution $\varphi(z)$ can be calculated with reasonable precision, is set out. Only the simplest methods are described; either the basic function $\varphi(z)$, or the reduced "characteristic" $T(z)$ can be used as a point of departure. $T(z)$, now defined by

$$T(z) = \frac{3}{16} \left(\frac{\varphi'}{\varphi} \right)^2 ,$$

is the function which appears in the reduced equation

$$R'' + T R = 0; \quad R = r \, \varphi^{1/4}.$$

(1) *The Störmer–Adams method*

This is a method of solving equations of the form

$$\frac{d^2 y}{d t^2} = f(x, y, t).$$

The first detailed discussions of the method were published by Störmer (1920) and Kryloff (1927); four points at the origin of the trajectory must be known, and as the trajectories of electron rays are rectilinear outside the lens, this method is most suitable for their calculation. It is less convenient for strongly convergent immersion objectives, where the beginning of the trajectory, close to the cathode, is a parabola.

Regenstreif (1947) has shown how the method may be adapted for electron mirrors. Time t is chosen as the parameter, and is measured in units of magnitude $\sqrt{m/2e}$. The Gaussian equations of motion are then of the form:

$$\frac{dz}{dt} = \varphi^{1/2},$$

$$\frac{d^2 r}{dt^2} = \frac{1}{2} \frac{\partial \Phi}{\partial r} = -\frac{r}{4} \varphi''.$$

Labelling the abscissa at the point of reflection z_i, the value of t at this point t_i, and the value of z at $t = 0$, z_0, we have

$$t(z) = \int_{z_0}^{z} \frac{dz}{\varphi^{1/2}} \quad \text{for} \quad t < t_i$$

and

$$t_i - t = \pm \int_{z}^{z_i} \frac{dz}{\{\varphi_i'(z - z_i)\}^{1/2}} = \pm 2 \left[\frac{z - z_i}{\varphi_i'} \right]^{1/2}.$$

In the neighbourhood of $t = t_i$, φ is replaced by an approximate expression, and

$$t = 2t_i - \int_{z_0}^{z} \frac{dz}{\varphi'^{1/2}} \quad \text{for} \quad t > t_i.$$

Having once obtained $t(z)$ and $\varphi(z)$ we can calculate $z(t)$ and $\varphi''(t)$; $r(t)$ is obtained by integrating

$$\frac{d^2 r}{dt^2} = -\frac{1}{4} r \varphi''(t)$$

by Störmer's method. $z(t)$ and $r(t)$ wholly define the trajectory.

(2) *The Runge–Kutta method (1901)* (see Scarborough, 1930)

This method requires only a knowledge of a single point on the trajectory and of the slope at this point. We use the reduced equation:

$$R'' = R T(z)$$

and an integration interval of length $\triangle z = h$. We simply quote the formulae with which the values of R and R' (the trajectory and its slope, in normalized coordinates) can be calculated at a point with abscissa $(z_i + h)$ when the values of these functions are known at the point with abscissa z_i, and also the values of the functions T, T_n and T_{n+1}, at z_i and z_{i+h} respectively (Lapeyre and Laudet, 1960). If an accuracy of one part in 10^3 is adequate, we use:

$$R_{n+1} = R_n \left\{ 1 - \frac{h^2}{6}(2T_n + T_{n+1}) \right\} + R_n' h \left\{ 1 - \frac{h^2}{6} T_{n+1} \right\},$$

$$R_{n+1}' = -R_n \frac{h}{2}(T_n + T_{n+1}) + R_n' \left\{ 1 - \frac{h^2}{2} T_n \right\}.$$

We can obtain an accuracy of better than one part in 10^6, however, if we use the following expressions which require the value of the function T

at the centre of the integration intervals to be known:

$$R_{n+1} = R_n \left\{ 1 - \frac{h^2}{6} \left(T_n + 2T_{n+1/2} - \frac{h^2}{4} T_n T_{n+1/2} \right) \right\}$$
$$+ R_n' h \left\{ 1 - \frac{h^2}{6} T_{n+1/2} \right\},$$

$$R_{n+1}' = - R_n \frac{h}{6} \left\{ T_n + 4T_{n+1/2} - \frac{h^2}{2} T_n T_{n+1/2} \right\}$$
$$+ R_n' \left\{ 1 - \frac{h^2}{3} T_{n+1/2} \right\} - \frac{h}{6} T_{n+1} R_{n+1}.$$

The accuracy of the method is discussed in an article by Lapeyre (1961). This method is particularly well adapted for calculations on magnetic lenses, which are described in Chapter 5, as the form of the equation of motion for a trajectory $r(z)$ resembles that of the Picht equation.

(3) Liebmann's method (1949)

The starting point here is the more complete equation (see §7.3.1):

$$r'' - \frac{1 + r'^2}{2U} \left(\frac{\partial U}{\partial r} - r' \frac{\partial U}{\partial z} \right) = 0$$

with

$$U = \Phi(r, z),$$

in which r is the radial coordinate of a paraxial trajectory. Using the Taylor series expansion, and integrating step-by-step, the method allows us both to pass from a point $P_n(z_n, r_n, r_n')$ to a point $P_{n+1}(z_{n+1}, r_{n+1}, r_{n+1}')$ and to calculate the error $\triangle r_n$ due the spherical aberration. The curves which properly represent U and its derivative are replaced by step functions, each "step" of constant value coinciding with an interval $\triangle z$. This interval must be chosen to be, at most, one-tenth of the radius of the smallest of the apertures, to attain a very high precision. Equations of the form

$$r_{n+1} = Q_1 r_n + Q_2 r_n',$$

$$r_{n+1}' = Q_3 r_n' + Q_4 r_n,$$

$$\triangle r_{n+1} = Q_1 r_n' + Q_2 r_n',$$

$$\triangle r_{n+1}' = Q_3 r_n' + Q_4 r_n + Q_5$$

appear, with coefficients—relatively simple for electrostatic lenses—which are functions of

$$\frac{\varphi'}{\varphi}, \quad \frac{\varphi''}{\varphi}, \quad \text{and} \quad \frac{\varphi'''}{\varphi}.$$

(4) *Gans' method* (*1937*)

Only $\varphi(z)$ need be known for this simple method, which is suitable for the electrostatic case. $\varphi(z)$ is represented as nearly as possible by a number of rectilinear segments. In each region z_j, the potential varies linearly and the equation of the Gaussian trajectories is simple to integrate; the length

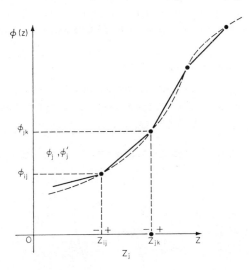

FIG. 30a. The polygonal approximation to the potential, used in Gans' method,

of each region may be varied, according to the shape of the curve $\varphi(z)$ in the vicinity, to increase the accuracy. The trajectories will be composed of a series of parabolic segments. Across each inter-regional boundary, ψ is continuous but φ', and hence r', is not. Integration of the trajectory equation over a limitingly small region around this discontinuity is equivalent to calculating the refraction at this point (see §8.4.1). Finally, with the notation of Fig. 30a, we obtain the formulae

$$r_j = r_{ij} + \frac{2C_j}{\varphi_j'}(\varphi_j - \varphi_{ij}) \quad \text{if } \varphi_j' \neq 0,$$

or

$$r_j = r_{ij} + r_{ij+}' \times (z_j - z_{ij}) \quad \text{if } \varphi_j' = 0,$$

in which

$$C_j = (r_j' \varphi_j)^{1/2} = (r_{ij+}' \cdot \varphi_{ij})^{1/2} \quad \text{and} \quad z_{ij} \leqq z_j \leqq z_{jk},$$

and also

$$r_{ij+}' = r_{ij-}' + \frac{\varphi_{ij-}' - \varphi_{ij+}'}{4\varphi_{ij}} r_{ij}.$$

We should also mention the methods due to Bertein (1952b), Burfoot (1952), and Laudet (1953).

4.5.3 Graphical Trajectory-tracing

A great deal of information about the properties of electrostatic lenses can be obtained by tracing out the trajectories on a large scale chart of the potential. Tracing in this way gives less precise trajectories than do the methods discussed in the last section; it is most useful, therefore, for rays far from the axis, for which numerical methods of calculation are extremely laborious (see for example Duchesne, 1953).

(1) *The radius of curvature procedure*

We consider two equipotentials, (Φ_1) and (Φ_2), sufficiently close that the potential in the region between them may be considered as effectively equal to the mean value $\dfrac{\Phi_1 + \Phi_2}{2}$. Within this region, the trajectory originating at some point A can be regarded as a circular arc, the radius, R, of which

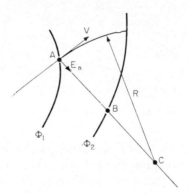

FIG. 30b. Tracing a trajectory by the radius of curvature method.

is obtained from the equation which expresses the fact that the normal forces balance:

$$\frac{m v^2}{R} = e E_n,$$

where E_n is the field component normal to the direction of motion. To a first approximation, we write

$$E_n = \frac{\Phi_1 - \Phi_2}{AB},$$

where the direction of AB is normal to that of the initial electron velocity (Fig. 30b).

(2) *The parabola method*

The space between the two equipotentials is now regarded as a region in which the potential varies linearly along the normals to these surfaces; the trajectories are therefore parabolae. If we consider a particle which arrives at a point I_1 on the equipotential (Φ_1) at an angle of incidence i_1

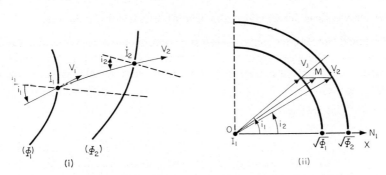

FIG. 30c. Tracing a trajectory by the parabola method.

with velocity V_1 (Fig. 30c), we have to determine the point I_2 at which it reaches (Φ_2), and its direction of motion and velocity V_2 at this point. Since E is normal both to (Φ_1) and to (Φ_2), the tangential component of V is constant, and we have

$$V_1 \sin i_1 = V_2 \sin i_2.$$

But

$$\tfrac{1}{2} m V_1^2 = e \Phi_1,$$

and

$$\tfrac{1}{2} m V_2^2 = e \Phi_2,$$

so that

$$\sqrt{\Phi_1} \sin i_1 = \sqrt{\Phi_2} \sin i_2.$$

Between the two equipotentials (Φ_1) and (Φ_2), the particles are uniformly accelerated, since the field is constant. The velocities, therefore, are related in the following way:

$$V = V_1 + \gamma t,$$
$$V_2 = V_1 + \gamma \tau.$$

The chord $I_1 I_2$ has the direction of the mean velocity V_m between the two equipotentials. V_m is given by

$$V_m = V_1 + \tfrac{1}{2}(V_2 - V_1).$$

It is now possible to devise a construction to give the directions of $I_1 I_2$ and V_2 (Fig. 30c). We draw two concentric circles, radii $\sqrt{\Phi_1}$ and $\sqrt{\Phi_2}$, with centre O; we draw $\overrightarrow{OV_1}$ at an angle i_1 to $\overrightarrow{OX}$ (the normal at O to the equipotential (Φ_1) and then $\overrightarrow{V_1V_2}$ parallel to $\overrightarrow{OX}$; $\overrightarrow{OM}$ provides us with the direction of I_1I_2 while $\overrightarrow{OV_2}$ gives that of V_2. It is convenient to draw $\overrightarrow{OX}$ and a family of circles with radii $\sqrt{\Phi_i}$ on transparent paper, which can then be moved along the trajectory which is being studied.

4.5.4 Mechanical Integration; the Use of Electrical Analogues

Electrical analogues have been successfully used to solve the Laplace equation for the potential; it is, however, also possible to find analogues by means of which the trajectory equation can be integrated. Consider,

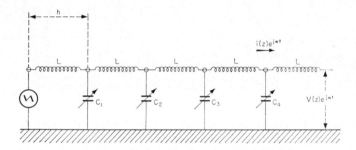

Fig. 31. The underlying principle of an analogue device to integrate the equation
$$y'' + f(x)\, y = 0.$$

for example, a transmission line of the form shown in Fig. 31 (Grivet and Rocard, 1949); the equations which describe the propagation of current and potential along the line are

$$\frac{\partial i}{\partial z} = j\frac{C}{h}\omega V, \qquad \frac{\partial V}{\partial z} = j\frac{L}{h}\omega i,$$

in which h is the length of each section of line.

Eliminating i, we obtain

$$\frac{\partial^2 V}{\partial z^2} + \frac{L C \omega^2}{h^2} V = 0$$

which is formally identical to the Picht equation provided

$$\frac{3}{16}\left(\frac{\varphi'}{\varphi}\right)^2 = \frac{L C \omega^2}{h^2} \qquad \text{and} \qquad V = R.$$

An integrating machine of this kind is in use at the Laboratoire de Radioélectricité (Faculty of Sciences, University of Paris), in which the characteristic function is simulated as nearly as possible with variable condensers. A low frequency generator supplies the potential, and the potential at each point is measured with a voltmeter. The trajectories can be constructed very rapidly (Hampikian, 1953).

The "trajectory-tracers" which have been current now for a decade (see, for example, Gabor and Langmuir) are not, however, particularly well adapted for electron optics, as the analogy upon which these devices rest depends upon angular relationships; as the angles in the Gaussian approximation are so slight, any attempt to obtain the point of intersection of trajectory and axis can only be imprecise.

Machines could be designed, however, which were specially adapted for the integration of this equation; we single out the work of Schiekel (1952) on this problem, using the mechanical integrator of Walther and Dreyer (1949), that of Barber and Sanders (1959), using the electrolytic tank described in Sanders and Yates (1953, 1956), and that of Verster (1961).

4.6 EXPERIMENTAL DETERMINATION OF THE STANDARD CARDINAL ELEMENTS

The positions of the asymptotic principal planes and foci (Fig. 25) can be rapidly calculated from a knowledge of particular trajectories in object and image space.

4.6.1 The Parallel Beam Method

In this, the simplest method, a parallel beam of electrons is incident upon the lens; the radius, r_0, of the beam is small in comparison with the radii of the apertures in the lens. After passing through the lens the beam

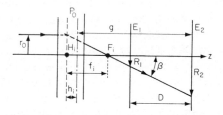

FIG. 32. The parallel incident beam method.

produces a spot of radius R_1 on a fluorescent screen at E_1. The screen is then moved to a new position E_2 a distance D away from E_1; the radius of the spot is here R_2 (see Fig. 32). We can now deduce the semi-aperture of the beam and the image focal length:

$$\tan \beta \sim \beta = \frac{R_2 - R_1}{D}, \qquad f_1 = \frac{r_0}{\beta} = \frac{r_0 D}{R_2 - R_1}.$$

The position of the image principal plane H_i is marked with respect to an arbitrary plane P_0 (which will be the symmetry plane when the lens is symmetrical). We then have

$$h_1 = H_i P_0 = H_i F_i - P_0 F_i,$$

or

$$h_1 = f - (P_0 E_2 - F_i E_2).$$

We know that the distance $g = P_0 E_2$ and $F_i E_2 = \dfrac{R_2}{\tan\beta}$; finally, therefore

$$h_1 = f - \left(g - \frac{R_2}{\tan\beta}\right).$$

If the lens is either electrically or geometrically unsymmetrical, a fresh set of measurements is made after turning the lens round without disturbing the plane P_0. In this way, the "object" elements are also determined (Septier, 1960).

The accuracy of the method depends upon the accuracy with which r_0, R_1 and R_2 are measured; it can only be used for large lenses, or for large-scale models of normal lenses.

We might equally well consider locating the position of the focus directly with a mobile screen, but this is a far less accurate method than the one described above.

4.6.2 Methods Using the Shadows of Grids

For microscope lenses, we can use the methods described by Spangenberg and Field (1942), Heise and Rang (1949), Everitt and Hanssen (1956) and Septier (1960). All of these methods are based upon the observation of the shadows of two grids G_1 and G_2 which are placed respectively in front of and behind the lens in question; the observations are made on a *fixed* fluorescent screen (see Fig. 33) which remains at a constant distance from a point electron source which is placed on the axis in front of the lens.

This method is particularly easy to apply to einzel lenses ($\Phi_{\text{object}} = \Phi_{\text{image}}$) and to symmetrical lenses (such as those illustrated in Figs. 20b, 26 and in § 8.2), as the object and image elements are symmetrically placed with respect to the geometrical centre of the lens.

The grids are composed of parallel wires; they are crossed, so that we shall easily be able to identify their respective shadows on the screen. We consider Figs. 33a and 33b in which the experimental arrangement is represented in the planes xOz and yOz. The image of the point source P is Q; an incident ray inclined at an angle α_0 produces an emergent ray with inclination α_i, which falls on G_2 at a distance x_i from the axis and on the screen at a distance X_i. An identical ray which passes through the grid G_1 at a distance y_0 from the axis meets the screen at a distance Y_0.

We can thus construct

$$OG_1 = g_0, \qquad OG_2 = g_i,$$
$$OP = p, \qquad OE = s,$$

where O, the centre of the lens, lies in the plane P_0.

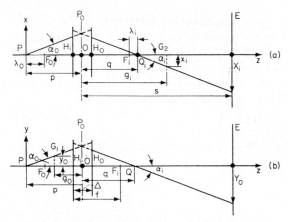

FIG. 33. The method using the shadow of a grid. (a) Ray-path in the xOz plane;
(b) Path of an identical ray in the yOz plane.

If f is the focal length and $\triangle$ is the distance between the two principal planes, we can write:

$$\frac{\tan\alpha_i}{\tan\alpha_0} = \frac{\lambda_0}{f} = \frac{f}{\lambda_i} \quad \text{with} \quad \begin{cases} \tan\alpha_i = \dfrac{X_i}{s - g_i}, \\[2mm] \tan\alpha_0 = \dfrac{Y_0}{p - g_0}, \end{cases}$$

and (see Fig. 33)

$$q = f - \frac{\triangle}{2} + \lambda_i,$$

$$p = f - \frac{\triangle}{2} + \lambda_0.$$

At the screen, we measure Y_0 and X_i; since we know the spacing between the wires of the grids, we can immediately calculate the quantities $V_i = X_i/x_i$ and $V_0 = Y_0/y_0$.

The distance q is given by

$$q = \frac{s - V_i g_i}{1 - V_i},$$

and combining all these relations, we arrive finally at the following expressions:

$$f = \frac{V_0(p - q)(s - q)(p - g_0)}{V_0^2(p - g_0^2) - (s - q)^2},$$

$$\frac{\triangle}{2} = p - f\left(1 + V_0 \frac{p - g_0}{s - q}\right).$$

For lenses which are either geometrically or electrically asymmetrical, the plane P_0 is selected arbitrarily. Two series of measurements are necessary: either the lens is reversed without disturbing the position of the plane P_0 with respect to P and E (Everitt and Hanssen) or it is displaced longitudinally a known distance without altering the distance between P and E (Spangenberg and Field, Septier). We can then determine:

(i) the focal lengths f_i and f_0 (if the lens is an einzel lens, f_i and f_0 are always equal), and

(ii) the distances h_i and h_0 between the principal planes and P_0.

CHAPTER 5

THE OPTICAL PROPERTIES OF
MAGNETIC LENSES

5.1 THE FORM OF THE TRAJECTORIES IN AN
AXIALLY SYMMETRIC FIELD

5.1.1 The Magnetic Force

The force which a magnetic field $\mathbf{B}$ exerts on an electron of charge $-e$ moving with velocity v is given by

$$\mathbf{F} = -e(v \times \mathbf{B}).$$

The relative orientation of the various vectors is shown in Fig. 34; it illustrates the sign convention which is described by such well-known mnemonics as Maxwell's right-hand rule, or the motion of a right-handed screw.

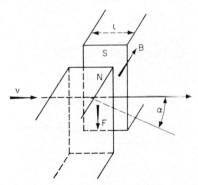

FIG. 34. The deflexion of a charged particle by a magnetic field.

In a uniform field, the particle follows a circular trajectory of radius $\varrho = \dfrac{mv}{eB}$; if this field acts only over a region of length l, the resulting deviation, α, is given by

$$\sin \alpha = \frac{eBl}{mv} \simeq \alpha.$$

We shall find subsequently that it is often convenient to characterize the electron velocity by the product $B\varrho$ to which it corresponds rather than by the potential Φ by which it was produced. These quantities are related by

$$v = \frac{e}{m} B\varrho = \sqrt{\frac{2e\Phi}{m}}, \tag{5.1}$$

or for a non-relativistic electron in particular:

$$v(\text{m sec}^{-1}) = 1\cdot77 \times 10^{11}\, B\rho \,(\text{webers m}^{-1}) = 5\cdot9 \times 10^5\, \sqrt{\Phi}\,(\text{volts}).$$

For relativistic electrons, we should have

$$p = mv = \sqrt{2em_0\Phi(1 + \varepsilon\phi)},$$

$$(B\rho) = \frac{p}{e} = \sqrt{\frac{2m_0\Phi}{e}(1 + \varepsilon\phi)}.$$

5.1.2 Trajectories Inside a Lens

In every round lens, the magnetic field B has a symmetry axis Oz, along the direction of the overall motion of the electron; two components, therefore,

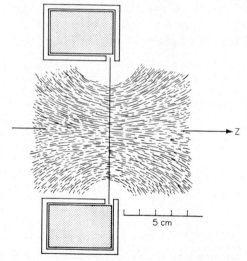

FIG. 35. Cross-section of a screened magnetic lens and the field distribution inside it.

suffice to define it: one radial $B_r(r, z)$, the other axial $B_z(r, z)$. The tangential component B_θ is zero and so the force on the electron has components

$$f_r = -e\, v_\theta\, B_z, \qquad f_\theta = -e(v_z\, B_r - B_z\, v_r), \tag{5.2}$$

$$f_z = +e\, v_\theta\, B_r.$$

Most magnetic lenses consist simply of short coils, either unshielded or, more often, provided with a magnetic circuit of iron which concentrates the field into a small region. Figure 35 represents a cross-section of a common design. The distribution of lines of force shows that the electrons will be affected simultaneously by the radial component B_r and equally strongly by the axial component B_z. An electron emitted at a point A follows a rectilinear path until it reaches the plane H in which the lens action effectively begins (Fig. 36); at this point, the field is predominantly radial (it is the region α, β on the frontal projection of the trajectory, where the curvature becomes large), and the trajectory is forced to rotate sideways out of the plane of the paper in Fig. 36. Therefore a transverse component of velocity v_t appears, perpendicular to the axis, which combines with the longitudinal field component to produce a radial force which draws the electron towards the axis. The particle then enters a region in which the field is effectively longitudinal and uniform—this is the region β, γ of the frontal projection which has become approximately circular. Finally, the electron passes through the field region in which B_r is once again dominant— conditions similar to those at the beginning of the lens obtain. The curvature increases again, but B_r has altered in direction and the resultant force draws the trajectory towards the axis. The relative intensities of these effects are such that the emergent ray always intersects the axis a second time. The radial component of the field as the electron leaves the lens "repels" the trajectory in a meridional plane, which is itself inclined at an angle θ_{oi} to the plane of incidence, and the image of an extended object will therefore appear to have been rotated relative to the object. This qualitative analysis shows that the trajectory is complex, and it seems a difficult task to give a simple description of the motion of an electron. The problem is, however, considerably simplified by taking into account *the existence of a new first integral*, while the conservation of energy relation remains simple.

We notice first that the integral of the conservation of energy equation remains valid, since the force due to the magnetic field is perpendicular

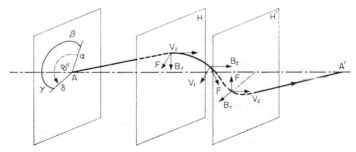

FIG. 36. The path of a ray through a magnetic lens. The field is effectively restricted to the region between the two planes H and H'. The projection of the trajectory back onto the object plane is also shown.

to the velocity, and there is therefore no tangential acceleration; the velocity and energy of the electron are constant. Equation (4.3) remains valid even if the lens uses a combination of electrostatic and magnetic fields. Further, there is another first order integral which Störmer discovered at the beginning of the century, but which was subsequently forgotten, until Busch rediscovered it in 1926. This Störmer integral is quite general, but we shall first consider only the Gaussian approximation, where the calculations are simpler; the full proof is to be found in § 6.2.2.

5.2 THE GAUSSIAN APPROXIMATION

5.2.1 The Equations of Motion in Fixed Axes

Just as the axial potential $\varphi(z)$ is sufficient to describe the electric field near the axis of an electrostatic system, so the axial field $B_z(z)$ is adequate to describe the radial field near the axis of a magnetic lens. The Maxwell equation div $\boldsymbol{B} = 0$ immediately implies

$$B_r(z, r) = -\frac{1}{2} r \frac{\mathrm{d} B_z(z)}{\mathrm{d} z}. \tag{5.3}$$

If we write out the general equation of motion for an electron in a magnetic field

$$\frac{\mathrm{d}(m\boldsymbol{v})}{\mathrm{d} t} = m \frac{\mathrm{d}\boldsymbol{v}}{\mathrm{d} t} = -e(\boldsymbol{v} \times \boldsymbol{B}) \tag{5.4}$$

in cylindrical polar coordinates, we obtain

$$m\left[\frac{\mathrm{d}^2 r}{\mathrm{d} t^2} - r\left(\frac{\mathrm{d}\theta}{\mathrm{d} t}\right)^2\right] = -e r \frac{\mathrm{d}\theta}{\mathrm{d} t} B,$$

$$m\left[r\frac{\mathrm{d}^2 \theta}{\mathrm{d} t^2} + 2\frac{\mathrm{d} r}{\mathrm{d} t}\frac{\mathrm{d}\theta}{\mathrm{d} t}\right] = e\left(B\frac{\mathrm{d} r}{\mathrm{d} t} + \frac{1}{2} r \frac{\mathrm{d} B}{\mathrm{d} z}\frac{\mathrm{d} z}{\mathrm{d} t}\right), \tag{5.5}$$

$$m z'' = 0.$$

The third equation expresses the fact that the velocity is constant in the Gaussian approximation—this is a direct consequence of conservation of energy. The second equation can be written

$$\frac{\mathrm{d}}{\mathrm{d} t}\left(r^2 \frac{\mathrm{d}\theta}{\mathrm{d} t}\right) = \frac{e}{m}\left(r B \frac{\mathrm{d} r}{\mathrm{d} t} + \frac{1}{2} r^2 \frac{\mathrm{d} B}{\mathrm{d} z}\frac{\mathrm{d} z}{\mathrm{d} t}\right),$$

which can be integrated directly, to give

$$\frac{\mathrm{d}\theta}{\mathrm{d} t} = \frac{e}{2m} B + \frac{C}{r^2}, \tag{5.6}$$

C being zero when the ray intersects the axis. This can occur in any of three ways; first, the ray, of arbitrary direction, may originate in an axial point object ($r_0 = 0$); alternatively, the object may have any position outside the lens, or may be a virtual immersed object, provided the rays lie in meridian planes $\left[r_0 \neq 0 \text{ but} \left(\dfrac{d\theta}{dt} \right)_0 = 0 \right]$; finally, for a real immersed object, rays which crossed the axis before they reached the lens will intersect it again after leaving the lens and entering free space once again. Such a case occurs in the magnetic microscope objective, where the rays are emitted by an axial point source in the electron gun.

In any of the cases for which $C = 0$, Störmer's equation reduces to the simple form:

$$\frac{d\theta}{dt} = \frac{e}{2m} B \qquad \left(\frac{e}{2m} = 8 \cdot 797 \times 10^{10} \text{ C kg}^{-1} \right). \qquad (5.7)$$

Any ray of this kind always leaves the lens as a straight line which intersects the axis (which lies, that is, in a meridian plane), since the condition $B = 0$ implies $\dfrac{d\theta}{dt} = 0$.

5.2.2 The Equations of Motion in the Rotating Coordinate System of Larmor

Larmor's theorem suggests that we should simplify our description of the motion still further by using a reference system which rotates about the axis with the Larmor angular velocity:

$$\omega = \frac{e}{2m} B.$$

For the family of rays for which $C = 0$, the simplification is obvious, as the various trajectories appear to be planar and meridional in the rotating system, for the new angular velocity of the particles is $\dfrac{d\theta'}{dt} = \dfrac{d\theta}{dt} - \omega = 0$. For the rays for which $C \neq 0$, however, the transformation is still more interesting, as we shall now show.

To apply Newton's laws of motion in a rotating system, we must take into account two artificial forces in addition to the force f. These are the Coriolis force, $-2m\omega \times v'$, and the drag (which corresponds to the "drag" acceleration) which has two components: the radial centrifugal force $m\omega^2 r$ and the tangential force $-m r \left(\dfrac{d\omega}{dt} \right)$. By a proper choice of ω, we can cancel all the force components save one. This can be seen from the following table in which the velocity and its transverse component are written v, v_θ, in the real motion, and v', v'_θ in the relative motion; ω is a vector of length $\dfrac{e}{2m} B$ along Oz (Fig. 37); the longitudinal force component is negligible in the Gaussian approximation, as it is proportional

	Magnetic force	Coriolis force	Inertial drag force	Resultant
Radial	$-ev_\theta B$	$e B v'_\theta$	$\dfrac{e^2}{4m}B^2 r$	$-\dfrac{e^2}{4m}rB^2$
Tangential	$-e\left(v_r B + \dfrac{1}{2}r\,\dfrac{dz}{dt}\,\dfrac{dB}{dz}\right)$	$e B v_r$	$\dfrac{e}{2}r\,\dfrac{dz}{dt}\,\dfrac{dB}{dz}$	0

to $r r'$ and is thus second order. The relation $v'_\theta = v_\theta - \omega r$ considerably simplifies the calculation of the resultant radial force:

$$f'_r = -e(v'_\theta + \omega r)B + e v'_\theta B + \frac{e^2}{4m}r B^2 = -e\omega r B + \frac{e^2}{4m}r B$$

$$= -\frac{e^2}{4m}r B.$$

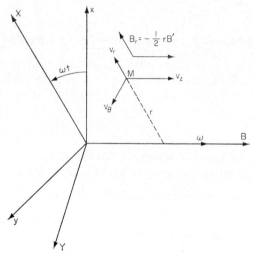

FIG. 37. Larmor's rotating coordinate system.

The equations of motion transform, in the frontal projection, into

$$\frac{d^2 r}{dt^2} + \frac{e^2}{4m^2}B^2 r = 0, \tag{5.8}$$

and along the axis Oz, into

$$\frac{dz}{dt} = \sqrt{\frac{2e\varphi_0}{m}}, \tag{5.8'}$$

where φ_0 is the constant electric potential which obtains throughout the whole lens. Provided φ'' is everywhere replaced by $(e/2m)\,B^2$, equation (5.8) is identical to (4.7) of § 4.2.2, which is its counterpart for electrostatic lenses.

(i) $C = 0$. For this family of rays, the trajectory is planar; the equation of motion in the rotating meridian plane is obtained by eliminating the time, thus:

$$\frac{d^2 r}{dz^2} + \frac{e}{8m\,\varphi_0}\,B^2\,r = 0. \tag{5.9}$$

It is useful to rewrite the quantity $(e/8m\,\varphi_0)$ in terms of the radius ϱ of the circle which the electrons would follow if they were introduced normally into a uniform magnetic field of intensity B with a velocity corresponding to an accelerating potential φ_0. We should have

$$\left(\frac{1}{\varrho}\right)^2 = \frac{e\,B^2}{2m\,\varphi_0},$$

whence

$$\varrho = 3\cdot371\,\sqrt{\varphi_0/B} \quad \text{(cm; V; gauss),}$$

or

$$\varrho = 3\cdot371 \times 10^6\,\sqrt{\varphi_0/B} \quad \text{(m; V; tesla),}$$

and equation (5.9) becomes:

$$\frac{d^2 r}{dz^2} + \frac{r}{4\varrho^2} = 0. \tag{5.10}$$

(ii) $C \neq 0$. Even in the rotating system, these rays remain skew. However, by using a complex coordinate to describe an off-axial position as in the parallel electrostatic situation, it is not difficult to show that these oblique rays also coincide, at the same point as the meridian rays, to form a point image. This involves writing $u = r\,e^{j\theta_1}$, in which θ_1 is the angle between the Larmor rotation and the actual rotation, defined by

$$\theta_1 = \int_{z_0}^{z_i} \frac{d\theta}{dt}\,\frac{dt}{dz}\,dz - \int_{z_0}^{z_i} \frac{e}{2m}\,B(z)\,dz.$$

After a certain amount of calculation, which is set out in detail in the book by de Broglie, u is found to satisfy equation (5.9) for all values of C.

(iii) *The image rotation.* The rotation of the meridian plane between object and image is given by

$$\theta_{0i} = \theta_i - \theta_0 = \sqrt{\frac{e}{8m\,\varphi_0}} \int_{z_0}^{z_i} B(z)\,dz = \int_{z_0}^{z_i} \frac{1}{2\varrho}\,dr. \tag{5.11}$$

(iv) The stages in the geometrical argument which the table summarizes can be interpreted very straightforwardly by using a complex notation for the forces, as in § 4.2.2,

$$f_u = j\,e\left(B\,u' + \frac{1}{2}\,u\,\frac{dB}{dz}\,\frac{dz}{dt}\right).$$

Writing

$$W = ue^{j\Omega}, \quad \Omega = \int_{t_0}^{t} \omega \, dt,$$

we obtain the trajectory directly.

If the system combines electrostatic and magnetic fields, we obtain an equation of motion

$$\frac{d}{dz}\left(\sqrt{\varphi} \, \frac{dr}{dz}\right) + \frac{1}{4\sqrt{\varphi}}\left(\varphi'' + \frac{e}{2m}B^2\right)r = 0. \tag{5.12}$$

Making the simplifying assumptions listed in § 4.2.2, we find that the focal length of a thin weak electromagnetic lens is given approximately by

$$\frac{1}{f} = \frac{1}{4\sqrt{\varphi_0}} \int_{-\infty}^{\infty} \left(\frac{\varphi''}{\sqrt{\varphi}} + \frac{e}{2m}\frac{B^2}{\sqrt{\varphi}}\right) dz, \tag{5.13}$$

in which φ'', φ and B are functions of z. For a purely magnetic lens, $\varphi(z) = \varphi_0$, so that

$$\frac{1}{f} = \frac{e}{8m\,\varphi_0} \int_{-\infty}^{\infty} B^2(z) \, dz. \tag{5.13'}$$

5.3 PROPERTIES OF MAGNETIC LENSES

5.3.1 The Optical Formalism

All the optical formalism of electrostatic lenses arises from the fact that the equations of motion are linear. As equation (5.12) is linear, therefore, all the reasoning presented in § 4.3 can be carried over directly into the magnetic or electromagnetic lens situation. The cardinal points, for example, are defined perfectly straightforwardly by a pair of rays in a rotating meridian plane.

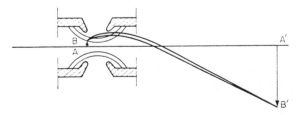

FIG. 38. An immersion object on the axis of a magnetic lens (the image rotation is not shown).

The only important difference between the electrostatic and electro-magnetic situations lies in the Larmor rotation of the image which appears in the latter, and which can be evaluated with the aid of formula (5.11). It is interesting to notice that immersion elements are now of considerable importance, as a magnetic field will not be perturbed by the intrusion of an object, or a screen with which to inspect the image. In Figure 38, two of the trajectories which form the real image, outside the lens, of an object immersed within it, are shown.

5.3.2 Physical Properties

Unlike the optical formalism, however, the physical properties are very definitely different, as we might have guessed from the elementary deviation formula quoted earlier,

$$\alpha = \frac{l}{\rho} = \sqrt{\frac{e}{2m\,\varphi_0}}\,Bl.$$

The value of e/m now effects the deviation, and the nature of the particle affects the trajectory it follows—the focal length of a magnetic lens is much longer for ions, with their greater weight, than for electrons, and a magnetic lens is of little use for fast heavy ions.

No longer are the coefficients homogeneous with respect to the electric potential $\varphi(z)$. The positions of the cardinal elements now vary, therefore, with the applied potential and with the lens currents.

In particular, accidental fluctuations of the potential will mar the image, and very careful stabilization of the accelerating high tension which is applied is necessary—similar precautions must be taken with the lens currents.

Yet another difference appears if we consider a beam of electrons moving away from an observer; the electrons will appear to rotate in a clockwise sense if the magnetic field points in the same direction as the motion of the electrons, but in an anticlockwise sense if it points in the opposite direction. The trajectory is dependent upon the direction of motion of the electrons relative to the field. Even though two trajectories pass through the same point with the same tangent, they will be different elsewhere if their velocities are in opposite directions (Fig. 39). The optical reversibility principle is no longer valid, and a close analogy with the optics of crystalline media is seen to be impossible.

In practice, however, it is the image itself in which we are interested, rather than the rays which produce it, so that this feature of magnetic lenses is not particularly disturbing.

The sense of the rotation is reversed by a change of sign of the magnetic field, while the convergence, depending as it does upon B^2, is unaffected. The image rotation, therefore, could be suppressed without interfering with the focusing action. All that would be required would be to couple together a pair of lenses in which the currents flow in opposite directions. This is in fact rarely done, however, as, for a given focal length, the

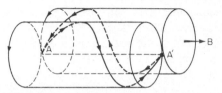

FIG. 39. The irreversibility of the trajectory in a magnetic field.

power expended in the magnetizing coils is greater, and some of the aberrations of the system are accentuated (Becker and Wallraff, 1940). It is much more satisfactory to break the system down into two separate lenses, the objective and projector in the microscope, for example.

5.4 RELATIVISTIC LENSES

The effect of relativity is to replace φ_0 or ϱ^2 by $\varphi_0(1 + \varepsilon_0)$ or $\varrho^2(1 + \varepsilon_0)$ respectively, in the equations of motion (5.9) and (5.10), where we have

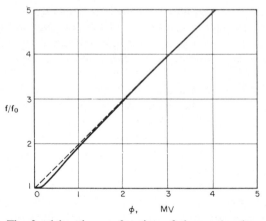

FIG. 40. The focal length as a function of the accelerating potential.

written ε_0 for $\dfrac{e}{2m_0c^2}\varphi_0$. The focal length increases steadily with φ_0 (Fig. 40). At velocities corresponding to energies in the neighbourhood of one MeV, the magnetic lens becomes impracticable, as saturation in the iron prevents B from exceeding some 26,000 gauss.

5.5 METHODS OF DETERMINING THE FUNDAMENTAL RAYS

Equation (5.12), which defines the rays in a rotating meridian plane, is formally identical with the Picht equation which gives the reduced rays of an electrostatic system. All the methods listed in § 4.5, therefore, can be invoked to help provide a solution of it. The equation can be integrated by using either a suitable mathematical model of the field—two such models, due to Glaser (1942) and to Grivet (1950b, 1952a, b) and Lenz (1951) are of particular importance—or one of the step-by-step methods already discussed (Störmer—Adams, Runge—Kutta, Liebmann, for example; see 4.5.2); or, again, by using an electrical analogue—the values of the condensers along the transmission line now being given by:

$$L\, C\, \omega^2 = \frac{e}{8\, m\, \varphi_0}\, B^2.$$

There are, however, other methods designed specifically for magnetic lenses; a very curious analogue method has been suggested by Vineyard(1952) in which the magnetic force is modelled by a Coriolis force. Unfortunately, the practical construction is very complicated, especially in a three-dimensional system, but we have already seen how useful this analogy is, permitting us as it does to integrate the equations of motion quite simply.

Another method, suggested by Loeb (1947), is known as the "hodoscope". This method rests on the analogy which exists between the trajectory of a particle and the equilibrium form of a conducting wire which is traversed by a current i and placed in the same magnetic field. (See Fig. 41.) From the equations

$$\frac{\mathrm{d}(mv)}{\mathrm{d}t} = \frac{\mathrm{d}\boldsymbol{p}}{\mathrm{d}t} = -e\,\boldsymbol{v} \times \boldsymbol{B}$$

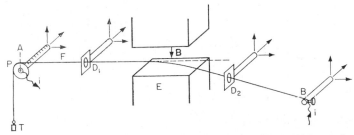

FIG. 41. The "Hodoscope" or "Floating Wire". The incident trajectory is fixed by the point B and the centre of the diaphragm D_2, for a given current i and induction B; the emergent trajectory is likewise determined by D_1 and A. The parts such as D_1 and D_2 which support the wire are mobile in every direction. The positions can be made precise microscopically. The tension is provided by the weight T; the pulley P must be frictionless.

we derive
$$dp + e(ds \times B) = 0$$

in which p denotes the momentum of the particle and ds an elementary vector, tangent to the trajectory.

If T is the tension of the wire and i is the intensity of the current flowing through it,
$$dT + i(ds \times B) = 0,$$

provided we neglect the other forces which may act upon the wire (such as its weight and rigidity); dT represents an elementary increase in the tension. The vectors p and T are parallel to ds, so that we can write:
$$p = p\frac{ds}{ds}, \qquad T = T\frac{ds}{ds},$$
and hence
$$dT = Td\left(\frac{ds}{ds}\right) + \frac{ds}{ds}dT.$$

Since $(ds \times B)$ is perpendicular to s, dT is perpendicular to s also; it follows therefore, that $dT = 0$, since $\frac{ds}{ds}$ is parallel to s—only the first term which represents a vector which is normal to s remains. Similarly, we show that $dp = 0$. In a magnetic field, both the momentum p of a particle and the tension T of a wire remain constant.

We thus obtain the following equations:
$$\frac{ds}{ds} \times B = -\frac{\varrho}{e}\frac{d}{ds}\left(\frac{ds}{ds}\right),$$
$$\frac{ds}{ds} \times B = -\frac{T}{i}\frac{d}{ds}\left(\frac{ds}{ds}\right).$$

If $p/e = T/i$ and the initial conditions are the same, these equations have identical solutions throughout the region.

The ratio $\frac{p}{e} = B\varrho$ is the "rigidity" of the particle. It can easily be expressed as a function of the accelerating voltage of the particles.
$$(B\varrho)^2 = \frac{2m_0\varphi_0}{e}\left(1 + \frac{e\varphi_0}{2m_0c^2}\right) = \frac{2m_0}{e}\Phi_0^*$$

in which $\Phi_0^* = \varphi_0\left(1 + \frac{e\varphi_0}{2m_0c^2}\right)$ as in all magnetic systems. For non-relativistic particles, $(B\varrho)^2 = \frac{2m_0\varphi_0}{e}$, where e denotes the absolute value of the charge. The condition to be satisfied, therefore, is
$$T = (B\varrho)i.$$

If, for example, $B\rho = 0\cdot3$ M.K.S. units (for 5 MeV protons) and $i = 1$ A, we find that $T = 0\cdot3$ newton, or about 30 gm weight. Since the tension is constant over the whole length of the wire, it is sufficient to apply this tension at a single point only, by attaching a stretched spring or a weight T to the end of the "trajectory", for example.

There are several possible sources of error, which arise from the action of perturbing forces on the wire:

(i) The force exerted by the image of the wire which is produced in the magnetic material, and hence in the four poles.

(ii) The force due to the rigidity of the wire.

In a well-annealed wire these first two causes are negligible provided T is sufficiently large.

(iii) The weight of the wire. This cannot be neglected, but its effect can be eliminated if we are looking only for plane trajectories, by arranging that the plane of the trajectories is horizontal. The friction of the system by means of which the force provided by a weight is converted into a horizontal tension should be eliminated as completely as possible.

Finally, for the given initial conditions, the wire may be in stable or unstable equilibrium in the magnetic field.

Detailed studies of these questions are to be found in the articles by Carlile (1957), by Citron *et al.* (1959) who have used this method at CERN to study the trajectories of protons and mesons within the magnet of the 600 MeV synchrocyclotron, and by Pinel (1959, 1960). Although this device was first suggested as a means of studying highly convergent, axially symmetric magnetic lenses, it has been used mainly for the determination of the cardinal elements and aberrations in magnetic prisms (see Chapter 22) and in the strong focusing lenses which are described in Chapter 10 (Carlile, 1957).

5.6 CALCULATION OF THE TRAJECTORIES AND OPTICAL ELEMENTS

5.6.1 Integration of the Equation of Motion

The form of the equation of motion:

$$\frac{\mathrm{d}^2 r}{\mathrm{d}z^2} + \frac{e}{8m\, \Phi_0^*} B^2(z)\, r = 0, \qquad (5.14)$$

in which $B(z)$ represents the induction on the axis and Φ_0^* the accelerating potential, relativistically corrected if necessary, is identical to that of the reduced (Picht) equation which we have examined in connexion with electrostatic lenses, namely:

$$\frac{\mathrm{d}^2 R}{\mathrm{d}z^2} + T^2(z)\, R = 0.$$

The methods of integration listed in the preceding chapter are all, in consequence, applicable to equation (5.14); in particular, the Runge–Kutta method has been used by Lapeyre and Laudet (1960) in the present case of magnetic lenses. If we refer back to the equations of § 4.5.2 we have simply to effect the following changes:

$$T(z) = \frac{e}{8m\,\Phi_0^*}\,B^2(z),$$

$$R = r,$$

$$R' = r'.$$

In addition, we should mention a method which has been described by Gautier (1953), which is quicker to use than the Runge–Kutta method, and which provides very fair results. We write

$$k^2 = \frac{e\,B_0^2}{8m\,\Phi_0^*}, \qquad x = \frac{z}{a}, \qquad y = \frac{r}{a}$$

(a is some arbitrary dimension of the lens), and

$$B(x) = B_m\,b(x),$$

where B_m is the maximum induction on the axis. This implies

$$T(x) = k^2\,b^2(x).$$

The values of $y_{n+1}, y_n \ldots$ at points with abscissae $x_{n+1}, x_n \ldots$ are obtained by a Taylor series expansion:

$$y_{n+1} = y_n + \Delta x \cdot y_n' + \frac{(\Delta x)^2}{2}\,y_n'' + \frac{(\Delta x)^3}{6}\,y_n''' + \cdots,$$

$$y_{n-1} = y_n - \Delta x \cdot y_n' + \frac{(\Delta x)^2}{2}\,y_n'' - \frac{(\Delta x)^3}{6}\,y_n''' + \cdots,$$

and from these expressions we obtain

$$y_n' = \frac{y_{n+1} - y_{n-1}}{2\Delta x} + \alpha',$$

$$y_n'' = \frac{y_{n+1} + y_{n-1} - 2y_n}{(\Delta x)^2} + \alpha'',$$

in which α' and α'' are second order in Δx. On replacing the differentials by finite differences, the trajectory equation $y'' + T(x)y = 0$ becomes

$$y_{n+1} = [2 - T_n(\Delta x)^2]\,y_n - y_{n-1} + \varepsilon,$$

in which ε is of the fourth order in Δx and can thus be neglected. We thus have a recurrence formula from which we can calculate y_{n+1}, once y_n and y_{n-1} are known. In practice, we evaluate the function $[2 - T_n(\Delta x)^2]$ throughout the useful region, after selecting the integration strip-width, Δx. At the beginning, the value of y must be known at x_0 and $x_1 = x_0 + \Delta x$. If $y = y_0$ and $y' = y_0'$ at x_0, we can write

$$y_1 = y_0 + y_0' \Delta x - y_0 T_0 (\Delta x)^2$$

and proceed with the calculation in this way. If we select trajectories incident parallel to the axis, we can easily establish the positions of the object and image cardinal points.

5.6.2 Direct Evaluation of the Optical Quantities

The convergence of any magnetic lens can be calculated quickly and simply (Durandeau and Fert, 1957), once the distribution of the induction on the axis, $B(z)$, has been measured. The real lens is replaced by an ideal lens of length L, such that $B(z) = 0$ outside and $B(z) = B_0 = $ constant within.

L is the "equivalent length" which is here defined by

$$L = \frac{1}{B_0} \int_{-\infty}^{\infty} B(z) \, dz,$$

or if $N I$ is the number of ampère-turns of the lens

$$L = \frac{\mu_0 N I}{B_0}.$$

The equation of motion in a rotating coordinate system can be integrated straightforwardly. Within the fictitious lens, of length L, we have

$$\frac{d^2 r}{dz^2} + \frac{e B_0^2}{8 m \Phi_0^*} r = 0,$$

and writing

$$\frac{e B_0^2}{8 m \Phi_0^*} = k^2,$$

this becomes

$$\frac{d^2 r}{dz^2} + k^2 r = 0.$$

If the lens lies between the planes $z = 0$ and $z = L$, the values of r and r' at the exit will be given by

$$r_s = r_0 \cos kL + \frac{r_0'}{k} \sin kL,$$

$$r_s' = - r_0 k \sin kL + r_0' \cos kL.$$

r_0 and r_0' are the initial conditions at $z = 0$. If $r_0' = 0$ we obtain

$$r_s = r_0 \cos kL,$$

$$r_s' = - r_0 k \sin kL.$$

When $kL < \dfrac{\pi}{2}$, the real focus lies outside the lens. If we calculate the expressions for the focal length f_i and the abscissa of the image focal point $z_{F'}$ with respect to the fictitious exit plane, we obtain

$$f_i = -\frac{r_0}{r_s'} = \frac{1}{k \sin kL},$$

$$z_{F'} = -\frac{r_s}{r_s'} = \frac{\cos kL}{k \sin kL}.$$

We can determine the position, $z_{H'}$, of the principal plane immediately: the object focus and principal plane are symmetrically placed with respect to their image counterparts about the point $z = \dfrac{L}{2}$, the centre of gravity of the curve which represents $B(z)$.

If $kL > \dfrac{\pi}{2}$, the preceding formulae give the normal cardinal elements. The immersion elements can be determined with the aid of the expression for the trajectory within the lens:

$$r = r_0 \cos kz + \frac{r_0'}{k} \sin kz,$$

by writing $r(z_1) = 0$ and calculating the slope $r'(z_1)$ at this point. In this way, we find

$$f_{min} = -\frac{r_0}{r'(z_1)}.$$

For a weak lens, we should write

$$\sin kL \simeq kL, \cdot$$

$$\cos kL \simeq 1 - \frac{k^2 L^2}{2},$$

so that

$$\frac{1}{f} \simeq k^2 L,$$

and

$$z_{F'} = \frac{1 - \dfrac{k^2 L^2}{2}}{k^2 L},$$

and hence

$$z_{H'} = z_{F'} - f = -\frac{L}{2}.$$

The principal planes coincide at the centre of the model lens.

The rotation of the coordinate system to which the trajectory is referred, and hence the rotation of the image with respect to the object, is given very simply by

$$\theta = kL$$

(see equation (5.11) of § 5.2.2).

CHAPTER 6

SIMILARITIES AND DIFFERENCES BETWEEN GLASS OPTICS AND ELECTRON OPTICS

6.1 ELECTROSTATIC LENSES

6.1.1 The Concept of Refractive Index in Electrostatic Electron Optics

Equation (4.21) and the Lagrange–Helmholtz relation for glass lenses become identical if we write $\sqrt{\varphi_0} = n_0$ and $\sqrt{\varphi_i} = n_i$. This is a special case of a very general formal analogy between the distribution of refractive index n in a glass lens and the distribution of the function $\sqrt{\varphi}$ in an electrostatic lens. If the two distributions are identical, the trajectory of an electron and that of a light ray, originating at the same point in their respective sources, are identical also.

The zero of potential is chosen to be the cathode potential; $n' = K n$ is equally suitable as refractive index, and a common choice is $K^2 = \dfrac{e}{mc^2}$, so that $n' = \dfrac{v}{c}$, where c is the velocity of light in free space. The refractive index, therefore, has become a dimensionless number, just as in glass optics. It is to be noted, however, that the ratio of the refractive index at the object to that at the image, $\dfrac{n_0}{n_i}$, can reach values far higher than those which can be achieved in glass optics.

6.1.2 Snell's Law and Electron Optics

In glass optics, the rays are bent straight lines, the sudden changes in gradient occurring when the ray crosses from one medium into another. The deviation is given by the elementary Snell law

$$n_1 \sin i_1 = n_2 \sin i_2,$$

and it is natural to enquire whether there is a similar law for electron optics. In this case, the boundary between two optical media becomes the boundary between two media of different potential φ_1 ($n_1 = \sqrt{\varphi_1}$) and φ_2 ($n_2 = \sqrt{\varphi_2}$). Such a boundary could be formed with the aid of two metal membranes distance l apart; between the membranes, the field is $E = \dfrac{\varphi_2 - \varphi_1}{l}$. If an electron arrives at the first membrane with velocity $\sqrt{\dfrac{2e\,\varphi_1}{m}}$ and at an angle of incidence i_1, it can proceed in either of two ways.

If $\sin i_1 \leqq \sqrt{\varphi_2/\varphi_1}$ the electron leaves the membrane at an angle of refraction i_2, such that $\sqrt{\varphi_1} \cdot \sin i_1 = \sqrt{\varphi_2} \cdot \sin i_2$, and thus obeys Snell's law exactly. Whatever the angle of incidence, the condition is satisfied

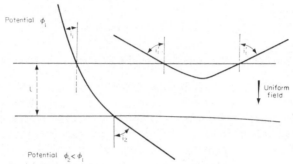

FIG. 42. A transmitted ray and a totally reflected ray at the boundaries between regions of different refractive index in electron optics.

provided $\sqrt{\varphi_2} > \sqrt{\varphi_1}$ (provided that the less refringent or "optically rarer" medium precedes the optically denser one); if the denser medium precedes the rarer, the condition is satisfied only for angles below a limiting value.

If, on the other hand, $\sin i_1 > \sqrt{\varphi_2/\varphi_1}$, which happens for large angles in the case in which $\sqrt{\varphi_2} < \sqrt{\varphi_1}$, total reflexion occurs; the rays turn back upon themselves, and leave the membrane on the same side as they were incident, at an angle of reflexion equal to their angle of incidence. Figure 42 illustrates the various possibilities.

It might seem reasonable, therefore, to try to build lenses from suitably curved membranes separating regions of different refractive index. In the early days of electron optics, Knoll and Ruska (1932) suggested that such a lens might be feasible, but in practice, it is very difficult to construct, as no membrane is available with which it might be built which is adequately transparent to electrons. One is forced to resort to grids, which produce marked optical defects as the holes perturb the potential and prevent a clear image from being formed.

Save for a suggestion by Ruska (1933) that they might be used to focus very wide beams, and as aids in teaching electron optics, therefore, lenses of this type have virtually no application.

6.1.3 The Refractive Index Applied to the Study of Lenses

If we examine a chart of the equipotential surfaces in an electrostatic lens (Fig. 20), a striking analogy is apparent between these surfaces (which become almost indistinguishable from spheres, near the axis) and the intermediate boundaries of a centred glass optical system. It is natural, therefore, to try to transfer the formulae for the cardinal elements in glass optics into electron optics.

As an example of how this can be done, we shall calculate the focal length of a thin lens. In glass optics, the power of a succession of regions of different refractive index, $n_0, n_1, n_2, \ldots, n_i, \ldots$ separated by surfaces of radius of curvature $R_1, R_2, R_3, \ldots, R_i, \ldots$ is given by

$$n_i C_i = \frac{n_1 - n_0}{R_1} + \frac{n_2 - n_1}{R_2} + \cdots + \frac{n_i - n_{i-1}}{R_i} = \sum \frac{\Delta n}{R}. \qquad (6.1)$$

To transfer this into electron optics, we have to integrate over an infinite number of regions instead of summing over a finite number, to replace the refractive index by $\sqrt{\varphi}$, and to replace the radii of curvature by those of the equipotentials, $R = 2\varphi'/\varphi''$, a standard result (φ is the axial potential). The power, therefore, is given by

$$C_i = \frac{1}{n_i} \int_{-\infty}^{\infty} \frac{dn}{R} = \frac{1}{4\sqrt{\varphi_i}} \int_{-\infty}^{\infty} \frac{\varphi'' \, dz}{\sqrt{\varphi}}. \qquad (6.2)$$

Other formulae are susceptible to this transition; the well-known Cotes formula, as Bernard (1951) has shown, can be translated in this way and thus be used for the step-by-step tracing of electron rays. The trajectory is represented not by a curve but by a bent straight line, a change of slope occurring at each of the equipotentials on the diagram. The magnitude of the change can easily be determined by a geometrical construction which stems from the analogous procedure (due to Huyghens) of glass optics. For further details, the reader is referred to the articles by Gans (1937), by Knoll (1935) and more recently, by Fert (1952).

A number of points are worthy of special emphasis. First, the converging power of an elementary intermediate boundary depends both upon the change in refractive index n across it and upon the radius of curvature R; in electron optics, R and n are no longer independent, but are related by the expression $R = \dfrac{2\varphi'}{\varphi''}$, and it is due to this lack of freedom that the correspondence between glass and electron optics is necessarily incomplete (so that, for example, divergent lenses are in general impossible to obtain in electron optics and spherical aberration cannot be eliminated).

In addition, there are no reflected electrons comparable to the finite proportion of light which is reflected at a glass optical boundary, as it is not possible to obtain an electric field strong enough to have a perceptible effect over a distance as short as the wavelength of an electron (some thousandths of a millimicron).

Further, the refractive index n in electron optics is proportional to the particle velocity v. In glass optics, on the other hand, $n = c/v'$ where v' is the phase velocity of the light wave. These two statements no longer seem contradictory, however, if we recall that the particle velocity and the phase velocity of the associated wave are related by $v \cdot w = c^2$. This relation is of extremely general validity, as it need no longer be based upon an analysis of the motion of wave-groups, such as that due to Rayleigh, Sommerfeld and L. Brillouin towards the beginning of the century, but can be derived directly from the principle of relativity (Ditchburn, 1948; Synge, 1952).

It is interesting to recall that even in the seventeenth century, a time when the distinction between v and w was still unknown, Fermat was aware of the disparity between the relations which related the index to the velocity in mechanics and in optics, and considered this sufficient reason to delay publication of the principle which bears his name until 1661, although he had discovered it in 1657. For more information on this point, and for a description of the discovery of the refractive index by Descartes, the reader is referred to the interesting historical study which comprises the first ten pages of *Geometrische Optik* by C. Carathéodory (1937).

6.1.4 Fermat's Principle and the Principle of Least Action

We can express the analogy between electron and glass optics most concisely and rigorously by a comparison of Fermat's principle and the principle of least action. We shall content ourselves here with a summary outline of the way in which this relation may be demonstrated; a more rigorous discussion is to be found in the book by de Broglie and in two specialized articles by Ehrenberg and Siday (1949) and by Glaser (1950).

It is well known that the path of a light ray is dictated by Fermat's principle, expressing the fact that the total optical path will have a stationary value

$$\delta \int_A^B n \, \mathrm{d}s = 0.$$

(6.3)

The motion of an electron, on the other hand, is governed by the principle of least action,

$$\delta \int_{t_0}^{t_i} (T + U) \, \mathrm{d}t = 0,$$

(6.4)

where U is the potential which defines the force applied to the particle and T the kinetic energy, $T = \frac{1}{2} m v^2$. As the conservation of energy relation,

$T - U = $ constant, is satisfied, we deduce immediately that

$$\int_{t_0}^{t_i} (T + U) \, dt = 2 \int_{t_0}^{t_i} T \, dt + \text{constant} = \int_A^B m v \, ds + \text{constant}.$$

Rewriting the velocity in terms of the corresponding electric potential, we obtain a modified version of the principle of least action:

$$\delta \int_A^B v \, ds = \delta \int_A^B K \sqrt{\Phi} \, ds = 0, \qquad (6.5)$$

which becomes exactly parallel to Fermat's principle if we write

$$n = K \sqrt{\Phi}(r, z), \qquad (6.6)$$

K being an arbitrary constant.

So far, we have restricted the argument to non-relativistic mechanics; if, however, we substitute the relativistic value of the function T, the refractive index becomes

$$n = K \sqrt{\Phi(1 + \varepsilon\Phi)}, \qquad \varepsilon = \frac{e\Phi}{2m_0 c^2}. \qquad (6.7)$$

Although it is possible to apply Fermat's principle directly, in certain numerical methods for example, it is more usual to work from the Euler equations which are derived from it, and which in fact prove to be the familiar differential equations for the electron trajectories. If we replace Φ in these equations by n^2, we regain the Bouguer formula with which it is possible to calculate the path of a light ray in an inhomogeneous medium (and to explain such phenomena as mirages). Here, the formula applicable to media with a symmetry axis is given. All these points are very fully dealt with in de Broglie's book on particle optics.

6.1.5 The Limits of Electron Optics

The preceding results represent only one particular aspect of a very general analogy, due to L. de Broglie, between wave-propagation and particle motion. We shall do no more than underline the fact that this analogy is maintained even when geometrical optics becomes invalid. In glass optics, the laws of geometrical optics (and Fermat's principle in particular) cease to be valid when the refractive index varies appreciably over a distance of the order of magnitude of the wavelength, and we have to resort to wave theories of light to resolve the problem.

Similarly, if the electric potential varies appreciably over a distance of the order of magnitude of the wavelength of the wave associated with the electrons, $\lambda = h/m v$, the laws of electron optics which we have enunciated are no longer valid, and the phenomena which appear can be described only in terms of the wave nature of the electron. This is precisely the case when an electron passes through a solid body—the object in an electron microscope, or the specimen in a diffraction camera, for example—and the attractive phenomenon of electron diffraction is observed. Diffraction patterns can be observed, too, in the geometrical shadow cast by a sharp edge illuminated by electrons, but in this case they are extremely tiny and can be distinguished only with the aid of a powerful microscope (Boersch, 1940). A similar effect is to be found in the very fine beams which are used in electron microscopes, and in fact, it is this which provides a limit to the resolution which can be obtained. All these questions are of cardinal importance in the electron microscope, to such an extent that when such an instrument is being designed using formulae such as those established in the present work, allowance must be made for wave mechanical effects.

6.2 MAGNETIC LENSES

6.2.1 The Refractive Index of the Fictitious Motion; the Gaussian Approximation

If we attempt to use the notion of refractive index to describe the focusing properties of a magnetic lens, a surprising feature of this focusing is revealed—the refractive index which characterizes the presence of a magnetic field is constant on the axis, and is of interest, therefore, only off the axis.

If we examine the ray equations of Chapters 4 and 5, we see that the transition from electrostatic focusing (equation 4.9) to mixed electromagnetic focusing (equation 5.12) is effected by replacing φ'' by $\varphi'' + \dfrac{e}{2m} B^2$.

This suggests that we should make the same substitution in the Gaussian electrostatic refractive index:

$$n^2 = \varphi(z) - \left(\frac{r^2}{4}\right)\varphi''(z), \tag{6.8}$$

and regard

$$n^2 = \varphi(z) - \left(\frac{r^2}{4}\right)\left[\varphi'' + \left(\frac{e}{2m}\right)B^2\right] \tag{6.8'}$$

as the general expression for the Gaussian refractive index.

It is easy to verify that the correct ray-path is obtained if we trace the refraction through a medium of fictitious refractive index n as given by (6.8′). In fact, applying the electrostatic equations to a medium with fictitious potential $\Phi_1(r, z) = n^2(r, z)$ gives the correct equations of motion in coordinates rotating with the Larmor precession velocity.

6.2.2 The Refractive Index of the Fictitious Motion: the General Case (Störmer, 1933)

Assuming that the observer too is rotating at the Larmor precession velocity, we attempt to establish a distribution of refractive index which will describe the motion in the rotating meridian plane, according to the rules of geometrical optics. To do this, we substitute for B in the general equations of motion in terms of the vector potential A, where $B = \text{curl } A$, so that

$$m\frac{dv}{dt} = e(v \times \text{curl } A) \tag{6.9}$$

(in which e is the algebraic magnitude of the charge). Oz is the symmetry axis of the magnetic field, so that A is reduced to single component A_θ. We find, then:

$$\frac{d^2 r}{dt^2} - r\left(\frac{d\theta}{dt}\right)^2 = -\frac{e}{m}\frac{1}{r}\frac{\partial(r\,A)}{\partial r}\,r\,\theta',$$

$$\frac{1}{r}\frac{d}{dt}(r^2\,\theta') = \frac{e}{m}\left[\frac{dr}{dt}\frac{1}{r}\frac{\partial(r\,A)}{\partial r} + \frac{dz}{dt}\frac{\partial A}{\partial z}\right], \tag{6.10}$$

$$\frac{d^2 z}{dt^2} = \frac{e}{m}\frac{\partial A}{\partial z}\,r\,\theta'\,.$$

As Störmer pointed out, the second equation integrates directly to

$$\theta' = \frac{e}{m}\left(\frac{A}{r} + \frac{C}{r^2}\right), \tag{6.11}$$

in which C is a constant of integration. Substituting this value of θ' into the other two equations, we find

$$\frac{d^2 r}{dt^2} = \frac{e}{m}\frac{\partial}{\partial r}\left[\frac{e}{2m}\left(A + \frac{C}{r}\right)^2\right],$$

$$\frac{dz^2}{dt^2} = \frac{e}{m}\frac{\partial}{\partial z}\left[\frac{e}{2m}\left(A + \frac{C}{r}\right)^2\right]. \tag{6.12}$$

Comparing equations (6.12) with

$$\frac{d^2 r}{dt^2} = -\frac{e}{m}\frac{\partial \Phi}{\partial r},$$

$$\frac{d^2 z}{dt^2} = -\frac{e}{m}\frac{\partial \Phi}{\partial z}, \tag{6.13}$$

which would describe the motion of the same particle in an electric field of potential Φ, the analogy is obvious.

The motion of the electrons with respect to a rotating observer is analogous to that of a light ray in a medium of refractive index (variable but axially symmetrical) n given by

$$n^2 = \Psi(z, r) = \Psi_0 - \frac{e}{2m}\left(A_\theta + \frac{C}{r}\right)^2. \tag{6.14}$$

The Gaussian approximation to this formula gives the expression found in the preceding paragraph; near the axis, $A_\theta = \dfrac{Br}{2}$, which gives the refractive index of (6.8') apart from a constant factor, in the case $C = 0$.

The trajectory *as seen by an observer in a rotating meridian plane*, therefore, is precisely the same as the real trajectory in the corresponding hypothetical

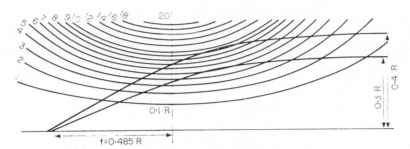

FIG. 43a. Curves of equal refractive index in a meridian plane rotating at the Larmor frequency — the field is created by a thin coil.

electrostatic case. If the constant C of equation (6.14) is not zero, however, the velocity of rotation—even to the Gaussian approximation—differs from the value given by equation (5.11).

The refraction of the ray can be traced through the combined electric and magnetic fields, step-by-step, by using a refractive index of the form

$$n^2 = \Phi(r, z) - \left(\frac{e}{2m}\right)\left[A_\theta + \frac{C}{r}\right]^2. \tag{6.15}$$

In Figure 43a, the general form of the surfaces $n = $ constant is shown, for a purely magnetic lens. The index defined by (6.15) has been used by Sandor (1941) to trace the trajectories through a magnetic lens; he also described a device to measure n^2 directly, but his work has not been followed up. The value of Störmer's theory lies in the light which it sheds upon the relation between electrostatic and magnetic systems; this will prove to be valuable in aberration theory, where the aberrations of magnetic and of electric systems can be related quantitatively by using Ψ.

6.2.3 The Refractive Index for a Stationary Observer

It is equally possible to use a refractive index to determine directly the motion relative to fixed axes. Schwarzschild (1905) discovered this representation in the course of his extension of the variational principle to motion in electric and magnetic fields. We shall consider a mixed system, in which the electrons experience the influence both of an electric field derived from a potential $\Phi(x, y, z)$ and of a magnetic field derived from a vector potential A [such that $B(x, y, z) = \text{curl}\, A(x, y, z)$].

As the electrons experience a force whose potential is not simply scalar, the principle of least action cannot be applied as straightforwardly as in the electrostatic case; Schwarzschild showed, however, that the principle is still applicable provided we take into account the fact that the kinetic energy T is no longer simply $m v^2/2$, but contains a further term due to the magnetic field.

The presence of this additional term can be made plausible by considering the kinetic energy in a system of coupled circuits. The electromagnetic energy stored in the field which is produced by two circuits of self-inductance L_1 and L_2 and mutual inductance M is

$$W = \tfrac{1}{2}L_1 i_1^2 + \tfrac{1}{2}L_2 i_2^2 + M i_1 i_2.$$

Let us now consider a very special case of such a pair of circuits, in which one circuit is the lens itself, and the other the beam of electrons which proceeds virtually along the axis. In this case, $\tfrac{1}{2} L_1 i_1^2$ is the electromagnetic energy stored in the self-inductance which constitutes the coil-windings; it is a constant, independent of the second circuit, which is formed by the electron beam. $\tfrac{1}{2} L_2 i_2^2$ represents the energy of an electron due to its motion, namely $\tfrac{1}{2} m v^2$. The term $M i_1 i_2$, however, represents energy which is present when an electron moves in a magnetic field. It is precisely the additional term for which we are searching—all we have to do is to transform it into a more suggestive form, in which it can be more easily manipulated.

For this purpose, we introduce the flux $\mathfrak{F}_{12}$ which threads the second circuit (the electron trajectory) as a consequence of the presence of the first (the coil). In terms of vector potential A:

$$\mathfrak{F}_{12} = \oint_{C_2} A . \, dl_2$$

(where dl_2 represents a line element of the second circuit), which allows us to define an energy of coupling for each element of the second circuit

$$dW = A . \, dl_2 . i_2,$$

but $i_2 \, dl_2 = e v$, so that the energy stored in the electron by virtue of its motion through a magnetic field is of the form

$$W = A . v e.$$

Returning to the principle of least action in the form (6.4), we find

$$\delta \int_{t_0}^{t_1} (\tfrac{1}{2} m v^2 + e v . A + U) \, dt = 0,$$

which can be simplified, using $\frac{1}{2}mv^2 = e\Phi$ and $U = e\Phi$, into

$$\delta \int_{t_0}^{t_1} (2\Phi + v.A)\,dt = 0,$$

or

$$\delta \int_{s_0}^{s_i} \left[\sqrt{\Phi} + \sqrt{\frac{e}{2m}} \frac{v \cdot A}{|v|} \right] ds = 0.$$

In order to render this integral identical to the integral which forms the basis of Fermat's principle, for refractive index we must use

$$n = K' \left[\sqrt{\Phi} + \sqrt{\frac{e}{2m}} A \cdot \cos\alpha \right],$$

in which α is the angle between the vector potential and the direction of the trajectory. K' is a constant, whose choice is arbitrary but which can with advantage be chosen to be $\sqrt{\dfrac{2e}{mc^2}}$, in order that the refractive index be dimensionless. We then find

$$n = \frac{1}{mc} \left\{ mv + e(A \cdot ds) \right\}$$

(ds being an element of length along the trajectory), in which c denotes the velocity of light *in vacuo*; e is the algebraic value of the charge on the particle (for electrons, $e = -|e|$) but we adopt the convenient convention that e is to be replaced by $|e|$ whenever it occurs under a square root, as in $\sqrt{(e/2m)}$.

Seen from this point of view, the magnetic field behaves like an anisotropic optical medium of a rather curious kind. The refractive index has different values for electrons moving in opposite directions, and the optical reversibility principle is no longer valid. n, on the other hand, can be adjusted at will by adding to A any function of the form grad $f(x, y, z)$ without the trajectories being affected at all, since any function of the form grad f vanishes identically if subjected to the curl operator. The refractive index has the sole simple property that it is constant for all rays lying in a plane perpendicular to A. As A is perpendicular to the meridian planes in a rotationally symmetrical system, the refractive index remains constant for rays moving in a meridian plane. It is upon this result that the success of the theory of magnetic lenses using a fictitious refractive index is based.

Because of these complications, the general expression for refractive index is rarely used. It is, however, possible to obtain a more general, relativistically exact expression of this form, and it is in this extended range of its validity that the interest of so general a refractive index resides. Every problem of electron motion can be attacked in the same way by using a standard procedure, namely, by obtaining the Lagrangian equations.

6.3 THE REFRACTIVE INDEX IN INTERFERENCE PHENOMENA AND THE "WAVE" MECHANICS OF THE ELECTRON

6.3.1 Relative and "Absolute" Definitions in Light Optics; Special Features of Electron Optics

In this chapter, we have hitherto been concerned with *geometrical* optics: the problem has always involved tracing a ray through a refracting medium, characterized by its refractive index $n(x, y, z)$. This never entailed more than a knowledge of the relative magnitude of the refractive index; in other words, if a "ray" or "trajectory" has been determined through a medium of index $n(x, y, z)$, we should obtain the same solution if we were to consider the distribution $n'(x, y, z) = kn(x, y, z)$, where k is a constant. This is clearly a consequence of Snell's law,

$$n_1 \sin i_1 = n_2 \sin i_2,$$

in which only the ratio, n_2/n_1, of the refractive indices of two successive elementary layers of the medium is involved. Since this law is valid in both light and electron optics, there seems at first sight to be a complete analogy between the laws of refraction in these two domains.

In the case of light, however, the possibility of transferring from n to kn (k being a constant) is never mentioned; a basic reason for this is that there is one medium with a privileged refractive index: the vacuum. The problem of changing from n to kn does not need to be considered because it seems perfectly natural to attribute to the vacuum—which is the simplest, best defined and most easily reproducible medium—the refractive index

$$n_{\text{vacuum}} = 1.$$

In electron optics, the vacuum is no longer such a simple medium. For, let us consider the common situation, which is also the simplest, in which the object scatters electrons from an electron gun and the focusing is of the magnetic type. The electrons emerge from the gun with velocity v_0 which is usually measured in terms of the equivalent potential, Φ_0; v_0 plays the role of initial velocity for the electron scattered by the object. The velocity v_0 is involved in calculations of the trajectory (in the coefficient of the second term of equation (5.9), for example) and hence, in the general expression for the refractive index which in this case takes the form

$$n = \left(\sqrt{\phi_0} + \sqrt{\frac{|e|}{2m}} A \cos \alpha \right) K'. \tag{6.16}$$

In the "vacuum", there is thus no unique reference medium; instead, we have a family of media, each element of which is defined by a value of

Φ_0. This leads to another convention for defining a dimensionless refractive index: K' is chosen to be $1/\sqrt{\Phi_0}$; in the commonest cases this leads to the following expressions for n:

electrostatic focusing

$$n = \sqrt{\Phi/\Phi_0}, \tag{6.17}$$

magnetic focusing

$$n = 1 + \sqrt{\frac{e}{2m\Phi_0} A_v}. \tag{6.18}$$

A_v denotes the component of vector potential in the direction of the particle velocity v.

6.3.2 The Refractive Index in Interference Phenomena; Electron Optics

For light, however, the refractive index is also involved in interference phenomena. Let us consider, for example, the classic experiment of Young's slits, in the improved form due to Lord Rayleigh (Fig. 43b). Thus if we select a as the reference tube and maintain the vacuum within it, we can measure the refractive index of a gas introduced into tube b, identical with tube a, by measuring the shift of the interference fringes caused by the presence of gas. This shift is a consequence of the phase difference between the two waves a and b, resulting from the gas, and this phase difference is given by

$$\delta\phi = 2\pi l\left(\frac{1}{\lambda_b} - \frac{1}{\lambda_a}\right) = 2\pi \frac{(n_b - 1)l}{\lambda_0}. \tag{6.19}$$

The length of tube b is denoted by l, λ_0 is the wavelength *in vacuo* of the monochromatic light employed, λ_b is the wavelength in the gas b, n_b is the refractive index of gas b and the refractive index of the vacuum is unity.

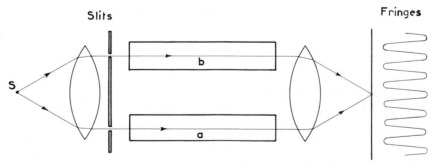

FIG. 43b. Interference between two coherent light beams, by means of which the refractive index of a gas can be measured. a: evacuated tube; b: gas-filled tube.

In this experiment, and in equation (6.19), therefore, we perceive the fundamental reason why the ordinary definition of the refractive index (with $n = 1$ for a vacuum) is regarded as an "absolute" definition.

We now know how to translate this experiment into electron optics, and several experimental workers have followed Chambers (1960) in attempting this. Möllenstedt and Bayh (1962) in particular have succeeded in producing fringes comparable in quality with those regularly obtained with light. This entailed overcoming the experimental problems caused by the extreme shortness of the electron wavelength (0·05 Å in a typical situation). Their experiment is described in §19.2.5 and Fig. 234g shows the fringe shift obtained in the electron analogue of Rayleigh's experiment; the thin graphite flake replaces the medium in tube b in the light optical situation.

This experiment can be analysed by transforming formula (6.19) into electron optical terms. Thus, if the de Broglie wavelength at a point M is λ, we can write $\lambda = \lambda_0/n$, if we introduce a refractive index n for the electrons, just as we do for light in the second part of equation (6.19). The wavelength is given by

$$\frac{1}{\lambda} = \frac{mv}{h} = \frac{\sqrt{2|e|m}}{h}\sqrt{\Phi} \qquad (6.20)$$

(where h is Planck's constant) in the simple case when the effects are wholly *electrostatic*. This is indeed the case in the experiment we are considering, since the graphite flake has the effect of modifying the electrical potential. The so-called "internal" potential is slightly different from that of the same point in space before the flake was inserted; this effect is adequately explained in the theory of solids. We obtain a general formula for the electrostatic case by supposing that the refractive index varies continuously (at the edge of the flake, we then assume that there are two "transition" zones in which the variation is rapid) and that the phase change $d\phi$ produced by an element of length ds is

$$d\phi = \frac{2\pi ds}{\lambda} = \frac{2\pi nds}{\lambda_0}, \qquad (6.21)$$

which leads to

$$n = \lambda_0/\lambda \qquad (6.22)$$

and, using (6.20), to

$$n = (\Phi/\Phi_0)^{1/2}. \qquad (6.23)$$

There are other experiments in which magnetic effects occur along the path b, and it is then interesting to consider the theory of the general case of mixed electric and magnetic fields. In this situation, the books on wave mechanics do not give a general expression for λ but instead, they derive a relation giving the phase change of the wave $(\Delta\phi)_A^B$ when we pass from a point A to a point B. This branch of wave mechanics is explained in a particularly careful fashion by Professor Feynman and his collaborators (1964) in §15.5 of the second volume of their *Lectures on Physics*: a simple discussion is also to be found in the opening chapters on quantum mechanics of Blokhintsev (1964), which is a good introduction to the fundamental article by Aharanov and Bohm (1959). The phase difference is given by

$$(\Delta\phi)_A^B = \frac{2\pi}{\lambda_0}\int_A^B\left\{\left(\frac{\Phi}{\Phi_0}\right)^{1/2} ds - \left(\frac{|e|}{2m\Phi_0}\right)^{1/2} A\cdot ds\right\}, \tag{6.24}$$

in which A is the vector potential of the magnetic field and λ_0 is the de Broglie wavelength which in the purely electrostatic case corresponds to the potential beyond the gun and is given by formula (6.20) with $\Phi = \Phi_0$.

A straightforward calculation shows that the wave mechanical expression (6.24) leads to the same result as we should obtain in light optics if the wavelength in the general mixed case is given by

$$\frac{1}{\lambda} = \frac{mv + eA_v}{h}. \tag{6.25}$$

For the refractive index, we then have

$$n = \frac{\Phi^{1/2} + eA_v}{\Phi_0^{1/2}}. \tag{6.26}$$

A_v denotes the component of vector potential A in the direction of the velocity v of the particle, the charge of which is equal to e in algebraic value (for the electron, $e = -|e|$).

In the geometrical optical approximation, for which the refractive index does not perceptibly vary over distances of the order of a wavelength, λ_0, we see that wave mechanics leads to the same expression for the refractive index as classical mechanics. The former is more exact, however, and the value of K' in equation (6.16) is determined: $K' = 1/\sqrt{\Phi_0}$. In addition, these wave mechanical arguments have the virtue of bringing out the character of the vector potential A as the *local physical quantity* representing magnetic effects in formulae (6.18) and (6.26).

This point has been clearly and exactly discussed by Feynman *et al.* (*loc. cit.*) and in the following section, we attempt no more than to describe the interference phenomena upon which this analysis reposes.

6.3.3 The "Reality" of the Vector Potential

In the foregoing formulae, the magnetic character is defined by the single condition $A \neq 0$. In other words, if in some region within our

experimental space a vector potential A is present, even though the magnetic field vanishes there ($B = 0$), we are confronted with a "magnetic"
case in which A_v must appear in the refractive index. In other words again,
an electron interference experiment may reveal the existence of A locally,
irrespective of the presence of $B =$ curl A, and hence even when curl $A =$
0. It is, moreover, very easy to create two regions with a solenoid, in
one of which $B = 0$, $A \neq 0$; this region is the space outside an idealized
solenoid of infinite length, or in practice, very long (there is no need to
adopt a toric winding). Inside the solenoid, both A and B exist. The experiment is thus in the form illustrated in Fig. 43c, and the sole complication
in comparison with Rayleigh's experiment is that A_v is not constant in
the region in which the electrons interact with A: in the space outside the

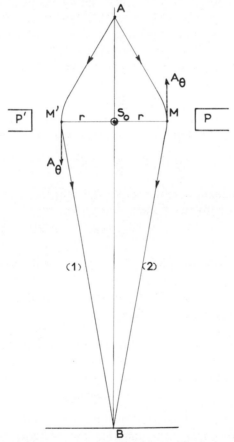

FIG. 43c. Interference between two electron beams, displaying the effect of the
vector potential A_θ of the solenoid S_0. The two beams (1) and (2) are emitted from
A, and deflected by an electric field between PP' and S_0. (This experiment is identical
with that using the Fresnel biprism.)

solenoid, the refractive index of the medium is variable. The dimensions of the solenoid must also be minute and A will affect both waves simultaneously but in opposite senses; the term A_v will be of opposite sign at two corresponding points of the two regions, and the fringe shift for a given current is thus doubled.

It is always extremely difficult to obtain an appreciable spatial separation between the two coherent beams that are to interfere. It was for this reason that the first experimentalists, Chambers (1960) and Fowler *et al.* (1961), employed not a solenoid but a very slender needle of magnetized iron (a whisker) and were only able to confirm the theory rather crudely because of uncertainty about the magnetization of the iron.

In 1962, Möllenstedt and Bayh succeeded in making a genuine solenoid of very small diameter ($2a = 10\ \mu$m), and by means of their biprism they were able to separate the beams sufficiently to insert this microscopic coil. With this, they obtained good quantitative confirmation of the theory. If Ψ is the magnetic flux inside the solenoid ($\Psi = B\pi a^2$), then at a point M distance r from the axis of the coil the vector potential has a component A_θ given by

$$A_\theta = \frac{\Psi}{2\pi r}.$$

There is no need to know this expression, however, since the fundamental property of the vector potential gives us the quantity of interest directly:

$$\oint_{(1)} A \cdot ds - \oint_{(2)} A \cdot ds = \Psi. \tag{6.27}$$

The wave mechanical formula (6.24) then shows directly that the phase difference between the two beams, $\delta\phi$, is given by

$$\delta\phi = \frac{2\pi e}{h}\Psi, \tag{6.28}$$

as Aharanov and Bohm showed for the first time in 1959 and as Professor Feynman explains in detail in chapter 15 of volume II of his *Lectures*. For $\delta\phi = 2\pi$, the interference pattern is shifted by one fringe-width, which in theory corresponds to flux Ψ_{th}:

$$\Psi_{th} = \frac{h}{e} = 4 \cdot 13 \times 10^{-15}\ \text{tesla} = 4 \cdot 13 \times 10^{-7}\ \text{maxwell.} \tag{6.29}$$

Experimentally, Möllenstedt and Bayh found

$$\Psi_{exp} = 4 \cdot 07 \times 10^{-7}\ \text{maxwell.}$$

The sources of error have been analysed in detail by Werner and Brill (1960), who showed in particular that the presence of time-dependent parasitic fields would not modify the theory in any way.

These interference effects were predicted as early as 1949 by Ehrenberg and Siday, who employed purely optical arguments.

CHAPTER 7

LENS DEFECTS

7.1 THE LIMITS OF THE GAUSSIAN APPROXIMATION

In practice, the Gaussian theory is of considerable importance, as there are a great many electron optical instruments which obey the laws to which it leads, and which form images which seem both well-defined and faithful. Nevertheless, a meticulous examination of the image always reveals defects, so that instead of a point image, we find a small spot which need not even be located at the Gaussian image point if the object is not situated on the axis. Whether, however, the origin of these defects is to be sought in the terms which we have neglected in establishing the Gaussian theory, or whether they are due to other causes which we have not as yet mentioned can only be decided by a careful study of each individual instrument, and is not known in every case.

We shall not expatiate here upon the effects caused by the mutual repulsion of the electrons, which are known as "space charge effects", as the effect of space charge—which has no glass optical parallel—on image formation is, as a general rule, negligible in the instruments which concern us here; the reason for this is simply that, since we can obtain very brilliant images and intense photographic effects with very low current densities, the electrons will be widely separated in space, and their interaction negligible. Space charge is of prime importance, however, in certain electron guns and in electron tubes, but this represents only a very specialized category of phenomena, and we shall simply refer the reader to § 8.5.3 and to the relevant books by Pierce (1949) and Klemperer (1953).

We shall, however, study in some detail the third order geometrical aberrations (which arise from terms in r^3, which we ignored in the Gaussian approximation), chromatic aberration, and the mechanical aberrations which arise from imperfections in the assembly and machining of the various parts of the instrument. It is the magnitude of these defects which decides the width of the tubular region around the axis within which the Gaussian approximation is valid; this, in practice, means the maximum diameters of the metal apertures which define this region.

A priori it would seem that the quality of an image can be improved to any desired extent simply by reducing the diameter of these apertures.

This cannot be continued indefinitely, however, owing to the corresponding reduction of luminous intensity (in a cathode ray tube, for example, we have to choose between spot size and luminous intensity), or more rarely, to diffraction; the latter is particularly marked in electron microscopy.

The angular aperture α of the elementary pencils which are emitted by a point on a microscope object is of the order of 5×10^{-4} radians. The image of a point is thus a small spot, of which the diameter, divided by the magnification, is

$$d = \frac{24}{\alpha \sqrt{\Phi}}$$

(in which d is measured in millimicrons, α in radians and Φ in kV).

The diffraction is a consequence of the wave-like qualities of the electron, which ceases to behave like a classical particle when its trajectory is too narrowly restricted by small diaphragms. This is an example of the existence in instrumental optics of the electron waves of L. de Broglie, which behave like light waves save in that their wavelengths are usually some 100,000 times smaller than those of light in the visible region.

7.2 THIRD ORDER ABERRATIONS

7.2.1 Definition of the Aberration Coefficients

We define x_0, y_0 to be the coordinates of some point A in the object plane $z = z_0$ and x_i, y_i to be the coordinates of A', the Gaussian image of A, in the plane $z = z_i$ (Fig. 44). The point N, at which a ray from A

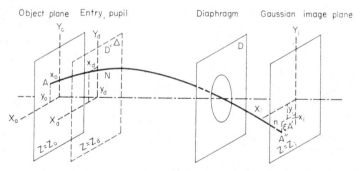

FIG. 44. The parameters which characterize a trajectory.

passes through the diaphragm which defines the extent of the beam, has coordinates x_d, y_d; if the actual aperture is situated within the lens, however, it is replaced by a hypothetical equivalent aperture, in the region in which the field is zero. We may for example, select the image of the real stop formed by the part of the lens which precedes it, which gives us the "entry pupil", the plane D' in Fig. 44. This ray through A and N intersects the image plane $z = z_i$ not at the Gaussian image point $A'(x_i, y_i)$ but at the point $A''(x_i', y_i')$. The aberration is characterized by the vector $A'A''$, thus:

$$A'A'' \begin{cases} \xi = x_i' - x_i = f(x_0, y_0, x_d, y_d), \\ \eta = y_i' - y_i = g(x_0, y_0, x_d, y_d), \end{cases} \tag{7.1a}$$

in which $x_i = G x_0$ and $y_i = G y_0$ (G is the linear magnification).

We now express f and g as series expansions up to terms of the third order. All the second order and other terms of even order drop out, as all the quantities x_0, y_0, x_d and y_d must change sign simultaneously in a rotation of 180°.

The general expression for the third order terms contains twenty coefficients for x_i and twenty for y_i; for example:

$$\begin{aligned}
x_i' - G x_0 = {} & a_1 x_d^3 + a_2 x_d^2 y_d + a_3 y_d^2 x_d + a_4 y_d^3 \\
& + b_1 x_0 x_d^2 + b_2 x_0 x_d y_d + b_3 x_0 y_d^2 + b_4 y_0 x_d^2 \\
& + b_5 y_0 x_d y_d + b_6 y_0 y_d^2 \\
& + c_1 x_0^2 x_d + c_2 x_0^2 y_d + c_3 x_0 y_0 x_d + c_4 x_0 y_0 y_d \\
& + c_5 y_0^2 x_d + c_6 y_0^2 y_d \\
& + d_1 x_0^3 + d_2 x_0^2 y_0 + d_3 x_0 y_0^2 + d_4 y_0^3.
\end{aligned} \tag{7.1b}$$

We shall show that the presence of axial symmetry implies a large reduction in the number of terms; in the general case, only twelve aberration terms remain.

If we write

$$\begin{aligned}
u &= x_0 + j y_0, \\
v &= x_d + j y_d, \\
w &= \xi + j \eta,
\end{aligned} \tag{7.2}$$

then w is a function of u, v, and their complex conjugates

$$\begin{aligned}
\bar{u} &= x_0 - j y_0, \\
\bar{v} &= x_d - j y_d
\end{aligned} \tag{7.3}$$

($\bar{u}$ and $\bar{v}$ are necessary, since the relation between w and x_0, y_0, x_d and y_d is too general for u and v alone to be sufficient). An arbitrary rotation of the coordinates about the axis transforms u and v into $u' = u e^{j\theta}$ and $v' = v e^{j\theta}$ respectively (with complex conjugates $\bar{u}' = \bar{u} e^{-j\theta}$, $v' = v e^{-j\theta}$) and w into $w' = w e^{j\theta}$. The relation between w' and u', v' must be identical to the relation between w and u, v. This condition is satisfied if we write the third order terms in the form

$$w = (a u + b v) \cdot P_2(u, \bar{u}, v, \bar{v}),$$

in which P_2 is a polynomial of the second degree which is invariant to the transformation above, and which is therefore of the form

$$P_2 = c\,u\,\bar{u} + d\,u\,\bar{v} + e\,v\,\bar{u} + f\,v\,\bar{v}.$$

w, therefore, is of the form

$$w = (A + jA')\,v^2\,\bar{v} + (B + jB')\,u^2\,\bar{v} + (C + jC')\,u\,\bar{u}\,v$$
$$+ (D + jD')\,\bar{u}\,v^2 + (E + jE')\,u\,v\,\bar{v} + (F + jF')\,u^2\,\bar{u}.$$

This function contains twelve real coefficients, but we can simplify it still further by taking an important physical property of the system into account; according to the theorem of Malus and Dupin, every ray is a normal to each member of a family of surfaces (a statement and proof of this theorem is to be found in books on geometrical optics, starting from the laws of refraction). This physical property is contained in the statement

$$\mathrm{Im}\left(\frac{\delta w}{\delta v}\right) = 0$$

(cf. *Optique Electronique et Corpusculaire*, by Louis de Broglie).

Applying this condition, we find

$$A' = C' = 0,$$
$$2D = E,$$
$$2D' = -E'.$$

In the general case, therefore, we have only eight aberration coefficients, and w becomes:

$$w = A\,v^2\,\bar{v} + (B + jB')\,u^2\,\bar{v} + C\,u\,\bar{u}\,v + (D + jD')\bar{u}\,v^2$$
$$+ 2(D - jD')\,u\,v\,\bar{v} + (F + jF')\,u^2\,\bar{u}. \tag{7.4}$$

This formula is in the form suitable for magnetic lenses; in electrostatic lenses, however, there is no force which would make an electron leave its meridian plane, which behaves therefore like a symmetry plane. w, in consequence, must be unaffected when u, $\bar{u}$, v and $\bar{v}$ are replaced by u, u, v and v respectively, which implies

$$B' = D' = F' = 0. \tag{7.5}$$

The aberration coefficients are now, as in glass optics, only five in number.

7.2.2 The Form of the Aberration Figures

With the aid of the formulae of the preceding section, we can determine both the shape and the position of the small spot which plays the role of image point. If all the terms are of comparable importance, the result is extremely complicated (Bricka and Bruck, 1948). It is, however, possible to choose the experimental conditions in such a way that all the terms except one are negligible. We shall, in fact, select these conditions in such a way that the spots have shapes already familiar in glass optics, made slightly more complicated by the effect of a magnetic field.

Aberrations produced by the aperture

The lens is assumed to be operating at maximum aperture, with the whole surface of the diaphragm illuminated, but the object is to remain in the neighbourhood of the axis. Polar coordinates in the aperture plane are denoted by r, φ ($v = r\,e^{j\varphi}$) and those in the object plane by ρ, θ ($u = \rho\,e^{j\theta}$); we shall always select $\theta = 0$ to simplify the analysis.

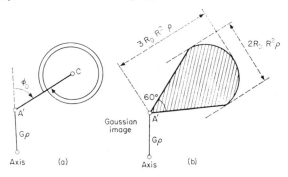

FIG. 45. Coma. (a) The circle described by those rays which pass through the diaphragm, at the same distance from the axis. (b) The appearance of the aberration figure.

(1) *Spherical aberration.* If we consider a point object on the lens axis, $u = 0$, equation (7.4) is reduced to

$$w = Ar^3\,e^{j\varphi}. \tag{7.6}$$

If we consider the rays which pass through the entry pupil along a circle of radius r, the angle φ varies between 0 and 2π; the points A'' in which these rays intersect the Gaussian image plane lie on a circle whose centre is the Gaussian image point and whose radius is Ar^3. If the whole diaphragm is irradiated, the family of circles will coalesce into a small circular spot, of radius AR^3, where R is the radius of the diaphragm, and A is the *spherical aberration coefficient*.

(2) *Coma.* Let us suppose that the spherical aberration is negligible for a pair of conjugate planes z_0 and z_i; we consider an object sufficiently close to the axis for u^2 and u^3 to be negligible. The aberration, therefore, is given by

$$w = R_D\,r^2\,\varrho\,[e^{j(2\varphi + \varphi_D)} + 2e^{-j\varphi_D}], \tag{7.7}$$

where the substitution

$$D + j\,D' = R_D\,e^{j\varphi_D}$$

has been made.

Once again, we consider those rays which pass through the entry diaphragm around a circle of radius r, over which φ varies from 0 to 2π; the points A'' lie on a circle, with radius $R_D\,r^2\,\varrho$ from the Gaussian image point at an azimuthal angle φ_D from the radius vector OA'. It is to be noticed that as φ varies from 0 to 2π, this circle is described twice (because of the presence of $2\,\varphi$ in the exponent). This is illustrated in Fig. 45 a.

If we consider the rays which pass through the remainder of the diaphragm, we obtain a family of superimposed circles, giving an aberration figure of the form shown in Fig. 45 b.

D is known as the *coma coefficient*, while D', which is responsible for the rotation and is peculiar to magnetic lenses, is the *coefficient of anisotropic coma*.

In general, the aberration figure which is observed will not be of this simple form, as the spherical aberration, which cannot be eliminated from electron lenses, will be superimposed onto the coma; preponderant as the spherical aberration often is, the coma is quite swamped. In glass optics, on the contrary, objectives rigorously corrected for spherical aberration can be obtained, and it is then possible to observe pure coma.

Field aberrations

We suppose now that the object is far from the axis, so that all the powers of u have to be taken into consideration, but that the system is stopped, on the other hand, in such a way that powers of v higher than unity can be neglected. The general formula (7.4) becomes

$$w = R_B\varrho^2 r e^{j(\varphi_B-\varphi)} + C\varrho^2 r e^{j\varphi} + R_F\varrho^3 e^{j\varphi_F}, \tag{7.8}$$

in which we have rendered the notation more consistent by writing

$$B + jB' = R_B e^{j\varphi_B}; \quad F + jF' = R_F e^{j\varphi_F}.$$

(1) *Distortion:* The last term is independent of the point in which the ray passes through the diaphragm—it produces, therefore, not a spot but an overall displacement of the image, which is subjected to a translation. F, from now on, will be known as the *distortion coefficient*, while F' which produces a rotational effect as well, will be called *the coefficient of anisotropic distortion*. (See Fig. 46.)

If the object is a square mesh, it is easy to verify that the image is deformed into a "cushion" or a "barrel" according to the sign of F. In a magnetic lens, the presence of F' produces, in addition, strain across the image. These results are illustrated in Fig. 47.

(2) *Astigmatism and field curvature:* If we consider the aberration from the point of view of the new image which has undergone translation as a result of distortion, the remaining term of equation (7.4) is simply

$$w = \varrho^2 r [R_B e^{j(\varphi_B - \varphi)} + C e^{j\varphi}].$$ (7.9)

Once again we consider the family of rays which pass through the entry pupil along a circle of radius r. It is easy to show that the points A'' define

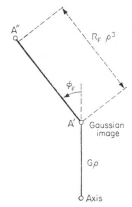

FIG. 46. The image displacement which distortion produces.

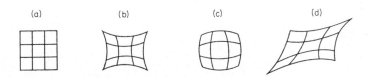

FIG. 47. The appearance of an object (a), in the presence of distortion: (b) cushion, (c) barrel and (d) anisotropic cushion.

an ellipse, the characteristics and position of which are set out in Fig. 48. When the whole diaphragm is illuminated, the spot is found to be elliptical, with major axis $2\varrho^2 R(R_B + C)$ and minor axis $2\varrho^2 R(R_B - C)$, and with the major axis inclined to the radius vector OA' at an angle $\varphi_B/2$.

If we calculate the shape of the pencil in image space, we find that it collapses into two focal lines—the lens is said to be essentially astigmatic. The standard procedures of geometrical optics show that the distance between the line foci is proportional to half the sum of the axes of the ellipse; the astigmatic difference depends, therefore, only upon $R_B e^{j\varphi_B}$. B, then, is the *coefficient of astigmatism*, while B', peculiar to magnetic lenses, which produces a rotation of the directions of the line foci, is the *coefficient of anisotropic astigmatism.* The line foci associated with a magnetic lens are no longer "sagittal" and "tangential", but are inclined to the reference direction at an arbitrary angle (Fig. 49).

Except in the special case of a rectilinear object, the best image which is obtainable when astigmatism is present is the circle of least confusion

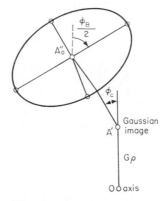

FIG. 48. The ellipse which is traced out by rays which pass through the diaphragm at the same distance from the axis.

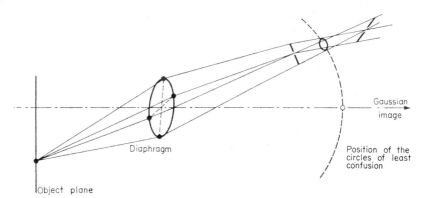

FIG. 49. The structure of an astigmatic pencil.

of the beam. It is of some interest to calculate the distance between this circle and the Gaussian image plane; this distance depends only upon the difference between the axes of the ellipse, upon C that is, as can be verified without difficulty.

C is the *field curvature coefficient.* Its presence in the formula implies that the best image is to be obtained not on a plane screen passing through the Gaussian image plane, but on a curved screen, like the ones used in cathode ray tubes.

7.2.3 A Comparison of Magnetic and Electrostatic Lenses

Both types of lens display the same basic defects: spherical aberration, coma, distortion, astigmatism and field curvature.

In addition to these, three further aberrations, which are said to be "anisotropic", appear in magnetic lenses as a result of the difference between the exact rotation of the meridian plane and the rotation of the Gaussian image, which are equal only in the first order. The signs of these aberrations depend upon that of $B(z)$, and on reversing the direction of the current, they take up a configuration which is symmetrical to the previous one with respect to the meridian plane. The coefficients B', D' and F' are peculiar to magnetic lenses.

With the aid of the Störmer potential, Ψ, we can compare the coefficients A, B, C, D and F which are common to both kinds of lens, and pass from the coefficients calculated for the electric lens to the corresponding quantities for a magnetic lens. We make a distinction between the two cases in which the integration constant appears and does not appear.

The constant of integration C does not affect three of the aberrations: spherical aberration, distortion and coma. For spherical aberration, $C = 0$ for all the rays when the point object is situated on the axis. Distortion only deforms the image, and does not blur it; rays for which $C = 0$ therefore converge onto the same point as those for which $C \neq 0$, and hence C cannot appear in the expression for this aberration.

For coma, finally, we know that two rays whose directions are initially symmetrical with respect to the meridian plane intersect again in the image plane—since these rays correspond to different values of C, however, the latter cannot appear in the general expression.

In these three cases, the transfer from the expressions for the coefficients A, D and F which are valid for an electrostatic lens over to those which are valid for a mixed lens, which produces a combined electric and magnetic field, is effected by replacing

$$\Phi(r, z) \quad \text{by} \quad \Phi(r, z) - \frac{e}{2m} A_\theta^2$$

(see § 6.2.2).

Since

$$A_\theta = \frac{1}{2} r B - \frac{1}{16} r^3 B'',$$

we have

$$A_\theta^2 = \frac{r^2}{4} B^2 - \frac{r^4}{16} B B''. \tag{7.10}$$

Hence

$$\Phi(r, z) = \varphi(z) - \frac{r^2}{4}\left(\varphi'' + \frac{e}{2m} B^2\right) + \frac{r^4}{64}\left(\varphi^{(iv)} + B B'' \frac{2e}{m}\right), \tag{7.11}$$

so that in practice the following substitutions have to be made:

$$\varphi(z) \rightarrow \varphi(z),$$

$$\varphi''(z) \rightarrow \varphi''(z) + \frac{2e}{m} B^2,$$

$$\varphi^{(iv)}(z) \rightarrow \varphi^{(iv)}(z) + \frac{2e}{m} B B''.$$

In the other situation, in which C has an effect, it is no longer possible to pass easily from the expressions for the aberrations in an electrostatic lens to those in a magnetic lens. This is the case for field curvature and astigmatism, where the corresponding coefficients contain supplementary terms in the magnetic case due to the presence of the constant C.

7.3 CALCULATION OF THE ABERRATION COEFFICIENTS

7.3.1 The Trajectory Method

The aberration coefficients can be fully evaluated by calculating the electron trajectories when the terms in r^3 which we neglected in the Gaussian approximation are taken into account. First, however, we must determine the terms up to and including the third order in the series expansion of the electric or magnetic field.

We shall show in detail how the spherical aberration is calculated for the simpler case of an electrostatic field. This means that we still have cylindrical symmetry about the axis, and it is profitless to concern ourselves with rays which do not lie in a meridian plane. We can, therefore, perform the whole calculation, as in the Gaussian approximation, simply by determining a meridian ray.

For the electrostatic potential, we find

$$\Phi(r, z) = \varphi(z) - \frac{r^2}{4} \varphi''(z) + \frac{r^4}{64} \varphi^{(iv)}(z). \tag{7.12}$$

The equations of motion, therefore, must be written in the following form, in which terms in r^3 and r'^2, hitherto neglected, have been retained:

$$m \frac{d^2 r}{dt^2} = +e\left[\frac{r}{2}\varphi'' - \frac{r^3}{16}\varphi^{(iv)}\right]; \quad m \frac{d^2 z}{dt^2} = -e\left[\varphi' - \frac{r^2}{4}\varphi'''\right]. \tag{7.13}$$

Replacing the second of these equations by the conservation of energy relation:

$$\frac{1}{2} m(z_t'^2 + r_t'^2) = e\left[\varphi(z) - \frac{r^2}{2}\varphi''\right], \tag{7.14}$$

we obtain the third order approximation to dz/dt:

$$\frac{d}{dt} = \sqrt{\frac{2e\varphi}{m}} \left(1 - \frac{r^2}{4} \frac{\varphi''}{\varphi}\right)\left(1 - \frac{r'^2}{2}\right)\frac{d}{dz}.$$

Finally, we obtain the trajectory equation for an electrostatic lens:

$$\frac{d^2r}{dz^2} + \frac{1}{2}\frac{\varphi'}{\varphi}\frac{dr}{dz} + \frac{1}{4}\frac{\varphi''}{\varphi}r = \left[\frac{1}{32}\frac{\varphi^{(iv)}}{\varphi} - \frac{1}{16}\left(\frac{\varphi''}{\varphi}\right)^2\right]r^3$$
$$+ \frac{1}{8}\left[\frac{\varphi'''}{\varphi} - \frac{\varphi''\varphi'}{\varphi^2}\right]r^2r' - \frac{1}{4}\frac{\varphi''}{\varphi}rr'^2 - \frac{1}{2}\frac{\varphi'}{\varphi}r'^3\cdots$$

$$(7.15)$$

The reader can establish without difficulty the trajectory equation for an axially symmetric system containing both an electric field and a magnetic field. The full expression is to be found in the book by de Broglie (1950).

Since the aberration terms are small, we shall integrate this equation by a perturbation method. We write

$$r = r_0 + \varepsilon$$

in which r_0 represents the Gaussian trajectory, originating in a point object, in some chosen direction. As was shown in § 4.3.1, it is possible to find the equation of this trajectory if we know two particular trajectories $X(z)$ and $Y(z)$. The perturbation ε is sufficiently small for powers greater than unity to be negligible.

The equation for the perturbation, therefore, is

$$\frac{d^2\varepsilon}{dz^2} + \frac{1}{2}\frac{\varphi'}{\varphi}\frac{d\varepsilon}{dz} + \frac{1}{4}\frac{\varphi''}{\varphi}\varepsilon = \left[\frac{1}{32}\frac{\varphi^{(iv)}}{\varphi} - \frac{1}{16}\left(\frac{\varphi''}{\varphi}\right)^2\right]r_0^3$$
$$+ \frac{1}{8}\left[\frac{\varphi'''}{\varphi} - \frac{\varphi''\varphi'}{\varphi^2}\right]r_0^2r_0' - \frac{1}{4}\frac{\varphi''}{\varphi}r_0r_0'^2 - \frac{1}{2}\frac{\varphi'}{\varphi}r_0'^3,$$

$$(7.16)$$

a linear second order differential equation with a term independent of ε on the right-hand side. Provided two solutions of the corresponding homogeneous equation are known, this equation can be solved by the method of "variation of parameters". These two solutions are, however, already known, since the homogeneous equation is none other than the equation for the Gaussian trajectories—Fig. 24 illustrates their behaviour.

The full calculations are lengthy, but not particularly difficult—we shall only give them in outline.

We consider a Gaussian ray which leaves an axial point object which has abscissa z_0, at an inclination α, so that

$$r_0 = \alpha X(z), \qquad (7.17)$$

in which $X(z)$ is the particular solution shown in Fig. 24. Both the perturbation and its derivative must be zero at z_0, since the third order trajectory certainly coincides with the Gaussian trajectory at the object. $\varepsilon(z_0) = \varepsilon'(z_0) = 0$, therefore. Elsewhere, we write

$$\varepsilon = \alpha(z) X(z) + \beta(z) Y(z) \qquad (7.18)$$

and apply the method of variations of parameters. We have

$$\alpha'(z)\,X(z) + \beta'(z)\,Y(z) = 0,$$
$$\alpha'(z)\,X'(z) + \beta'(z)\,Y'(z) = \alpha^3\,P(X, X', \varphi),$$

in which P represents the inhomogeneous term of the differential equation for ε into which the value of r_0 has been substituted.

The spherical aberration represents the value of ε in the Gaussian image plane, $z = z_i$. But $X(z_i) = 0$, and the radius of the aberration disc is given, therefore, by the expression:

$$\varepsilon(z_i) = |w| = \beta(z_i)\,Y(z_i).$$

Solving the pair of simultaneous equations for α' and β' we obtain

$$\beta'(z) = \frac{\alpha^3\,X\,P(X, X', \varphi)}{X'Y - XY'}.$$

But according to (4.16''')

$$X'Y - XY' = \sqrt{\frac{\varphi_0}{\varphi}},$$

so that

$$\beta'(z) = \frac{\alpha^3}{\sqrt{\varphi_0}}\,X\,\sqrt{\varphi}\cdot P(X, X', \varphi),$$

and integrating, this gives

$$\beta(z_i) - \beta(z_0) = \frac{\alpha^3}{\sqrt{\varphi_0}}\int_{z_0}^{z_i} X\,\sqrt{\varphi}\,P(X, X', \varphi)\,\mathrm{d}z.$$

Since $\varepsilon(z_0) = 0$, however, $\beta(z_0)$ is necessarily zero, so that finally we obtain

$$|w| = \frac{\alpha^3}{\sqrt{\varphi_0}}\int_{z_0}^{z_i} X\,\sqrt{\varphi}\cdot P(X, X', \varphi)\,\mathrm{d}z. \tag{7.19}$$

In order to express this result in the notation which we use in the general study, we introduce an entry pupil, radius R, which is placed a distance δ from the object. R and α are related by the equation $R = \alpha\,\delta$ (we have in fact two points in object space, between which the rays are straight lines). The transverse linear magnification, G, is given by $Y(z_i)$, as Fig. 24 shows. We know that

$$|w| = A\,R^3,$$

and hence

$$A = \frac{G}{\delta^3\,\sqrt{\varphi_0}}\int_{z_0}^{z_i} X^4\,\sqrt{\varphi}\left\{\frac{1}{32}\,\frac{\varphi^{(iv)}}{\varphi} - \frac{1}{16}\left(\frac{\varphi''}{\varphi}\right)^2 + \frac{1}{8}\left(\frac{\varphi'''}{\varphi} - \frac{\varphi'\,\varphi''}{\varphi^2}\right)\frac{X'}{X}\right.$$
$$\left. - \frac{1}{4}\,\frac{\varphi''}{\varphi}\left(\frac{X'}{X}\right)^2 - \frac{1}{2}\,\frac{\varphi'}{\varphi}\left(\frac{X'}{X}\right)^3\right\}\mathrm{d}z.$$

Integration by parts removes φ''' and $\varphi'\,\varphi''$; for example,

$$\int_{z_0}^{z_i} \frac{\varphi'''}{\sqrt{\varphi}} X^3 X \,\mathrm{d}z = \left[\frac{\varphi''}{\sqrt{\varphi}} X' X^3\right]_{z_0}^{z_i} + \int_{z_0}^{z_i} X^4 \sqrt{\varphi}\left[\frac{1}{4}\left(\frac{\varphi''}{\varphi}\right)^2\right.$$

$$\left. + \frac{\varphi'\,\varphi''}{\varphi^2}\frac{X'}{X} - 3\frac{\varphi''}{\varphi}\left(\frac{X'}{X}\right)^2\right]\mathrm{d}z,$$

and finally

$$A = \frac{G}{\delta^3\sqrt{\varphi}}\int_{z_0}^{z_i} X^4 \sqrt{\varphi}\left[\frac{1}{32}\frac{\varphi^{(iv)}}{\varphi} - \frac{1}{32}\left(\frac{\varphi''}{\varphi}\right)^2 - \frac{1}{4}\frac{\varphi''}{\varphi}\left(\frac{X'}{X}\right)^2 - \frac{1}{2}\left(\frac{X'}{X}\right)^4\right]\mathrm{d}z.$$

$$(7.20)$$

A different aberration constant is frequently used, which is defined by the relation

$$|w| = G\,C_3\alpha^3,$$

C_3 and A being connected by the relation

$$C_3 = \frac{A\,R^3}{G\,\alpha^3} = \frac{A\,\delta^3}{G}.$$

C_3 can therefore be calculated directly from (7.20) by numerical integration; it is, however, better to use the formula below, which has been taken from the list given in § 7.3.2 as it is equivalent to (7.20) but no longer contains terms in $\varphi^{(iv)}$.

$$C_3 = \frac{1}{16\sqrt{\varphi_0}}\int_{z_0}^{z_i} X^4 \sqrt{\varphi}\cdot\left[\frac{5}{4}\left(\frac{\varphi''}{\varphi}\right)^2 + \frac{5}{24}\left(\frac{\varphi'}{\varphi}\right)^4\right.$$

$$(7.20')$$

$$\left. + \frac{14}{3}\left(\frac{\varphi'}{\varphi}\right)^3\frac{X'}{X} - \frac{3}{2}\left(\frac{\varphi'}{\varphi}\right)^2\left(\frac{X'}{X}\right)^2\right]\mathrm{d}z.$$

We could equally well use yet another equivalent formula in which only the function $T(z) = \dfrac{\varphi'}{\varphi}$ and its derivative $T'(z)$ appear (see Glaser, 1956, p. 229).

$$C_3 = \frac{1}{64\sqrt{\varphi_0}}\int_{z_0}^{z_i} X^4 \sqrt{\varphi}\cdot\left[\left(3T^4 - \frac{9}{2}T^2T' + 5T'^2\right) + 4TT'\frac{X'}{X}\right]\mathrm{d}z.\quad(7.20'')$$

Using Störmer's analogy we can transfer this to the magnetic lens situation, to give:

$$A = \frac{G}{\delta^3\sqrt{\varphi}}\int_{z_0}^{z_i} X^4 \sqrt{\varphi}\left\{\frac{1}{32}\frac{\eta^{(iv)}}{\varphi} + \frac{1}{16}\frac{e}{m}\frac{B\,B''}{\varphi}\right.$$

$$\left. - \frac{1}{32}\left(\frac{\varphi'' + \dfrac{e}{2m}B^2}{\varphi}\right)^2 - \frac{1}{4}\left(\frac{\varphi'' + \dfrac{e}{2m}B^2}{\varphi}\right)\left(\frac{X'}{X}\right)^2 - \frac{1}{2}\left(\frac{X'}{X}\right)^4\right\}\mathrm{d}z.$$

$$(7.21)$$

For a purely magnetic lens, in which $\varphi = \varphi_0$, the expression for C_3 will become

$$C_3 = \frac{e}{m} \int_{z_0}^{z_i} X^4 \left[\frac{1}{16} \frac{B\,B''}{\varphi_0^*} - \frac{1}{32} \left(\frac{e}{2m} \frac{B}{\varphi_0^*} \right)^2 - \frac{1}{8} \frac{B^2}{\varphi_0^*} \frac{X'^2}{X^2} - \frac{1}{2} \frac{X'^4}{X^4} \right] dz,$$

$$(7.21')$$

and this can be transformed into

$$C_3 = \frac{e}{96m\,\varphi_0^*} \int_{z_0}^{z_i} X^4 \left(\frac{2e}{m\,\varphi_0^*} B^4 + 5B'^2 - B\,B'' \right) dz, \qquad (7.21'')$$

and hence into

$$C_3 = \frac{e}{128\,m\,\varphi_0^*} \int_{z_0}^{z_i} X^4 \left[\frac{3e}{m\,\varphi_0^*} B^4 + 8B'^2 - 8B^2 \frac{X'^2}{X^2} \right] dz.$$

In all these formulae, the potential φ_0^* represents the relativistically corrected value of the acceleration potential. It is (7.21'') which is used above all for the numerical calculation of C_3.

The complete calculation of the coefficients has been carried out by Glaser (1933) and by Scherzer (1936). The latter had the idea of eliminating all the terms containing third or fourth derivatives by a series of partial integrations. The integrated terms disappear, a consequence of the Lagrange–Helmholtz relation, and the formulae listed in the next section result.

Recknagel (1940) attempted to obtain the aberration coefficients of an electron lens, starting from the reduced equations (§ 4.2.2), which give a simpler Gaussian equation. Later, however, Sturrock (1951) established these formulae by a more direct method, the principle of which we shall explain later.

7.3.2 List of the Aberration Coefficients

We give below the formulae obtained by Scherzer, in a notation which is in conformity with that which we have used earlier; $X(z)$ and $Y(z)$ represent the two Gaussian trajectories defined in § 4.3.1, $\varphi(z)$, $B(z)$ and C represent the axial potential, the axial magnetic field, and the integration constant (not to be confused with the field curvature coefficient), and G the magnification; δ is the distance between the object plane and the entry diaphragm.

We obtain

$$A = \frac{G}{16\delta^3 \sqrt{\varphi_0}} \int_{z_0}^{z_i} X^4 \sqrt{\varphi} \left[f + 4g\,\frac{X'}{X} + 2h\left(\frac{X'}{X}\right)^2 \right] dz,$$

$$B = \frac{G}{16\delta \sqrt{\varphi_0}} \int_{z_0}^{z_i} X^2 Y^2 \sqrt{\varphi} \cdot \left[f + 2g\left(\frac{X'}{X} + \frac{Y'}{Y}\right) \right.$$
$$\left. + \left(3h + \frac{\varphi'^2}{\varphi^2}\right)\left(\frac{Y'}{Y} - \frac{X'}{X}\right)^2 + 2h\,\frac{Y' X'}{Y X} \right] dz + \frac{G}{4\delta} [X'\,Y']_{z_0}^{z_i},$$

$$B' = \frac{G}{8\delta} \sqrt{\frac{2e}{m}} \int_{z_0}^{z_i} \frac{X Y B}{\varphi^2} \sqrt{\varphi} \cdot \left[k - \frac{\varphi'}{4}\left(\frac{Y'}{Y} + \frac{X'}{X}\right) + \varphi\,\frac{X' Y'}{X Y} \right] dz$$
$$+ \frac{G}{16\delta} \sqrt{\frac{2e}{m}} \,\varphi_0 \left[\frac{B}{\varphi}\right]_{z_0}^{z_i},$$

$$C = \frac{G}{8\delta \sqrt{\varphi_0}} \int_{z_0}^{z_i} Y^2 X \sqrt{\varphi} \cdot \left[f + 2g\left(\frac{X'}{X} + \frac{Y'}{Y}\right) \right.$$
$$\left. - \left(h + \frac{\varphi'^2}{2\varphi^2}\right)\left(\frac{X'}{X} - \frac{Y'}{Y}\right)^2 + 2h\,\frac{X' Y'}{X Y} \right] dz + \frac{G}{8\delta} \left[\sqrt{\varphi_0}\,\frac{\varphi'}{\varphi^{3/2}} 4 X'\,Y' \right]_{z_0}^{z_i},$$

$$D = \frac{G}{16\delta^2 \sqrt{\varphi_0}} \int_{z_0}^{z_i} Y X^3 \sqrt{\varphi} \cdot \left[f + g\left(3\,\frac{X'}{X} + \frac{Y'}{Y}\right) \right.$$
$$\left. + h\left(\frac{X' Y'}{X Y} + \left(\frac{X'}{X}\right)^2\right) \right] dz + \frac{G}{8\delta^2} [X'^2]_{z_0}^{z_i},$$

$$D' = \frac{G}{16\delta^2} \sqrt{\frac{2e}{m}} \int_{z_0}^{z_i} \frac{B X^2}{\varphi^2} \sqrt{\varphi} \cdot \left[h - \frac{\varphi'}{2}\,\frac{X'}{X} + \varphi\left(\frac{X'}{X}\right)^2 \right] dz,$$

$$F = \frac{G}{16 \sqrt{\varphi_0}} \int_{z_0}^{z_i} Y^3 X \sqrt{\varphi} \cdot \left[f + g\left(3\,\frac{Y'}{Y} + \frac{X'}{X}\right) + h\left(\left(\frac{Y'}{Y}\right)^2 + \frac{Y' X'}{Y X}\right) \right] dz$$
$$+ \frac{G}{4} \left[Y^2 \left(\frac{\varphi''}{8\varphi} + \frac{5}{32}\left(\frac{\varphi'}{\psi}\right)^2 + \frac{\varphi'}{2\varphi}\,\frac{Y'}{Y} + \frac{3}{2}\left(\frac{Y'}{Y}\right)^2 + \frac{e}{8m}\,\frac{B^2}{\varphi}\right) \right]_{z_0}^{z_i},$$

$$F' = \frac{G}{16} \sqrt{\frac{2e}{m}} \int_{z_0}^{z_i} \frac{B Y^2}{\varphi^2} \sqrt{\varphi} \cdot \left[k - \frac{\varphi'}{2}\,\frac{Y'}{Y} + \varphi\left(\frac{Y'}{Y}\right)^2 \right] dz$$
$$+ \frac{G}{32} \sqrt{\frac{2e}{m}} \left[\frac{B Y^2}{\sqrt{\varphi}} \left(\frac{3}{2}\,\frac{\varphi'}{\varphi} + 2\,\frac{Y'}{Y} - \frac{B'}{B}\right) \right]_{z_0}^{z_i},$$

in which

$$f = \frac{5}{4}\left(\frac{\varphi''}{\varphi}\right)^2 + \frac{5}{24}\left(\frac{\varphi'}{\varphi}\right)^4 + \frac{e}{m}\frac{B'^2}{\varphi} + \frac{3}{8}\left(\frac{e}{m}\right)^2\frac{B^4}{\varphi^2} + \frac{35}{16}\frac{e}{m}\left(\frac{\varphi'}{\varphi}\right)^2\frac{B^2}{\varphi}$$
$$- \frac{3e}{m}\frac{\varphi'}{\varphi}\frac{B\,B'}{\varphi},$$

$$g = \frac{7}{6}\left(\frac{\varphi'}{\varphi}\right)^3 - \frac{e}{2m}\frac{\varphi'}{\varphi}\frac{B^2}{\varphi},$$

$$h = -\frac{3}{4}\left(\frac{\varphi'}{\varphi}\right)^2 - \frac{e}{2m}\frac{B^2}{\varphi},$$

$$k = \frac{3}{8}\frac{e}{m}B^2 + \frac{9}{8}\frac{\varphi'^2}{\varphi} - \frac{\varphi'\,B'}{B}.$$

7.3.3 The Eikonal Method

The essence of the method

An electrical system is wholly defined if its potential function is known; likewise, the relevant thermodynamic potential is sufficient for all the principal characteristics of a chemical or physical system to be calculable. Hamilton had the idea of trying to find a function from which all the properties of an optical system could be simply derived. Such functions (for many equivalent definitions have been suggested since that due to Hamilton) bear the generic title of "eikonals". This word is surrounded by an air of mystery, which conjures up, quite wrongly, pictures of bizarre calculations. It is important, therefore, to convince oneself that the eikonal is, in principle, no more complicated a concept than the potential in electricity.

Practical calculations, however, will be more difficult, since there is, in optics, no simplifying principle analogous to the conservation of energy in electrostatics, for example. In consequence, *the difference between the values of the eikonal function at two points will depend upon the actual trajectory along which the particle travels* from one point to the other, unlike electric potential, which is independent of path. It seems, therefore, that the eikonal function can be calculated only if we already know the light rays or corpuscular trajectories. Since the whole point of the function is precisely to calculate these rays, its introduction seems rather to beg the question than to answer it. It is doubtless due to this aspect of the eikonal method that it has had little success as a basis for teaching geometrical optics.

In fact, we shall show that it is possible to surmount this difficulty by taking into account the fact that the optical trajectory between two points A and B has a "stationary" value—if a ray is displaced slightly from its real position, only a second order change in the length of the trajectory ensues. This property is sufficient for us to be able to extract a great deal of information from the eikonal, and in particular, to calculate the aberration coefficients. The results can be obtained much more easily and quickly than by using the trajectory method. It is, however, to be emphasized that the eikonal method does not involve any numerical calculation, so that the final formulae are no less complicated irrespective of the method with which they have been established.

The Hamilton eikonal and the Schwarzschild eikonal

We consider two points, M_0, situated in object space, with coordinates x_0, y_0, z_0, and M_i, lying in image space, with coordinates x_i, y_i and z_i. In general, only one light ray, which we shall assume to be known, links these points. Along this ray, we calculate the "optical distance"

$$\int_{M_0}^{M_i} n \cdot ds = V, \qquad (7.22)$$

in which ds is the element of arc-length along the ray. It is possible to express this quantity in terms of the coordinates of M_0 and M_i; we shall denote it, therefore, by

$$V(x_0, y_0, z_0; x_i, y_i, z_i)$$

and refer to it as the "Hamilton eikonal".

We consider now two neighbouring points M_0' and M_i', lying on the same ray, with coordinates $x_0 + dx_0$, $y_0 + dy_0$, $z_0 + dz_0$, and $x_i + dx_i$, $y_i + dy_i$, $z_i + dz_i$. The difference between the optical paths $M_0 M_i$ and $M_0' M_i'$ is given by

$$dV = \frac{\partial V}{\partial x_0} dx_0 + \frac{\partial V}{\partial y_0} dy_0 + \frac{\partial V}{\partial z_0} dz_0 + \frac{\partial V}{\partial x_i} dx_i + \frac{\partial V}{\partial y_i} dy_i + \frac{\partial V}{\partial z_i} dz_i ,$$

or alternatively, if we introduce the direction cosines of the ray — in object space α_0, β_0, γ_0, and in image space α_i, β_i, γ_i — by

$$dV = -(\alpha_0 dx_0 + \beta_0 dy_0 + \gamma_0 dz_0) n_0 + (\alpha_i dx_i + \beta_i dy_i + \gamma_i dz_i) n_i$$

(see Fig. 50).

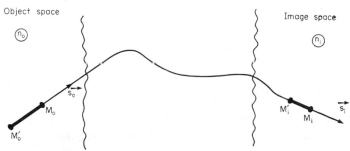

FIG. 50. The relationship between the portion of the trajectory which lies in image space and the portion which lies in object space.

Finally, comparing these two expressions for dV, we obtain

$$\frac{\partial V}{\partial x_0} = -n_0 \alpha_0, \qquad \frac{\partial V}{\partial y_0} = -n_0 \beta_0, \qquad \frac{\partial V}{\partial z_0} = -n_0 \gamma_0,$$

$$\frac{\partial V}{\partial x_i} = +n_i \alpha_i, \qquad \frac{\partial V}{\partial y_i} = +n_i \beta_i, \qquad \frac{\partial V}{\partial z_i} = +n_i \gamma_i. \qquad (7.23)$$

Given a point in object space and a point in image space, therefore, the Hamilton eikonal allows us to determine the ray which joins them, by simple operations. Unfortunately, this is neither the kind of data which is available in practical problems, nor the kind of information of which we are particularly appreciative. In consequence, many nineteenth-century mathematicians (Seidel, 1856 and Schwarzschild, 1906 in particular) strove to modify the structure of the eikonal in such a way that it would be better adapted to practical problems, without affecting the ease with which it could be manipulated. Seidel's "mixed eikonal" seems the best adapted to our needs, and it is the eikonal which we shall consider as an example (cf. the general discussion of eikonals in the books by Synge, 1936, by Born, 1933, and by Born and Wolf, 1959).

We select a point A in image space, in principle arbitrary, although it is often instructive to examine particular cases; A, for example, might be the centre of the exit pupil, with coordinates x_a, y_a, z_a. We examine the function

$$S = V(M_0, M_i) + \overrightarrow{AM_i} \cdot s_i n_i, \qquad (7.24)$$

in which s_i is the unit vector tangential to the emergent ray. In the expression for S, the coordinates x_i, y_i, z_i are replaced by the slope α_i, β_i, γ_i of the emergent rays, with the aid of the relations (7.23). Thus

$$dS = dV + n_i[(x_A - x_i)\, d\alpha_i + (y_A - y_i)\, d\beta_i + (z_A - z_i)\, d\gamma_i]$$
$$- n_i[\alpha_i\, dx_i + \beta_i\, dy_i + \gamma_i\, dz_i],$$

and substituting the expression for dV already obtained, we arrive at

$$dS = -n_0(\alpha_0\, dx_0 + \beta_0\, d y_0 + \gamma_0\, d z_0) + n_i[(x_A - x_i)\, d\alpha_i$$
$$+ (y_A - y_i)\, d\beta_i + (z_A - z_i)\, d\gamma_i],$$

whence

$$\frac{\partial S}{\partial x_0} = -n_0 \alpha_0, \qquad \frac{\partial S}{\partial y_0} = -n_0 \beta_0, \qquad \frac{\partial S}{\partial z_0} = -n_0 \gamma_0,$$

$$\frac{\partial S}{\partial \alpha_i} = n_i(x_A - x_i), \qquad \frac{\partial S}{\partial \beta_i} = n_i(y_A - y_i), \qquad \frac{\partial S}{\partial \gamma_i} = n_i(z_A - z_i). \qquad (7.25)$$

Given an object point, (x_0, y_0, z_0), and the gradient of a ray in image space, $(\alpha_i, \beta_i, \gamma_i)$, therefore, it is possible to obtain, by straightforward differentiation, the coordinates of a point on the ray. It is with this kind of data that we approach the problem of calculating aberrations, where we express the variation of these coordinates as a function of the emergent direction. Undeniably, this is an ideal procedure for calculating the aberration coefficients.

From what we have said so far, however, it still seems necessary that we should know the ray path in order to calculate the eikonal. We shall now show, with the aid of an example, that the fact that S is stationary allows us to make certain approximations, which has the effect of resolving this apparent contradiction.

The principle behind the calculation of spherical aberration

We shall attempt to determine the spherical aberration of an electrostatic lens, using the eikonal method. This permits us to consider only meri-

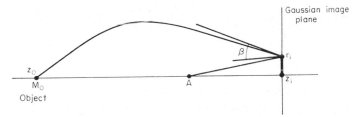

FIG. 51. The rotationally symmetric system.

dian rays. The point M_0 is situated on the axis, at abscissa z_0. The exact ray $r(z)$ will be accurately defined if we know its slope β in image space (Fig. 51). First, we calculate the Hamilton eikonal:

$$V = \int_{z_0}^{z_i} \sqrt{\Phi(r, z)} \sqrt{1 + r'^2}\, dz. \tag{7.26}$$

The integrand can be expanded in powers of r and r', thus

$$V = \int_{z_0}^{z_i} \sqrt{\varphi} \left[1 + \frac{r'^2}{2} - \frac{1}{8} \frac{\varphi''}{\varphi} r^2 + \frac{1}{64} \left(\frac{\varphi^{(iv)}}{\varphi} - \frac{1}{2} \left(\frac{\varphi''}{\varphi} \right)^2 \right) r^2 \right.$$
$$\left. - \frac{1}{16} \frac{\Phi''}{\Phi} r^2 r'^2 - \frac{1}{8} r'^4 \right] dz.$$

Interested as we are only in points lying in the Gaussian image plane with abscissa z_i, it is easy to pass over to the mixed eikonal by adding the term $\overrightarrow{AM_i} \cdot s_i\, n_i$ which takes the form

$$\sqrt{\varphi_i}\,[(z_A - z_i)\cos\beta - r_i \sin\beta]$$

in this particular case.

Much more difficult, however, would be the problem of obtaining an explicit expression for S, in which z_i and r_i were eliminated and replaced by $\cos\beta$ and $\sin\beta$, but we have no need to attempt this. For, let us consider first of all only terms of second order in r and r', that is to say, let us make the Gaussian approximation. Since the object lies on the axis, the Gaussian image is also axial, and we are led unequivocally to the result,

$$\frac{\partial S}{\partial (\sin \beta)} = n_i \, r_i = 0,$$

if we consider only second order terms.

Now, we include terms up to the fourth order, and evaluate the ordinate of the point at which the ray intersects the Gaussian image plane, *which is precisely the radius of the spherical aberration disc.* About this radius w, the general eikonal relations state that

$$n_i(w) = \sqrt{\overline{\varphi_i}} \cdot (w) = - \frac{\partial S}{\partial \beta_i} = \frac{\partial S}{\partial (\sin \beta)},$$

so that

$$\sqrt{\overline{\varphi_i}} \cdot (w) = \frac{\partial}{\partial (\sin \beta)} \int_{z_0}^{z_i} \sqrt{\overline{\varphi}} \cdot \left[\frac{1}{64} \left\{ \frac{\varphi^{(iv)}}{\varphi} - \left(\frac{\varphi''}{\varphi} \right)^2 \right\} r^4 - \frac{1}{16} \frac{\varphi''}{\varphi} r^2 \, r'^2 \right.$$
$$\left. - \frac{1}{8} r'^4 \ldots \right] dz.$$

This formula, however, is essentially the expression of a correction term; we can make a good approximation to it, therefore, by replacing the exact trajectory by the Gaussian approximation $r = \beta \, G \sqrt{\left(\dfrac{\varphi_i}{\varphi_0} \right)} X(z)$, following the notation already introduced, which gives

$$\sqrt{\overline{\varphi_i}} \cdot (w) \simeq \frac{d}{d\beta} \int_{z_0}^{z_i} \beta^4 \frac{G^4 \, \varphi_i^2}{\varphi_0^2} \sqrt{\overline{\varphi}} \left[\frac{1}{64} \left\{ \frac{\varphi^{(iv)}}{\varphi} - \left(\frac{\varphi''}{\varphi} \right)^2 \right\} X^4 \right.$$
$$\left. - \frac{1}{16} \frac{\varphi''}{\varphi} X^2 X'^2 - \frac{1}{8} X'^4 \right] dz.$$

We introduce the spherical aberration coefficient A, and relate the angular aperture of the beam in object space, α, to its value in image space, β, with the aid of the Lagrange–Helmholtz relation, and arrive finally at

$$A = \frac{G}{\delta^3 \sqrt{\overline{\varphi_0}}} \int_{z_e}^{z_i} \sqrt{\overline{\varphi}} \cdot X^4 \left[\frac{1}{32} \frac{\varphi^{(iv)}}{\varphi} \right.$$
$$\left. - \frac{1}{32} \left(\frac{\varphi''}{\varphi} \right)^2 - \frac{1}{4} \frac{\varphi''}{\varphi} \left(\frac{X'}{X} \right)^2 - \frac{1}{2} \left(\frac{X'}{X} \right)^4 \right] dz, \qquad (7.27)$$

using the notation of the first paragraph of this section; δ is the distance between the object and the plane of the entry pupil. This expression is identical to the one obtained in the preceding section. The analogous approximations which are necessary in the general case have been discussed in detail by Glaser (1933), who demonstrated that the method is in every case valid.

The complete results

By introducing non-meridian rays, the calculations for the general case can be effected and the expressions for the other aberration coefficients obtained. For a magnetic lens, the starting point is the optical path-length, which is obtained from the generalized refractive index in the form

$$V = \int_{z_0}^{z_i} \left[\sqrt{\Phi(x, y, z)} - \sqrt{\frac{e}{2m}} A(x, y, z) \cos\alpha \right] ds. \tag{7.28}$$

The calculation is set out in detail in the books by Picht (1939), Cosslett (1946) and Zworykin (1945).

In conclusion, we point out one further application of the Hamilton eikonal function. If we leave the coordinates x, y, z of M_i unspecified, then the equation

$$V(x_0, y_0, z_0, x, y, z) = \text{const.}$$

defines a family of wave fronts originating in the point M_0; since we know that each light ray is a normal to the wave front (Huyghens' principle) it is possible to calculate the effects which will be seen in the image plane from the shape of the wave front. Nijboer (1942) has used this method in glass optics, where he was able to classify in a very logical manner the various aberrations of different orders. In electron optics, however, where only third order aberrations are of appreciable importance, we arrive once again at the results which were obtained in the first paragraph by direct study.

7.3.4 Sturrock's Formulae

Another set of formulae has been obtained by Sturrock (1951a), who started from the reduced equation and the corresponding Gaussian rays. The main feature of interest in the use of the reduced rays arises from the fact that if we consider the two familiar basic Gaussian rays, differentiation is replaced, as it were, by integration since (cf. § 4.2.3) $R'(z)$ represents the operation†

$$R'(z) = R'(z_0) - \frac{3}{16} \int_{z_0}^{z} \left(\frac{\psi'}{\varphi} \right)^2 R(z) \, dz.$$

An integration is far more simple and precise an operation than graphical differentiation.

---

† It is because this unusual notation is used by Sturrock that we employ it here.

The Sturrock formulae are listed below; an electrostatic lens is described by its "characteristic function" $T = \varphi'/\varphi$,† and by two "reduced" Gaussian rays $R(z)$ and $S(z)$, defined in an identical fashion to X and Y.

The method with which Sturrock establishes these formulae is of extreme elegance; a Lagrangian, adapted to the reduced variables, is constructed, and after some calculation, for details of which the reader is referred to the original paper, the following expressions are obtained:

$$A = \sqrt{\varphi_0} \int_{z_0}^{z_i} \frac{R^4}{\sqrt{\varphi}} \left[g + f \left(\frac{R'}{R} \right)^2 \right] dz,$$

$$B = \sqrt{\varphi_0} \left\{ \int_{z_0}^{z_i} \frac{S^2 R^2}{\sqrt{\varphi}} \left[g + f \frac{S' R'}{S R} \right] dz + \int_{z_0}^{z_i} \frac{h}{\sqrt{\varphi}} dz - \left[\frac{a S' R' + d}{\sqrt{\varphi}} \right]_{z_0}^{z_i} \right\},$$

$$C = \sqrt{\varphi_0} \left\{ \int_{z_0}^{z_i} \frac{S^2 R^2}{\sqrt{\varphi}} \left[g + f \frac{S' R'}{S R} \right] dz + \int_{z_0}^{z_i} \frac{f + 2h}{\sqrt{\varphi}} dz - \left[\frac{4 a R' S'}{\sqrt{\varphi}} \right]_{z_0}^{z_i} \right\},$$

$$D = \sqrt{\varphi_0} \left\{ \int_{z_0}^{z_i} \frac{S R^3}{\sqrt{\varphi}} \left[g + f \left(\frac{R'}{R} \right)^2 \right] dz + \int_{z_0}^{z_i} \frac{f R R'}{\sqrt{\varphi}} dz - \left[\frac{a R'^2}{\sqrt{\varphi}} \right]_{z_0}^{z_i} \right\},$$

$$F = \sqrt{\varphi_0} \left\{ \int_{z_0}^{z_i} \frac{S^3 R}{\sqrt{\varphi}} \left[g + f \frac{S' R'}{S R} \right] dz + \int_{z_0}^{z_i} \frac{f S S'}{\sqrt{\varphi}} dz \right.$$
$$\left. + \left[\frac{3 a S'^2 + b S S' + c S^2}{\sqrt{\varphi}} \right]_{z_0}^{z_i} \right\},$$

in which

$$f = \frac{3}{64} T^2, \quad g = \frac{1}{512} (15 T^4 - 25 T^2 T' + 20 T'^2), \quad h = - \frac{5}{128} T^2,$$

$$a = - \frac{1}{8}, \quad b = - \frac{1}{16} T, \quad c = - \frac{1}{32} (2 T^2 + T'), \quad d = - \frac{1}{32} T.$$

7.4 CHROMATIC ABERRATION

7.4.1 Phenomena which Produce Velocity Dispersion

In electron optics, the refractive index n depends upon the electrostatic potential Φ or the magnetic field B. It is a constant of the apparatus, therefore, and it may at first sight seem strange that we should be able to observe

---

† Certain changes have been made in Sturrock's notation to make it accord with the notation of the present work. In his article Sturrock refers to $T = \frac{3}{16} \left(\frac{\varphi'}{\varphi} \right)^2$ as the "characteristic function".

a phenomenon analogous to the variation with colour of the refractive index of a glass lens.

The function Φ, however, includes the constant Φ_c, the mean kinetic energy of the electrons as they leave the cathode. In fact, the electron energies, like the corresponding velocities, are distributed continuously over a narrow band, and to consider this band as a "line" is only a first approximation. This then is one of the sources of variation of n, the effect of which is to replace the image point by a small spot—by analogy with the similar effect in glass optics, this defect is known as chromatic aberration. The fundamental characteristic of this aberration—the arbitrary reduction (as Φ_c/Φ_0) of its magnitude which can be obtained by increasing the accelerating potential Φ_0—is immediately apparent.

The importance of this effect depends essentially upon the nature of the cathode. When the cathode emits thermal electrons, the velocity spread is very narrow, not departing from a mean value Φ_c by more than a few tenths of a volt either way. Φ_c is given accurately enough by $\Phi_c = \dfrac{T}{11,600}$, so that for oxide-coated cathodes ($T \simeq 1100°$), $\Phi_c = 0.1$ V while for tungsten cathodes ($T \simeq 2500°$), $\Phi_c = 0.2$ V.

The breadth $\delta\Phi$ of the energy spread, which is the energy such that 50 per cent of the electrons have an energy which lies in the region between $\Phi_c \pm \frac{1}{2}\delta\Phi$, is correspondingly less as the temperature is lower.

Since lens potentials range between 1 and 60 kV, in order of magnitude, the chromatic aberration is in fact slight. For photoelectric emission, it is much larger, for if ν represents the frequency of the light being used, and ν_0 is the photoelectric threshold frequency of the cathode, the relation $e\Phi_c = h(\nu - \nu_0)$ shows that Φ_c can be as large as a few volts. The aberration becomes most troublesome in secondary emission sources, where the electrons can leave the cathode with any velocity up to that of the primary electrons, and where the energy band may be 25 V wide. The reason why thermal sources are usually employed is vividly illustrated by this comparison.

In the important special case of the microscope, the object itself is a secondary source, since it is simply a thin film through which an electron beam passes, and hence a second way in which dispersion of the electron velocities can be caused appears. On passing through the specimen, the electrons are retarded, not continuously since the phenomenon is quantum mechanical, but in such a way that the electron velocities fall into discrete groups, tracing out a line spectum; the first "line" corresponds to the primary value, and the subsequent lines appear at intervals of about ten volts (Ruthemann, 1941, 1942, 1948; Möllenstedt, 1949, 1952). The overall spectrum, which is still imperfectly understood despite a very promising recent theory, appears to be characteristic of the substance, and reaches

about a hundred volts for the usual thickness of collodion film which
provides the specimen support.

Fortunately, it is usual to work at a high operating potential (between
30 and 100 kV), and a very small contrast diaphragm placed in the focal
plane of the objective will eliminate some of the electrons which have
suffered scattering. Experiment shows that the overall effect is not too harmful, though why this should be so is still uncertain. Recently, a method of
filtering off almost all the electrons which have been retarded has been
tested (Boersch, 1953; Möllenstedt, 1955), and found to produce a marked
improvement of the image.

7.4.2 The Magnitude of the Aberration

We can calculate the size of this defect by a perturbation method, not
dissimilar to the method used to calculate spherical aberration. If, in an
electrostatic lens, φ increases by a small amount, ε, the Gaussian equation
becomes

$$r'' + \frac{1}{2} \frac{\varphi'}{\varphi + \varepsilon} r' + \frac{1}{4} \frac{\varphi''}{\varphi + \varepsilon} r = 0.$$

ε being small, we collect up the terms in which it appears onto the right-
hand side of the equation:

$$r'' + \frac{1}{2} \frac{\varphi'}{\varphi} r' + \frac{1}{4} \frac{\varphi''}{\varphi} r = \varepsilon \left[\frac{1}{2} \frac{\varphi'}{\varphi^2} r' + \frac{1}{4} \frac{\varphi''}{\varphi^2} r \right]. \tag{7.29}$$

We make the substitution $r = r_0 + \varrho$, in which ϱ is a correction sufficiently
small for the product of ϱ and ε to be negligible, and r_0 is the Gaussian
trajectory which we can express in terms of the familiar functions $X(z)$
and $Y(z)$. The correction term ϱ, therefore, satisfies the equation

$$\varrho'' + \frac{1}{2} \frac{\varphi'}{\varphi} \varrho' + \frac{1}{4} \frac{\varphi''}{\varphi} \varrho = \varepsilon \left[\frac{1}{2} \frac{\varphi'}{\varphi^2} r_0' + \frac{1}{4} \frac{\varphi''}{\varphi^2} r_0 \right], \tag{7.30}$$

which is a linear, inhomogeneous, second order differential equation. Since
we know two independent solutions of the corresponding homogeneous
equation, it can be solved immediately.

We shall not pursue the calculation further, but simply refer the reader to
such original publications as Glaser (1940), where the full calculation,
both for the electrostatic and for the magnetic case, is to be found. The
results of these calculations are given below: the effect of a perturbation
ε is to displace the Gaussian image to a plane with abscissa $z_i + \triangle z_i$, to
alter its magnification from G to $G + \triangle G$, and to modify the Larmor

rotation in magnetic lenses by $\Delta\,\theta$, thus:

$$\Delta z_i = \frac{G^2\,\varepsilon}{\sqrt{\varphi_0}}\,\sqrt{\left(\frac{\varphi_1}{\varphi_0}\right)}\int_{z_0}^{z_i}\left[\frac{1}{2}\,\frac{\varphi'}{\varphi^2}\,\frac{X'}{X} + \frac{1}{4}\,\frac{\varphi''}{\varphi^2} + \frac{e\,B^2}{8\,m\,\varphi^2}\right] X^2\,\sqrt{\varphi}\cdot\mathrm{d}z,$$

$$\Delta G = \frac{G\,\varepsilon}{\sqrt{\varphi_0}}\int_{z_0}^{z_i}\left[\frac{1}{2}\,\frac{\varphi'}{\varphi^2}\,\frac{Y'}{Y} + \frac{1}{4}\,\frac{\varphi''}{\varphi^2} + \frac{e\,B^2}{8\,m\,\varphi^2}\right] X\,Y\,\sqrt{\varphi}\cdot\mathrm{d}z,\qquad(7.31)$$

$$\Delta\theta = \frac{\varepsilon}{4}\,\sqrt{\left(\frac{e}{2m}\right)}\int_{z_0}^{z_i}\frac{B}{\varphi^{3/2}}\,\mathrm{d}z.$$

The meaning of these corrections is illustrated in Fig. 52.

In reality, ε has a continuous range of values, so that a whole series of images is superimposed; the overall effect is very complicated, although in

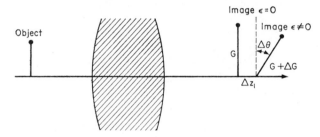

FIG. 52. The perturbation produced by a variation ε of the potential.

practice the fluorescent screen lies in the image plane which corresponds to $\varepsilon = 0$, and the effect of ε is to create an aberration disc of radius r_c around this image. The radius r_c of this disc is given by

$$r_c = \frac{1}{G}\,\sqrt{\left(\frac{\varphi_0}{\varphi_i}\right)}\,\Delta z_i\,\alpha = C_c\,\alpha\,\frac{\varepsilon}{\varphi_0},\qquad(7.32)$$

in which α is the angular aperture of the beam in object space, straight-forwardly connected by the Lagrange–Helmholtz relation to the angular aperture in image space which actually appears in the formula. For ε, a mean value is selected; we shall return later to the coefficient C_c, introduced here.

7.4.3 The Chromatic Aberration of Magnetic Lenses Used for Beta-ray Spectrography

In the β-ray spectrograph, we take advantage of the extreme dispersion of electron lenses, that is, of the fact that the position of the image depends strongly upon the emission velocity of the particles. This velocity is inversely proportional to the de Broglie wavelength of the particles, hence the use of the term "dispersion". The energies which we have in mind range over a very wide interval, between some tens of keV up to 25 MeV. Electrostatic lenses, too feebly convergent in the megavolt region, must therefore be rejected, in favour of magnetic lenses which are effective over the whole energy range.

The diagram below, simplified in the extreme, represents the principle of the arrangement which Kapitza suggested in 1924, and which recalls the optical method of focal separation which R. W. Wood used to study

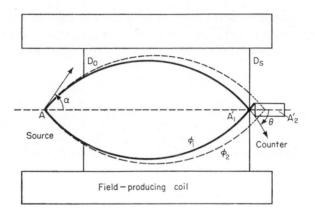

FIG. 53. The principle of the β-ray spectrograph in which the foci are isolated independently.

the infra-red spectrum. A point source at A (Fig. 53) emits β-rays in all directions; the diaphragm with aperture D_0 permits only a beam of angular aperture α to fall upon a lens which focuses the family of rays of energy φ_1 into a point image A_1'. Rays corresponding to an energy $\varphi_2 > \varphi_1$ intersect at a different point A_2', beyond A_1', since the convergence of the lens diminishes as the particles are faster. A small diaphragm D_S with a very small hole placed at A_1', therefore, will separate off the rays of energy φ_1 from all the others, and with the aid of a Geiger counter placed behind the diaphragm the intensity of these rays can be measured.

If the image A_1' were a point image, the selection could be made rigorous, but in fact the lens cannot be perfect, and the image is a small circular spot, of radius ϱ_s, given by

$$\varrho_s = A \, \delta^3 \alpha^3,$$

in which A is the aberration coefficient defined in § 7.2.2, and δ is the distance between the object and the entry pupil.

The rays which correspond to energy φ_2 will be well separated from those of energy φ_1 only if the former are scattered over a circle of radius ϱ_c at D_S, such that ϱ_c is considerably larger than ϱ_s.

Writing $\varphi_2 = \varphi_1 + \delta\varphi$ and $\varphi_1 = \varphi$, ϱ_c will represent the radius of the chromatic aberration spot given by equation (7.32),

$$\varrho_c = C_c \alpha \frac{\delta\varphi}{\varphi}.$$

An idea of the purity of the spectrum which is obtained is given by the quotient ϱ_c/ϱ_s:

$$\frac{\varrho_c}{\varrho_s} = \frac{C_c}{A\,\delta^3}\frac{1}{\alpha^3}\frac{\delta\varphi}{\varphi},$$

although in fact, too small a value for ϱ_c has been used. We should rather substitute $\varrho_c' = \varrho_c + \varrho_s$, since the spherical aberration is produced both at A_2' and A_1', so that

$$\frac{\varrho_c'}{\varrho_s} = 1 + \frac{C_c}{A\,\delta^3}\frac{1}{\alpha^2}\frac{\delta\varphi}{\varphi}. \tag{7.33}$$

Two monochromatic lines of the same intensity with wavelengths λ_1 and λ_2 will be separated, therefore, if $\varrho_c'/\varrho_s = 2$, since the transmitted intensity which is proportional to $\left(\dfrac{\varrho_c'}{\varrho_s}\right)^2$ will fall in relative value from 1, when the spectrograph is adjusted perfectly for one of the lines λ_1 or λ_2, to 0.71 when it is adjusted for their mean wavelength $(\lambda_1 + \lambda_2)/2$.

Such a calculation is admittedly over-simplified, but in fact it proves to be sufficient to justify selecting the value of the ratio $\varphi/\delta\varphi$—or better

$$P = \frac{\lambda}{\delta\lambda} = 2\frac{\varphi}{\delta\varphi}$$

(since this makes the notation conform with the usual optical notation)—which corresponds to $\varrho_c/\varrho = 1$ to characterize in the usual way the resolving power of the spectrograph. For P, we have

$$P = \frac{2C_c}{A\,\delta^3\,\alpha^2}. \tag{7.34}$$

This theory can be considerably improved by taking into account the various different designs of instrument (Grivet, 1950b, 1951a). Abundant information on the chromatic aberration of magnetic lenses is naturally available in the literature of β-ray spectrography.

The chromatic aberration of electrostatic lenses too has found a useful application, in the hands of Möllenstedt (1949, 1952); the three-electrode lens combines strong dispersion in the transgaussian region with a high stability in the presence of variations of the source potential. Möllenstedt took advantage of this to obtain a velocity spectrograph with very high dispersion, which he used to study the velocity loss of electrons as they passed through thin films such as those which are used as specimen supports in transmission microscopy, and to examine the distribution of initial energies at hot cathodes.

7.5 MECHANICAL DEFECTS

7.5.1 Classification

So far we have been presuming that the electrodes are perfect, in the sense that they are of exactly the shape and dimensions which we have ascribed to them in the theory. In practice, however, they are machined with tools of limited precision and the exactness with which the electrodes are assembled, too, may leave something to be desired—the planes in which they lie may be imperfectly parallel, and their symmetry axes may not quite coincide.

These imperfections have the effect of degrading the definition of the image; even though the lenses be machined with the most precise tools available, mechanical defects still produce an appreciable effect. The electron microscope is the extreme case in this respect, for its optical qualities depend largely upon the precision with which it has been constructed; the image quality is an excellent test of the accuracy with which the electrodes have been constructed and assembled. The optical effect of an ellipticity of a fraction of a micron on the rim of a microscope objective—with a radius of some millimetres—is perceptible, and the precision has still further to be improved upon.

In glass optics, the situation is quite different, for this kind of defect is normally quite negligible. This is due to the ease with which a glass surface can be ground with extreme precision, and to the simplicity of the shapes of the lens surfaces, which are always either spherical or planar. If necessary, the error can be reduced below a tenth of a micron.

We shall segregate these mechanical defects into two groups. On the one hand, those caused by imperfect alignment, a lack of parallelism, or of eccentricity in the assembly; and on the other, the defects which are due to a lack of symmetry of the electrodes themselves.

The theory of defects of the first group is well developed (Cotte, 1938; Wendt, 1943; and Hachenberg, 1948), and seems to go beyond the needs of most practical cases. Generally speaking, an adequate and rapid calculation can be made by considering the system as a set of slightly eccentric lenses, and tracing the rays from one lens to its successor (Bruck, 1947). We conclude from such a calculation that no appreciable aberration results from an eccentricity of a hundredth of a millimeter, and that the electrodes can always be set sufficiently far apart for any lack of parallelism of the cross-sectional planes to be significant.

7.5.2 Astigmatism due to Ellipticity

The defects which result from asymmetry in the electrodes themselves are both more important and more difficult to calculate. The simplest method is to consider a model, and we shall suppose that the aperture of the central

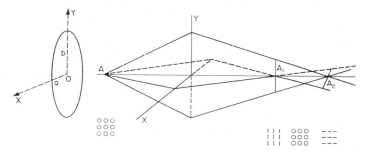

FIG. 54. Ellipticity astigmatism.

electrode is slightly elliptical (Fig. 54). The potential Φ, therefore, is no longer axially symmetric but possesses two symmetry planes which contain the axes of the ellipse; it can be expanded in the form

$$\Phi(r, \theta, z) = \varphi(z) - \frac{1}{4}\varphi''(z)\{1 - \varepsilon(z)\cos 2\theta\}r^2 + \cdots, \qquad (7.35)$$

where the ellipticity of the electrodes is characterized by the function $\varepsilon(z)$. For an axial point object, therefore, we obtain two slightly different families of rays (to the Gaussian approximation) in the symmetry planes xOz and yOz of the elliptical cross-section. Each family of rays gives a clear sharp image, at one of the two points A_1 and A_2, while the other rays are such that their projections onto one or other of the two symmetry planes intersect the axis at A_1 or A_2. The overall effect, therefore, is of two short straight lines, lying in the symmetry planes perpendicular to the axis, at the points A_1 and A_2.

Every axial point object corresponds to two line foci (Glaser, 1943a), and the best image is to be found at the circle of least confusion midway between them; the radius d of this circle provides a measure of this defect, which is known as "ellipticity astigmatism"; d is given by

$$d = C_e \alpha, \qquad (7.36)$$

where α is the angular semi-aperture of the pencil and C_e is related to $\epsilon(z)$ by the expression

$$C_e = \frac{1}{4 r_0^2 \sqrt{\varphi_0}} \int_{z_0}^{z_i} \frac{\varphi'' \, \varepsilon \, r^3}{\sqrt{\varphi}} \, dz. \tag{7.37}$$

If the object is near the focus, an approximate expression for C_e is

$$C_e = 1 \cdot 2 f \bar{\varepsilon}$$

($\bar{\varepsilon}$ being the mean value of ε, presumed to be independent of z_0 and z_i).

All that remains to be done now is to relate ε to the geometrical ellipticity of the opening in the electrode, a rather difficult task. The necessary calculation is to be found in the publications of Bertein (1947 a, b, c; 1948 a, b) and of Bertein, Bruck and Grivet (1947); confirmation of these results has been obtained by Bruck, Remillon and Romani (1948), who made careful measurements in an electrolytic tank, and by Cotte (1949), who followed an alternative line of reasoning. The result is that ε is related to the ellipticity η (if a is the mean radius of the opening and δa is the difference between the extreme radii, then $\eta = \dfrac{\delta a}{2a}$) by the equation

$$\varepsilon \, \varphi'' = \frac{0 \cdot 88 \, E_r \, \eta}{a} \frac{1}{\left[1 + 0 \cdot 7 \left(\dfrac{z}{a} \right)^2 \right]^{5/2}} \sim \eta \, \varphi_c'' \tag{7.38}$$

(E_r, in this equation, is the radial field at the inside edge of the opening, and φ_c'' is the value of $\varphi''(z)$ at the centre of the lens). The relation between cause δa and defect C_e can therefore be summarized thus:

$$C_e \sim f \eta \sim \delta a. \tag{7.39}$$

This theory provides a qualitative explanation of certain observations in the microscope—a careful scrutiny of the image reveals that it is marred by an astigmatism originating in asymmetries of the electrodes (Hillier and Ramberg, 1947; Bertein and Regenstreif, 1947). Combining this image defect with the effect of diffraction, we obtain the following formulae for the resolving power d and for the optimum semi-aperture α of the incident beam (Bruck and Grivet, 1947; Bruck, 1947a):

$$d = 0 \cdot 85 \, \lambda \sqrt{\frac{C_e}{\lambda}},$$

$$\alpha = 0 \cdot 71 \sqrt{\frac{\lambda}{C_e}}. \tag{7.40}$$

λ, measured in millimicrons, is the electron wavelength, which is related to the accelerating potential Φ_0 (V) by the expression

$$\lambda^2 = \frac{1 \cdot 24}{\Phi_0}. \tag{7.41}$$

A result which is important for the construction of pole-pieces or of electrodes is that the resolving power depends upon the absolute value of δa, and thus upon the absolute precision of the machining. It does not depend upon the diameter of the aperture, which can therefore be as small as is necessary to produce the required focal length.

Good agreement is found between the values of d and α given by the formulae above, and their experimental values.

An analogous theory for magnetic lenses has been developed by Hillier and Ramberg (1947) and by Sturrock (1951c), and successfully applied to the magnetic microscope. In this case, however, the defects perhaps originate rather in magnetic inhomogeneities in the pole-pieces than in their mechanical imprecision. Since the methods which we have used to combine mechanical defects and diffraction are rather primitive, this theory only gives the order of magnitude of the various quantities involved. We can appreciate the difficulty of the exact calculation by considering a different problem: the problem of calculating the combined effect of diffraction and spherical aberration. The theory, though it has been to some extent explored, is by no means complete, as the phenomena can only be calculated in the Gaussian image plane, which does not correspond to be best focus (Glaser, 1943a; Bruck, 1947a; Maréchal, 1944, 1947, 1948; Bremmer, 1950).

7.6 MEASUREMENT OF THE ABERRATION COEFFICIENTS, AND THEIR ORDERS OF MAGNITUDE

The aberration coefficients which we have defined earlier are in a convenient form for theoretical calculation. The quantities which are obtainable experimentally, however, are: the angular apertures of the beams in object and image space, α_o and α_i, and the off-axial distances of the various rays; the longitudinal and transverse aberrations in image space, $\triangle z$ and $\triangle r$, which are connected by $\triangle z = \alpha_i \triangle r$; and the Gaussian focal length f_0. It is convenient, therefore, to represent the aberration discs of radius $\triangle r$ by relations of the form

$$\triangle r = k \alpha_i^n,$$

where

$$1 \leqq n \leqq 3.$$

For third order spherical aberration, for example, we have

$$\Delta r = C_s \alpha_i^3. \tag{7.42}$$

If only the angles α_o are known, α_i can be deduced by a form of the Lagrange–Helmholtz relation modified to include third order terms (Grivet, 1950a).

Experimentally, spherical aberration has received the most attention, for it is present even for axial objects, and restricts both the resolving power, an important quantity in microscopy, and the definition of the spot in cathode ray oscillographs. Rather than introduce a correcting device as complicated as the one designed by Scherzer (see § 7.7.6), experimental workers have tended to reduce C_s to the minimum by choosing the best shapes for their lenses.

The astigmatism which results from ellipticity gives a term of the form $\Delta r = C_e \alpha_o$ (on the axis), which has only very rarely been measured, and which is nowadays easily corrected.

Another aberration is particularly important in projective lenses which are often required to focus rays far from the axis, namely distortion. The effect of this aberration appears not in the resolving power but in the image fidelity. If $\Delta r'$ represents the distance between the actual image point and the corresponding Gaussian image point, r the distance of the corresponding object point from the axis and G the linear magnification of the lens, then

$$\Delta r' = C_d G r^3. \tag{7.43}$$

In image-converters also, it is necessary to correct the distortion.

7.6.1 Measurement of the Spherical Aberration Constant

The methods which are used were originally modelled on those of geometrical optics, the Hartmann test and the shadow method using the edge of a screen.

The Hartmann test

This method has been used to study experimentally two-cylinder lenses (Epstein, 1936; Klemperer and Wright, 1939) and three-electrode lenses of uncommon design; it gives the optical constants of the lens in question as well as the spherical aberration. A diaphragm D, perforated by a number of small holes distributed concentrically about the axis (Fig. 55), is placed in the path of a parallel electron beam. The pencils which are defined by this diaphragm pass through the lens L at distances $r_1, r_2, \ldots, r_i$ from the axis, subsequently intersect it at points $F_1, F_2, \ldots, F_i$, and finally impinge on a mobile fluorescent screen, E, which can be moved along the axis; the points of intersection lie at distances $h_1, h_2, \ldots, h_i$ from the axis. With an annular stop we can intercept all but one of the pencils, in object space; knowing $h_1, h_2, \ldots, h_i$ for a range of positions of E, therefore, it is a simple matter to determine the positions of the foci $F_1, F_2, \ldots, F_i$ (and to extrapolate to F_0, the Gaussian focus), and of the principal planes.

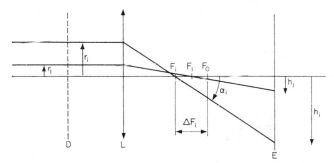

FIG. 55. The Hartmann test.

If α_i represents the inclination of the i-th beam, in image space, and $\triangle F_i$ the distance $F_i F_0$, C_s is given by

$$C_s = \frac{\triangle F_i}{\alpha_i^2}. \tag{7.44}$$

The diffraction diagram method

The preceding method is only feasible if the lens being studied has a large aperture, as its accuracy depends upon the holes in the diaphragm D being small relative to the off-axial distance.

To investigate lenses with a small opening, Mahl and Recknagel (1944) used beams of known inclination, which were produced by diffraction in a thin film of aluminium placed on the specimen support of an electron microscope. In the focal plane of the objective which is being studied, an image

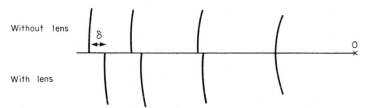

FIG. 56. The diffraction diagram method.

of the Debye–Scherrer diagram is formed, the outer rings of which are contracted by spherical aberration. This diagram is magnified by a second lens, with a large opening, whose own spherical aberration is negligible since the beams which emerge from the objective are so feebly inclined. This final diagram is compared with the diagram which is obtained if the objective is suppressed (the specimen lying beneath the second lens only) and the image magnified optically until the first rings of the two diagrams coincide (Fig. 56). The discrepancy δ between subsequent rings is a function of the various Debye angles at which the beam emerges from the object. In the paper quoted above, $C_s = 80$ mm for one of the three-diaphragm objectives considered, and 65 mm for the other; the focal lengths were respectively 7 mm and 6·1 mm.

Method using the edge of a screen

Before arriving at the lens, a parallel beam is partially intercepted by the straight edge of a screen, placed at some minimum distance δ_1 from the axis. Were the lens perfect, the shadow on the screen E would have a straight

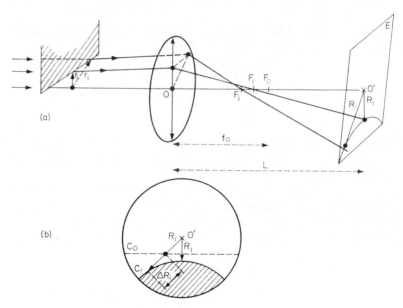

Fig. 57. The method using the edge of a screen. (a) The overall appearance. (b) The shadow which is seen on the fluorescent screen.

edge. The various points on the edge (Figs. 57a and b) lie at different distances from the axis, however, and the edge of the shadow is convex. If the line C_0 represents the ideal Gaussian shadow, and the curve C_1 represents the real shadow, then (Liebmann, 1949)

$$\triangle R_i = L \left(\frac{r_i}{f_0} \right)^3 \frac{C_s}{f_0}. \tag{7.45}$$

Methods using the shadows cast by a wire or a grid

A variant upon the preceding method consists of examining the shadow on a screen of a very fine wire (about a micron in diameter) which is inserted into the path of the beam. Dosse (1941 b) has studied strongly convergent magnetic objectives in this way, placing the wire before the objec-

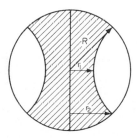

FIG. 58. The shadow of a wire situated at the focus.

tive. Castaing (1950 and 1951) has measured the spherical aberration constant of an objective with long focal length by placing the wire in the Gaussian focal plane. The appearance of the shadow is shown in Fig. 58; if L represents the distance between the wire and the screen, r_1 and r_2 the breadth of the image corresponding to the angular apertures r_1/L and R/L, and r the radius of the wire, then

$$C_s = L^3 r \frac{1/r_1 - 1/r_2}{R^2 - r_1^2}. \tag{7.46}$$

Castaing, using a lens with three plane electrodes, found that for $r_1 = 4\cdot5$ mm, $r_2 = 7\cdot35$ mm, $R = 9\cdot75$ mm, $L = 560$ mm and $r = 10^{-3}$ mm, the value of C_s was 200 mm with $f_0 = 14$ mm. Should the objective not be free of ellipticity astigmatism, however, the appearance of the image is very different (see § 7.6.2).

Spangenberg and Field (1942) and Heise and Rang (1949) use two grids: the first, G_1, is placed in front of the lens L, and defines either the height of the incident pencils if they are parallel, or their inclination if they stem from a point source S; the second, G_2, has much coarser meshes and is placed just behind the lens. The shadows of the two grids are observed on a fluorescent screen, and the origin for the distances h_i and H_i (see Fig. 59) is fixed by marking the bars of G_1 and G_2 which pass through the axis in some distinctive way. With L earthed, S and the various values of α_0 are determined; with L at the desired potential, the values of h_i and α_i are measured, and the position of the image S_i of S is calculated. If S is sufficiently distant, ΔS_i is effectively the longitudinal aberration. By placing S in a different position, we have sufficient data to calculate the focal length and the positions of the principal planes; the curvature of the principal surfaces can also be calculated—they are planar only in the immediate vicinity of the axis.

If the lens is illuminated with a parallel beam, the grid G_2 is sufficient to determine C_s, and the method is equally applicable to an immersion

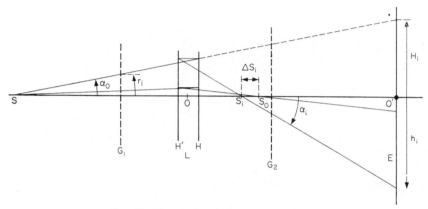

FIG. 59. The method using two gratings.

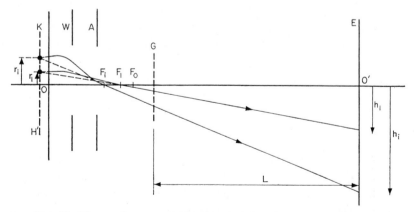

FIG. 60. The grating as a device for studying an immersion objective
(Septier, 1954b).

objective (Septier, 1954b) as Fig. 60 shows in a diagrammatic form. The pencils which are emitted at the surface of the cathode K, with median ray parallel to the axis, are focused onto a screen E which is placed at a distance great in comparison with the focal length f_0 of the objective. Despite the shadow of the grid, the image of K can be distinguished on E, the magnification G can be measured, and hence f_0 is known. Given L and the periodicity of the grid, we can deduce the longitudinal aberration F_i.

For openings for which $R = 1$ mm, W is 0.7 mm in thickness, $WA = 2$ mm and $KW = 0.4$ mm, C_s is found to be 250 mm. Klemperer and Klinger (1951), using identical objectives but operating them as guns, found that 100 mm $\leq C_s \leq 600$ mm.

The halo test

For certain operating conditions an emission system with an exit diaphragm produces a small spot surrounded by a brilliant ring. This phenomenon can be described only in terms of the spherical aberration of the

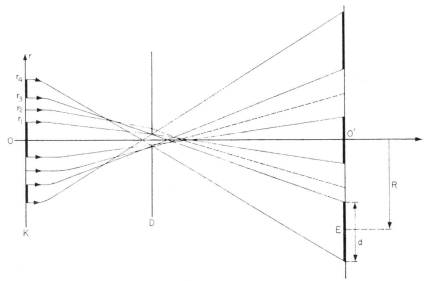

FIG. 61. The halo test.

objective (Fig. 61); only rays which have originated either in the region of the cathode surface between 0 and r_1 or between r_3 and r_4 can pass through the diaphragm D—the former produce the central spot, while the latter create the bright ring. The mean radius R and thickness d of the bright ring are the only quantities which need be measured, for a known value of L, in order to calculate C_s. This method can be extended quite straightforwardly to cover any lens illuminated by parallel radiation.

7.6.2 Measurement of the Other Aberration Coefficients

The coefficient of ellipticity astigmatism

If the lens is weak, the constant C_e can be obtained rapidly with the shadow method which has already been described (Castaing, 1950). C_e has a simple geometrical significance, for it represents one-half of the distance between the two line foci (Fig. 62a) f_1 and f_2. The form of the shadow of the wire D varies according as the wire lies beneath, above, or between the line foci; more precisely, the central part of the shadow near the axis rotates through 90° as the wire moves from f_1 to f_2. We can obtain $2C_e$ directly, therefore, and it is found to be of the order of a few hundredths of a millimetre with $f_0 = 14$ mm (and $C_s = 200$ mm).

Alternatively, C_e can be deduced from C_s for one special position of the wire. If the latter is inclined at 45° to the directions of the focal lines (Fig. 62b), and intersects the axis at the Gaussian focus f_0 (midway between f_1 and f_2), and if we write u_0 for the angular aperture which corresponds to the point on the image at which the latter intersects the axis X', we find

$$C_e = C_s u_0^2.$$

Bertein, Grivet and Regenstreif (1949) have suggested yet another method, which depends upon the use of rays far from the axis, which are styled "transgaussian".

The search for the minimum distortion coefficient

The constant C_d, or the constant D_d which is derived from it by

$$D_d = C_d r^2 = \frac{\triangle G}{G},$$

is measured by forming the images of grids of known dimensions with the lens in question. The quantity

$$\triangle r' = C_d \cdot G \cdot r^2$$

is positive or negative according as the distortion is "cushion" or "barrel"—G is the linear magnification.

The validity of the relation:

$$\frac{\triangle G}{G} = C_d r^2,$$

has been verified experimentally (Jacob and Mulvey, 1949) for an electrostatic unipotential lens with three equally large openings, and at the same time, the influence upon C_d of variations in the thickness T of the central electrode and in the distance S between the electrodes was studied; the distortion is zero for various sets of values of the three quantitites ($S = 2D = 2T$ is a particular case).

Heise and Rang (1949) have studied the variation of C_d with the polarization of the central electrode, and find that the distortion is cancelled when the focal length of the lens is a minimum; this is a most important result, for this is the situation in which both the spherical and the chromatic aber-

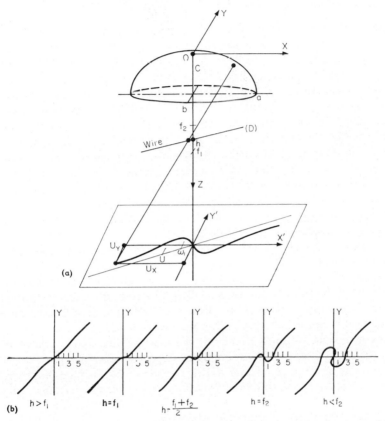

(a)

(b) $h > f_1$ $h = f_1$ $h = \frac{f_1 + f_2}{2}$ $h = f_2$ $h < f_2$

Fig. 62. The shadow of a wire lying perpendicular to the axis of a lens with ellipticity astigmatism. (a) The wave surface and general appearance. (b) The form of the shadow as the wire is moved along the axis.

ration are minimized also. In practice, perfectly corrected lenses can even be achieved without polarization of the central electrode, by altering the thickness or the diameter of this latter in such a way that the lens is operating at its minimum focal length (Regenstreif, 1951a). It is equally possible to reduce the distortion of an immersion objective which is being used in an image converter, by employing a cathode which is concave towards the image side (Morton and Ramberg, 1936).

As for magnetic lenses, Liebmann (1952 and 1953) has shown that the distortion starts off as cushion distortion for feeble excitations, is annulled at a well-defined value of B^2/Φ, and then becomes barrel-shaped—the critical value is dependent upon the pole-piece geometry, but it is always possible to construct a projective lens free of distortion.

The study of field curvature and of astigmatism

Like distortion, field curvature is of interest only in projective lenses and immersion objectives. The field curvature of the objective of an image converter has been studied by Morton and Ramberg (1936) and by Duchesne (1953)—a concave cathode produces partial correction.

The field curvature of the immersion objectives with planar electrodes, operating at a high magnification, which are used in microscopy, has been explored by Johannson (1933) and Septier (1952, 1954b). Briefly, we can conclude that the curvature of the sagittal surface is less pronounced than that of the tangential surface, although both surfaces are always concave on the cathode side; that for any given geometry, the actual curvatures vary markedly with the polarization Φ_1/Φ_2 which is applied to the focusing electrode of the objective, and pass through a minimum near $\Phi_1/\Phi_2 = 0$; and that at this optimum adjustment, the aberration is virtually independent of the geometrical parameters of the lens.

The field curvature of unipotential three-tube lenses has been measured by Gobrecht (1942), using a point source which can be moved in a plane perpendicular to the axis—the focal surfaces have the same general properties as in the preceding case, and their curvature decreases with the focal length. Using the same method, Becker and Wallraff (1940) have obtained substantially equivalent results for magnetic lenses.

In conclusion, we must point out the fact that the magnitude of distortion, field curvature, and astigmatism depends upon the position of the aperture, as we shall show later in this chapter (§ 7.7.1); they can, therefore, only be specified for a rather fully defined ensemble of lens and diaphragm.

7.6.3 The Magnitude of Spherical Aberration

The values of C_s and C_e in some of the common kinds of electrostatic lens have already been mentioned—now, therefore, we shall indicate the laws which govern the behaviour of C_s in terms of the various parameters which characterize electrostatic and magnetic lenses, and quote the smallest values which have so far been attained.

Electrostatic lenses

Two-cylinder lenses (cf. § 8.1). Lenses with two equal cylinders have been studied by Klemperer and Wright (1939), who show that the spherical aberration, which varies with the ratio of the potentials applied to the two electrodes, Φ_1/Φ_2, is considerably smaller in retarding lenses than in accelerating lenses. Measured in terms of the tube radius as the unit of length, the coefficient C_s ranges between 300 and 2 when the electrons are being retarded, and Φ_1/Φ_2 is varied between 3 and 15, as opposed to 500 and 25 for the same Φ_1/Φ_2 range when they are being accelerated.

Lenses with two unequal cylinders invariably have a larger spherical aberration than those with two equal cylinders.

Unipotential lenses (cf. § 8.2). A number of strongly convergent symmetrical three-electrode lenses, of somewhat original forms, have been studied by Liebmann (1949); for the majority, C_s obeys the approximate formula:

$$C_s = K\left(\frac{f_0}{R_i}\right)^2 f_0,$$

in which R_i is the radius of the central opening. C_s decreases as the radius of the outer openings decreases, when the divergent region of the field extends less and less outside the lens. Again using the radius of the apertures as our length unit, the constant C_s is of the order of $30\,R$ for a symmetrical lens with three equal openings, although it falls to $4\,R$ if the outer electrodes are closed off with grids. Modifications of this kind unfortunately make it virtually impossible to use the lens in practice.

The experimental work of Heise and Rang (1949) has confirmed the existence of a minimum value of C_s which corresponds to a minimum value of focal length for a given ratio between the potentials; this ratio varies with the geometry of the lens, but in all types of three-diaphragm lens in common use, C_s is of the order of ten times the focal length.

Immersion objectives (cf. § 8.3). Only a little work has been done on such microscope objectives. Septier's measurements (1954b) show that in many cases,

$$C_s \simeq 50-70 f_0$$

for an objective of standard type with plane electrodes. For this species of lens, however, spherical aberration is usually a negligible factor in calculating the resolving power.

Finally, we would point out that spherical aberration exists equally prominently in the emission systems from which guns are built, but is quite swamped by the aberration which results from space charge effects.

Mirrors. In 1948, Regenstreif (unpublished) calculated the spherical aberration coefficients of a number of unipotential lenses used as mirrors; they proved to be some four times greater than the values in the corresponding lenses.

Magnetic lenses

C_s is affected by a host of different parameters—the length of the field, of which the half-width a of the $B(z)$ curve (using the bell-shaped distribution) is a convenient measure; the extent to which it is symmetrical; and the value of the maximum field B_0, or better, of the coefficient

$$k^2 = \frac{e}{8m} \frac{B_0^2}{\Phi} a^2.$$

Generally speaking, C_s decreases rapidly as k^2 increases (Fig. 63); C_s cannot be decreased indefinitely in this way, however, as B_0 possesses an upper limit of about 2·6 weber m^{-2} (26,000 gauss) beyond which the pole-pieces become saturated.

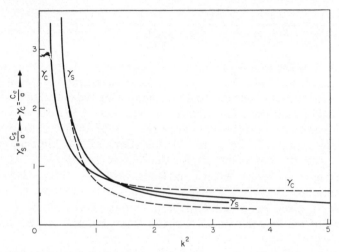

FIG. 63. The values of C_s/a and C_c/a as functions of k^2 in a symmetrical magnetic lens (Dosse, 1941a). The broken curves represent the theoretical values given by a Glaser bell-shaped field distribution (§ 9.2.1).

For given values of B_0^2/Φ, however, C_s passes through a minimum as a is increased (Fig. 64).

Finally, the spherical aberration is diminished if the field distribution is unsymmetrical in the sense that it increases more slowly on the object side (Fig. 65).

Examples of these kinds of variation has been published by Dosse (1941a). For more exact data about these phenomena, the reader is referred to Ruska (1934), von Ardenne (1939) and Liebmann (1950a, b; 1951).

The smallest value of C_s which can be obtained is of the order of

$$C_s = 0.8 f_0$$

for $f_0 \simeq 1$ or 2 mm. Objectives are considered very good if

$$C_s = 2 \quad \text{or} \quad 3 f_0.$$

As far as the smallness of C_s is concerned, it is clear that magnetic lenses are superior to electrostatic ones, the more so in that the risk of break-down between the electrodes prevents the focal lengths of electrostatic lenses from being as short as those of magnetic lenses. At voltages of some tens of kilovolts, the interelectrode spacing cannot be reduced below about 3 mm, and f_0 is then of the order of 3 mm.

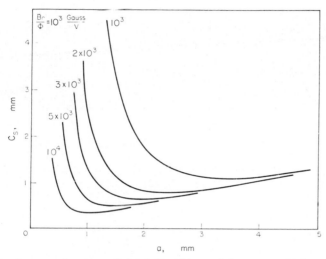

FIG. 64. C_s as a function of the half-width a of the curve which represents the axial field (cf. Fig. 12; Dosse, 1941a).

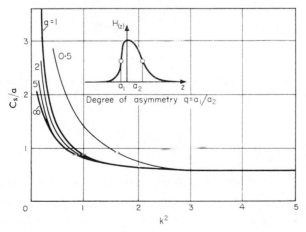

FIG. 65. C_s/a as a function of k^2 for various values of $q = a_1/a_2$; $a = a_1 + a_2$ (Dosse, 1941b).

In Fig. 66, the relative values of the spherical aberration coefficient in a magnetic and a three-electrode electrostatic lens, each of width $2a$, are shown, as calculated by Dosse (1941 b). Magnetic lenses also prove to be more satisfactory than electrostatic lenses as regards chromatic aberration C_c.

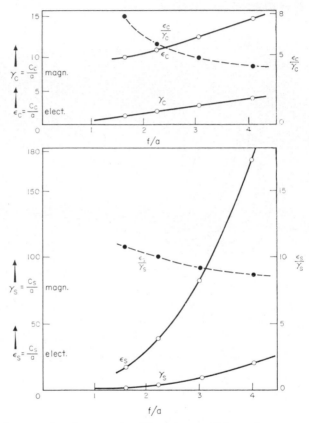

FIG. 66. A comparison between the abberration coefficients C_s and C_c of electric and magnetic lenses of comparable dimensions (Dosse, 1941 b).

7.7 CORRECTION OF THE ABERRATIONS

7.7.1 The Influence of the Diaphragm Position on the Field Aberrations

The aberrations of electron optical lenses are far less amenable to correction than their counterparts in glass optics, as the only parameter which is easily varied is the position of the diaphragm. The position of this diaphragm has an effect upon the aberrations which affect only an off-axial object point, the coma, distortion, field curvature and astigmatism, but from considerations of symmetry, it affects neither spherical aberration nor the aberrations peculiar to magnetic lenses. The reason why the diaphragm position does affect certain of the aberrations is that the pencil of rays by which the image is produced varies according to where the aperture is placed, and the paths which the rays will have followed through the lens therefore varies too (Fig. 67).

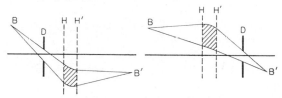

FIG. 67. The variation of the pencil of rays which produces the image, with the position of the diaphragm.

A detailed quantitative study of magnetic lenses, using the Glaser model, is available (Glaser and Lammel, 1941, 1943) and a parallel study of electrostatic lenses might well be made—this has not as yet been done, however, as the calculation is rather more complicated. In particular, the suppression of distortion has been explored; experimentally, Hillier (1946) has verified the results for a magnetic microscope projective, and Le Rutte (1948) has done the same for an electrostatic projective and for electron mirrors.

7.7.2 Spherical Aberration; the Impossibility of Complete Correction

Spherical aberration cannot be perfectly eliminated from electron lenses as we can see if we search for a distribution of refractive index such that all the rays which leave a point object are concurrent in a point image. Gabor (1942) established the requisite distribution, which is identical to that known as "Maxwell's fish-eye", but it seems to be impossible to create it, even roughly, with charges on metal electrodes (Glaser, 1948).

Much earlier, however, than the references just quoted and more straightforwardly, Scherzer (1936) had shown that the integrand of the expression for the spherical aberration coefficient can be written as the sum of a number of squared terms, and hence can never be zero. For magnetic lenses, the result is

$$\frac{|w|}{G} = \frac{1}{16} \frac{e}{m \Phi_0} \alpha^3 \int_{z_0}^{z_i} X^4 \left[\left(B' + B \frac{X'}{X} \right)^2 + B^2 \left(\frac{X'}{X} \right)^2 + \frac{e}{4m} \frac{B^4}{\Phi_0} \right] dz, \quad (7.47)$$

which is zero only if each of the terms of the integrand is zero, which only happens if there is no lens. In glass optics, the situation is quite different, for the sign of the spherical aberration is different according as the lens is convergent or divergent; by combining two lenses, therefore, the aberration can be cancelled without the overall converging action being lost. In electron optics, however, the spherical aberration is positive whether the lenses be convergent or divergent.

There was a period during which Glaser (1940a) believed Scherzer's conclusion could be invalidated: by applying certain algebraic transformations, he was able to put equation (7.47) into the form

$$\frac{|w|}{G} = \frac{e}{96m\Phi_0} \alpha^3 \int_{z_0}^{z_i} \left[\frac{2e}{m\Phi_0} B^4 + 5B'^2 - BB''\right] X^4 \, dz \qquad (7.48)$$

and hence to deduce that a lens whose magnetic field $B(z)$ is a solution of

$$\frac{2e}{m\Phi_0} B^4 + 5B'^2 - BB'' = 0, \qquad (7.49)$$

will be strictly free of spherical aberration. Unfortunately, equations (7.47) and (7.48) are equivalent only if the lens which the field describes is very convergent—a lens described by the solution of (7.49), however, is invariably negligibly convergent (Recknagel, 1941; Rebsch, 1940), and the two descriptions are contradictory. Nevertheless, formula (7.48) is of interest, as it can be used to calculate the spherical aberration of the ordinary types of lens which are always powerfully convergent.

7.7.3 Partial Correction of Spherical Aberration

It is possible to select the field in such a way as to reduce the spherical aberration; the physical reason for this aberration is that marginal rays are converged more strongly than paraxial rays. This in turn is due to the fact that the longitudinal component of the velocity is less for marginal rays—about this we can do nothing—and to the fact that the longitudinal component of the magnetic field,

$$B_z(r, z) = B(z) - \frac{r^2}{4} B''(z),$$

depends upon the sign of B''. The marginal rays pass through regions where B'' is negative, where the spherical aberration is accentuated, and regions where B'' is positive, in which it is diminished.

A means of producing a magnetic field of the second type (in which the intensity of $B(z)$ is greater in the neighbourhood of the object and the image than at the centre of the lens) has been devised by Siegbahn (1946); using this field, he has built a β-ray spectrograph of markedly improved brilliance. Marton and Bol (1947) attempted to extend this method to lenses with the kind of magnification that would be useful in electron microscopy, since Siegbahn's apparatus has unit magnification only; no worthwhile advantage was gained, however, as the clarity of the image depends upon the fourth root of the spherical aberration—a reduction of the latter, therefore, barely affects the resolving power.

If we take into account the fact that Scherzer's formula cannot be applied to the case of mirrors (Recknagel, 1940), we have the possibility of cancelling the positive spherical aberration of a lens with the negative spherical aberration of a mirror, without losing the overall convergence. In a microscope, however, it is difficult to get the mirror sufficiently close to the lens—no feasible arrangement has yet been suggested; if the mirror is not close to the lens, the rays are insufficiently steeply inclined to the axis for the correction to be at all adequate. Ramberg has clarified this situation (1949).

As we shall see later, the attempt to correct mechanical defects has led to the introduction of systems formed from a stigmatic combination of cylindrical lenses.

7.7.4 Correction of Chromatic Aberration

Chromatic aberration likewise cannot be completely corrected, for as Scherzer (1936) has shown, C_c too can be written as the integral of a sum of squared terms, and cannot, therefore, be made zero. In electron optics it is not possible to compensate for the dispersion of the velocities with which the electrons leave the cathode by a judicious choice of lens, unlike the analogous situation in glass optics. As before, we might attempt to correct the aberration by combining lens and mirror action, as mirrors do have a negative chromatic aberration, but the correction is almost always insignificant (Ramberg, 1949). It is interesting, however, to note that the introduction of a composite lens-mirror unit has been envisaged in emission systems (the image-converter).

A closely related defect is the sensitivity of the cardinal elements (the focal lengths, and the position of the principal planes) to accidental variations of the potential $\delta\Phi$ or of the current δi.

Electrostatic lenses

The harmful effect is a consequence of relativity, when all the potentials are drawn from a single source by means of potentiometers. It can, however, be suppressed, as Le Rutte (1948) discovered, by introducing a comparatively small, fixed polarization potential—provided by a thousand-volt battery, say—into the lead of the central electrode of a three-electrode lens.

Magnetic lenses

If the lens is operated under the conditions which correspond to the minimum of the curve which describes f_i as a function of k^2 (the parameter which summarizes the form of B and Φ), variations of B and Φ have little effect, and only a residual effect, upon the distance between the two principal planes, remains.

Alternatively, the construction parameters—the diameter D of the opening between the pole-pieces, the interval s between the poles—can be chosen in such a way as to minimize the chromatic aberration C_c. This is particularly desirable in the gun, where it is no easy problem to regulate the potential with precision. Experimentally, Le Poole and van Dorsten (1951) have studied this problem, and, theoretically, Liebmann (1952) has also. Liebmann's results are summarized in the following table:

	D mm	s mm	f_{min} mm	C_c mm	C_s mm	z_o mm	NI ampère turns
Experiment (L.P. and v.D.)	0·5	2	0·85				4·000
Theor. $\quad$ { f_{min}	0·48	1·9	0·83	0·66	1·2	0·43	3·680
values giving { C_{cmin}	0·54	2·2	0·84	0·66	1·15	0·32	4·210

The lenses which were studied were the objectives of a microscope operating at 80 kV; between the parallel faces of the pole-pieces, the field may reach some 24,000 gauss, although at the centre it is weaker. From the point of view of constructing microscope objectives, it is very fortunate that the conditions under which C_s and C_c are minimized are so very similar to those under which f is least.

7.7.5 Correction of the Mechanical Aberrations

The choice of an elliptic model for these aberrations which we made in §7.5.2 is arbitrary, and at first sight seems highly schematic. In a general theoretical study of this class of aberrations, however, Bertein (1947a, b, c; 1948a, b) has shown to what extent it is valid; the causes of the mechanical aberrations are elucidated in detail, and a great deal of exact information about the whole question is provided. However, the really interesting feature of his theory is that it suggests possible ways by which the aberrations can be corrected. Independently, Hillier and Ramberg (1947) and Scherzer (1947) suggested similar methods. To produce the correction, a weakly convergent cylindrical lens is used which is mechanically extremely asymmetrical (Rang, 1949). This weak lens is placed either just in front of or just behind the lens which is being corrected in such a way that it produces an aberration of the same nature as that of the lens, but of opposite sign. The problem is to find a weak lens which cancels the aberration to be corrected without creating other aberrations—to correct astigmatism without introducing distortion, for example, requires the assembly to be exceedingly accurate. The very simple system of Fig. 68, which is called a stigmator, possesses this interesting property.

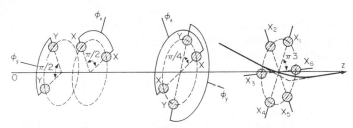

FIG. 68. The stigmator (Bertein, 1947c).

7.7.6 Cylindrical Lenses and the Correction of Spherical Aberration

However large it may be, the astigmatism produced by ellipticity can always be corrected—we are familiar with the analogous situation in glass optics where two crossed cylindrical lenses have the same effect as a spherical lens, although in practice, three cylindrical lenses have to be used if the image is not to be intolerably distorted.

Scherzer (1947, 1949, 1950) has shown that by placing such a combination of lenses behind an ordinary round lens it should be possible to correct the spherical aberration of the latter. We shall briefly outline the principle upon which this correction is based.

We consider the expression for the potential $\Phi(x, y, z)$ in an optical system with four radial symmetry planes which intersect along the Oz axis and are mutually inclined at an angle $\theta = 45°$. Φ can be written in the form:

$$\Phi(x, y, z) = \Phi_0(x, y, z) + \Phi_2(x, y, z) + \Phi_4(x, y, z) + \cdots,$$

in which

$$\Phi_0(x, y, z) = \varphi(0, 0, z) - \frac{1}{4}\varphi''(0, 0, z)(x^2 + y^2)$$

$$+ \frac{1}{64}\varphi^{(iv)}(0, 0, z)(x^2 + y^2)^2 - \cdots,$$

$$\Phi_2(x, y, z) = \varphi_2(0, 0, z)(x^2 - y^2) - \frac{1}{12}\varphi_2''(0, 0, z)(x^2 - y^2)(x^2 + y^2) + \cdots$$

$$\Phi_4(x, y, z) = \varphi_4(0, 0, z)(x^4 - 6x^2y^2 + y^4) - \cdots.$$

We shall not go beyond terms of the fourth degree in x and y. $\Phi_0(x, y, z)$ we recognize as the familiar term associated with axial symmetry; $\Phi_2(x, y, z)$ represents a term with quadrupole symmetry (see Chapter 10) and $\Phi_4(x, y, z)$ a term with octupolar symmetry. We could obtain this latter by means of eight identical electrodes lying parallel to Oz, distributed regularly round a circle centred on Oz, and supplied with potentials $+\Phi_4$ and $-\Phi_4$ alternately.

φ'' and $\varphi^{(iv)}$ denote various derivatives of $\varphi = \Phi(0, 0, z)$ with respect to z. In an axially symmetric electrostatic lens, the radial force is proportional to r only over an extremely slender axial zone; in reality, this force, which is proportional to the radial field E_r:

$$E_r = -\frac{1}{2} r \frac{\partial E_z}{\partial r} = +\frac{r}{2} \left\{ \varphi(z) - \frac{r^2}{4} \varphi''(z) + \frac{r^4}{64} \varphi^{(iv)}(z) - \cdots \right\},$$

consists of terms in r^{2n+1}. In the third order approximation, the essential term varies as r^3, the mean effect of which is a reinforcement of the convergence of the lens for the outer rays. We might hope to be able to introduce a defocusing force proportional to r^3, with the aid of the term $\Phi_4(x, y, z)$, but (for the Laplace equation to be satisfied), this potential obligatorily varies periodically with θ; the force is certainly defocusing in the Ox and Oy directions, say, but has then necessarily a focusing effect in the directions at 45° to these. If, therefore, we annul the aberration in the Ox and Oy directions, we automatically double it along the bisectrices of the angle $\widehat{xOy}$.

If, as Scherzer (1947) suggested, we employ a combination of several different systems in which both Φ_2 and Φ_4 are present, it is effectively possible to correct the third order aberration in every radial direction. The field distributions $\varphi_2(z)$ and $\varphi_4(z)$ are not superimposed onto $\varphi_0(z)$, but succeed it along the optic axis. The potentials $\varphi_0(z)$ and $\varphi_2(z)$ determine the paths of the rays through the system to the first order. The beam is axially symmetric as it emerges from the round lens which is to be corrected; it then passes through a quadrupole lens $[\varphi_2(z)]$ which produces a strong astigmatism and results in the formation of two real focal lines, and after this, a second quadrupole lens and a round lens which render the beam again axially symmetric. Generally, we try to obtain the same convergence as that which would have resulted had the lens being corrected been used alone; this determines the potentials to be applied to the different elements.

Various octupole systems create the field distribution $\varphi_4(z)$ which is necessary for correction.

The equations of motion for the whole system are of the form:

$$\varphi x'' + \frac{1}{2} \varphi' x' - \left(\varphi_2 - \frac{1}{4} \varphi'' \right) x = 2\varphi_4(x^3 - 3xy^2) = S_x,$$

$$\varphi y'' + \frac{1}{2} \varphi' y' + \left(\varphi_2 + \frac{1}{4} \varphi'' \right) y = 2\varphi_4(y^3 - 3x^2 y) = S_y.$$

We calculate two particular trajectories: x_a, which lies in the plane zOx in such a way that $x_a = 0$ and $x_a' = 1$ in the object plane, and y_a, which lies in zOy and satisfies $y_a = 0$ and $y_a' = 1$ in this same object plane. A real trajectory which originates in an axial object point A at an angle r_0' (the projection of r_0' onto the plane zOx is α, and onto zOy, β) can always be represented by

$$x = \alpha\, x_a, \qquad y = \beta\, y_a.$$

If we consider all the rays which emerge from A with the same slope r_0' we should obtain a circular aberration figure defined by the equations:

$$\begin{cases} x_b = a\alpha(\alpha^2 + \beta^2) \\ y_b = a\beta(\alpha^2 + \beta^2) \end{cases}$$

and of radius

$$\varrho = a(\alpha^2 + \beta^2)^{3/2};$$

if the system were axially symmetrical the aberration would be completely described by a single coefficient, $a = C_s$.

If axial symmetry is abandoned, however, but nevertheless only quadrupole and octupole field distributions are considered [$\varphi_2(z)$ and $\varphi_4(z)$], we find

$$\begin{cases} x_b = a\alpha^3 + b\alpha\beta^2 \quad \text{(in a stigmatic system)} \\ y_b = c\beta^3 + b\alpha^2\beta \end{cases}$$

so that here the aberration is characterized by the three coefficients, a, b and c; a judicious choice of $\varphi_4(z)$ permits all of these to be cancelled.

To do this, we calculate first the trajectories x_a and y_a to the first order, and then the spherical aberration x_s and y_s without taking account of $\varphi_4(z)$. The equations are then completely integrated by considering S_x and S_y as small perturbations—S_x and S_y can be calculated everywhere on the Oz axis by writing $x = \alpha\, x_a$ and $y = \beta\, y_a$.

The contributions $(x_b)_c$ and $(y_b)_c$ of the correction term $\varphi_4(z)$ are then obtained in the form of integrals taken from z_a (the object) to z_b (the image):

$$\begin{cases} (x_b)_c = \dfrac{2G}{\sqrt{\varphi_a}} \displaystyle\int_{z_a}^{z_b} \dfrac{\varphi_4}{\sqrt{\varphi}} (3\alpha\,\beta^2\, x_a^2\, y_a^2 - \alpha^3\, x_a^4)\, dz, \\[4mm] (y_b)_c = \dfrac{2G}{\sqrt{\varphi_a}} \displaystyle\int_{z_a}^{z_b} \dfrac{\varphi_4}{\sqrt{\varphi}} (3\alpha^2\,\beta\, x_a^2\, y_a^2 - \beta^3\, y_a^4)\, dz. \end{cases}$$

We next break up $\varphi_4(z)$ into three distinct regions by using three octupole correctors—the first two lie in the planes P_1 and P_2, with abscissae z_1 and z_2; in these planes the trajectories pass respectively through the following pairs of points:

$$x_a = 0; \quad y_a \neq 0,$$
$$x_a \neq 0; \quad y_a = 0.$$

We can then choose the amplitudes of the potentials to be applied to the correctors in such a way as to cancel the aberration terms in β^3 and α^3, that is, the coefficients a and c.

By placing a third corrector at the exit of the system at a point in which $x_a = y_a$, we can cancel the aberration terms in $\alpha\beta^2$ and $\alpha^2\beta$ which have been aggravated by the two earlier correctors.

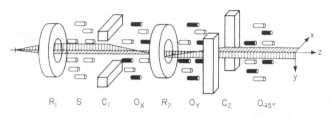

$$R_1 \quad S \quad C_1 \quad O_X \quad R_2 \quad O_Y \quad C_2 \quad O_{45°}$$

FIG. 69. The Scherzer correction system.
R_1: lens to be corrected. S: stigmator. C_1, C_2: cylindrical lenses (either slit lenses or quadrupoles). R_2: intermediate round lens. O_x, O_y, $O_{45°}$: octupole correcting elements, which act in the directions Ox, Oy, and along the bisector of xOy respectively. Note the two real line foci which are formed at the centres of O_x and O_y.

The complete corrector system, as employed by Seeliger (1949) and Möllenstedt (1956), is shown in Fig. 69. In practice, it is extremely difficult to manipulate as the problems of alignment and adjustment of the potentials of the various elements are most complex.

Seeliger (1951) showed that the third order aberration could indeed be annulled with such an assembly, and hence, for example, larger apertures could be used in microscopy. It is then the fifth order aberration which provides the limit (terms in α^5). Recently, Meyer (1961) has calculated the fifth order aberration of this corrector, set out practical alignment procedures, and discussed the residual aberrations which may result from poor alignment.

Many attempts have been made to simplify the assembly to the uttermost (Burfoot, 1953a, 1953b; Archard, 1954a, 1954b, 1955 and 1958; Archard, Mulvey and Petrie, 1960) by collecting the $\varphi_2(z)$ and $\varphi_4(z)$ terms into a single lens, for example. Recently, Deltrap and Cosslett (1962) and Deltrap (1964a, b) have studied, first theoretically and then experimentally, a system consisting of a round lens together with four complex magnetic lenses; the aberration coefficients were measured and Deltrap showed that the three coefficients a, b and c can be reduced to zero in practice. We might even hope to be able to suppress $\varphi_0(z)$ altogether—to remove the lens under correction that is—and to obtain instead an optical system equivalent to a thick round lens and wholly free of third order spherical aberration.

CHAPTER 8

SOME ELECTROSTATIC LENSES
AND TRIODE GUNS

8.1 TWO COAXIAL CYLINDERS

8.1.1 Principal Studies

Used as they are in cathode ray oscillographs, lenses of this kind received a great deal of attention in the earliest work on electron optics. Epstein (1936) calculated the rays and cardinal elements in such a lens from the potential distribution which he had obtained in an electrolytic tank, and compared these with his experimental results; Spangenberg and Field(1943) completed his experimental work, using the method which has been explained in the chapter on aberrations. We must mention too the work of Gundert (1941) on cylinders with different radii, and of Nicoll (1938), who examined the transition from lens to mirror using the same electrodes.

Not until much later, however, was the theory of this type of lens developed; Goddard (1944) calculated the cardinal elements, by integrating the Gaussian equation with a method of successive approximations. Bernard and Grivet (1951, 1952a) adapted the simple model which Hutter (1943) had already suggested for this form of lens; Glaser and Robl (1951) showed that the axial potential could be described by

$$\varphi = \varphi_0 + \triangle \varphi \cot^{-1} kz,$$

in which case the trajectories are described by $J_{1/4}(z)$, although tables of this Bessel function are not easily to be found.

We shall simply outline the method which Grivet and Bernard use; for the experimental results, the reader is referred to the bibliography.

8.1.2 The Grivet–Bernard Model

The two cylinders are assumed to be of the same radius, which we shall use as our unit of length; as a first approximation, we assume also that the gap between the two cylinders is small enough to be neglected (Fig. 70). The axial potential is given by

$$\varphi = \frac{\Phi_1 + \Phi_2}{2} \left[1 + \frac{1 - \gamma}{1 + \gamma} \left\{ 1 - 2 \sum_n \frac{e^{-\mu_n |z|}}{\mu_n J_1(\mu_n)} \right\} \right], \qquad (8.1)$$

in which μ_n is the nth root of $J_0(\mu) = 0$. This complicated function is, however, represented very closely by

$$\varphi = \frac{\Phi_1 + \Phi_2}{2}\left[1 + \frac{1-\gamma}{1+\gamma}\tanh \omega z\right], \qquad (8.2)$$

in which Φ_1 and Φ_2 are the potentials of the two cylinders of unit radius, $\gamma = \Phi_1/\Phi_2$ and $\omega = 1\cdot318$; Gray (1939) pointed out this happy coincidence.

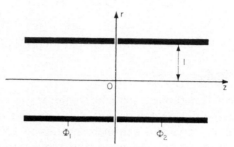

FIG. 70. The form of the electrodes in a two-cylinder lens.

The function $T(z)$, characteristic of the lens, which is defined by

$$T(z) = \frac{\varphi'(z)}{\varphi(z)}, \qquad (8.3)$$

is of the form (Fig. 71):

$$T(z) = \frac{\varphi'(z)}{\varphi(z)} = \frac{2\omega(1-\gamma)t}{(1+t)(\gamma+t)}; \qquad 2\omega z = \log_e t, \qquad (8.4)$$

and passes through a maximum, T_m, at $t_m = \sqrt{\gamma}$, which is given by

$$T_m = -2\omega \tanh(\omega z_m); \qquad z_m = \frac{1}{4\omega}\log_e t. \qquad (8.5)$$

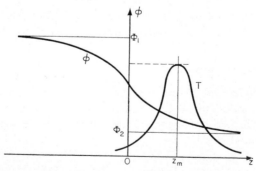

FIG. 71. The form of the functions $\varphi(z)$ and $T(z)$.

If we substitute $Z = z - z_m$, $s = e^{2\omega Z}$ and $T = y\,T_m$, it is clear that the two points on the bell-shaped function T which have reduced ordinate y are *symmetrical* about the abscissa of the maximum since the equation for the "coordinate" s of these points is

$$s^2 - 2s\left[\frac{(1 + \sqrt{\gamma})^2}{2y\sqrt{\gamma}} - \frac{1 + \gamma}{2\sqrt{\gamma}}\right] + 1 = 0, \tag{8.6}$$

which gives

$$\frac{(1 + \sqrt{\gamma})^2}{2y\sqrt{\gamma}} - \frac{1 + \gamma}{2\sqrt{\gamma}} = \cosh(2\omega Z), \tag{8.7}$$

for the abscissae Z. To each value of y, therefore, there correspond two values of Z, since the hyperbolic cosine is an even function.

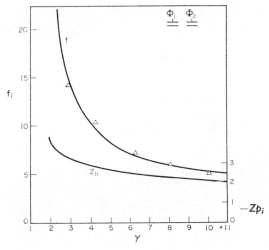

FIG. 72. The cardinal elements of lenses with two equal cylinders.
— —; the Grivet–Bernard model. △△△: Spangenberg's experimental results.

In terms of these reduced quantities, therefore, the lack of electrical symmetry has vanished, its only effect being to shift the axis of the model relative to the gap between the cylinders by an amount z_m. If F_0 and F_i are the focal lengths of the reduced lens, defined by:

$$F_0 = \frac{R(+\infty)}{R'(-\infty)}, \qquad F_i = \frac{R(-\infty)}{R'(+\infty)},$$

then the focal lengths of the real lens will be

$$f_0 = F_0\left[\frac{\Phi_2}{\Phi_1}\right]^{-\frac{1}{4}}, \qquad f_i = F_i\left[\frac{\Phi_2}{\Phi_1}\right]^{+\frac{1}{4}}. \tag{8.8}$$

We can now replace the actual bell-shaped function T by the Glaser expression for such functions, $T = \dfrac{T_m}{1 + (Z/a)^2}$, having the same maximum T_m, and a half-width a such that the T^2 curves, real and model, have the same area—using formula (4.15), we see that this ensures that the focal lengths are identical in the extreme case of weak convergence. For a we obtain the expression:

$$a = -\left(\frac{1}{\pi \omega}\right)\left(\frac{1 + \sqrt{\gamma}}{1 - \sqrt{\gamma}}\right)^2 \left[\left(\frac{1 + \gamma}{1 - \gamma}\log_e \gamma\right) + 2\right]. \tag{8.9}$$

The values of focal length which have been calculated in this way are accurate to within two per cent for very convergent lenses in the region which is used in practice: $0 \cdot 1 < \gamma < 10$; an excellent agreement. The numerical values are obtained from the formulae:

$$f_i = \frac{a\,K\,\gamma^{-1/4}}{\sin(K\pi)}, \qquad f_0 = \frac{a\,K\,\gamma^{1/4}}{\sin(K\pi)},$$

$$z_{F_i} = a\,K\cot(K\pi) - \frac{T_m\,a^2}{4},$$

$$z_{F_0} = -a\,K\cot(K\pi) - \frac{T_m\,a^2}{4},$$

$$K = \sqrt{1 + \left(\frac{3}{16}\right)a^2\,T_m^2}, \tag{8.10}$$

$$a = 0\cdot483\,\coth^2\left(\frac{x}{2}\right)(x\coth x - 1),$$

$$T_m = 2\cdot636\,\tanh\left(\frac{x}{2}\right), \qquad x = -\log_e \sqrt{\gamma}.$$

K is always very close to unity, and the formulae approximate to:

$$\sqrt{\gamma} = 1 + \varepsilon, \qquad K = 1 + 0\cdot0675\,\varepsilon^2\left(\frac{1}{\sqrt{\gamma}}\right), \tag{8.11}$$

$$a = 0\cdot644\left(1 + \frac{\varepsilon^2}{10}\right), \qquad f_0 = 3\cdot036\,\frac{\gamma^{3/4}}{\varepsilon^2}.$$

The reduced ray can be written in terms of sinusoidal functions.

In Fig. 72, the closeness of the agreement between the values of the focal length calculated in this way and the values obtained experimentally by Spangenberg and Field (1943) is displayed—the other curve in the same figure represents the behaviour of the abscissa of the image focal plane.

The curve $T(Z)$ and the cardinal elements are even more accurately represented if we write

$$T = T_0\,\mathrm{sech}\left(\frac{Z}{b}\right) \tag{8.12}$$

(Fig. 73); the reader is referred to the article by Bernard and Grivet (1952a) for further details. In Fig. 74, the results furnished by this improved model, the theoretical results of Goddard (1944), and the experimental results of Spangenberg and Field (1943) are all presented simultaneously for comparison.

So far we have been making the assumption that the gap between the cylinders is thin; if this gap is in fact of width $2d$, however, the axial potential is found to be well represented by

$$\varphi = \frac{\Phi_1 + \Phi_2}{2} \left[1 + \frac{1 - \gamma}{1 + \gamma} \frac{1}{2\omega d} \log_e \frac{\cosh \omega (z + d)}{\cosh \omega (z - d)} \right], \qquad (8.13)$$

so that the formulae listed earlier are still valid provided ω is replaced by ω_1, where

$$\omega_1 = \frac{\tanh (\omega d)}{d}, \qquad \omega = 1 \cdot 318. \qquad (8.14)$$

These same formulae also describe quite accurately (to within a few per cent) the lens which is formed by two diaphragms in which two equal circular holes have been cut.

8.2 LENSES FORMED FROM THREE DIAPHRAGMS

8.2.1 The Potential

The three-electrode, or "einzel" lens is frequently used as the second lens of a cathode ray oscillograph, and is also the standard type of lens for an electrostatic electron microscope. As the bibliography for this paragraph shows, it has often been studied; we shall describe the properties of the einzel lens in terms of the very complete, though analytically simple, theory which Regenstreif (1951a) has proposed. We have already stated that the potential around a single plate with a circular hole, lying between two distant parallel plates without holes, can be calculated; on the axis, it takes the form

$$\Psi(z) = a + bz + c\,z \tan^{-1}\left(\frac{z}{R}\right). \qquad (8.15)$$

Regenstreif, and later Bertein (1952b), have shown that the potential on the axis of the three-electrode lens can be represented approximately by a linear superposition of three potential functions of the form (8.15). We try, therefore, to use a function of the form

$$\varphi(z) = A\,\Psi(z - z_0) + B\,\Psi(z) + C\,\Psi(z + z_0). \qquad (8.16)$$

A function constructed in this way satisfies Laplace's equation exactly, but can only be approximately matched to the boundary conditions.

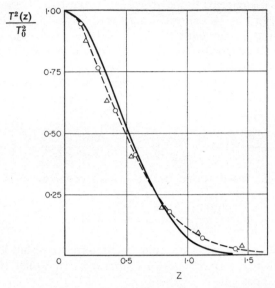

FIG. 73. Approximations to the characteristic function.

——— : the actual form of $T^2(Z)$. △ △ △ : the model $T^2(Z) = \dfrac{T_0^2}{1 + (Z/a)^2}$.

—∘—∘— : the model $T^2(Z) = T_0^2 \operatorname{sech}^2 (Z/b)$.

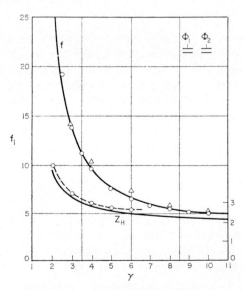

FIG. 74. The cardinal elements of lenses with two equal cylinders.

——— : results obtained with the second approximation (Grivet–Bernard).

△ △ △ : experimental results. —∘—∘— : Goddard's theoretical results.

Regenstreif selected the values of A, B, C, a, b, c in such a way that $\varphi(z)$ would represent the actual field most faithfully at the centre of the lens where the electrons move most slowly, and are hence most susceptible to deviation. This produces

$$\varphi(z) = a + b\left[(z + z_0)\tan^{-1}\left(\frac{z + z_0}{R_2}\right)\right.$$

$$\left. + (z - z_0)\tan^{-1}\left(\frac{z - z_0}{R_2}\right) - 2z\tan^{-1}\left(\frac{z}{R_1}\right)\right],$$

$$a = \Phi_1 - 2b\left[R_1 + z_0\tan^{-1}\left(\frac{z_0}{R_2}\right)\right],$$

$$b = \frac{\Phi_2 - \Phi_1}{2R_1 + 2z_0\tan^{-1}\left(\frac{z_0}{R_1}\right)},$$

(8.17)

in which R_1 is the radius of the opening in the central electrode, which is held at potential Φ_1; R_2 is the radius of the opening in each of the outer electrodes, which are held at potential Φ_2; and z_0 is the distance between the central electrode and each of the other two. This notation is shown in Fig. 75a; in Figs. 75b and 75c a cross-section of the actual lens and a photograph of the actual component parts are to be seen. A potential function of the form (8.17) is not difficult to handle, and a comparison between electrolytic tank measurements and this mathematical model shows that the latter provides a very good description of the lens. We should, however, point out that a considerably less accurate model is necessary

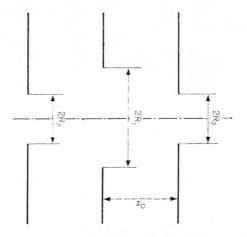

FIG. 75a. The lens formed from three thin diaphragms.

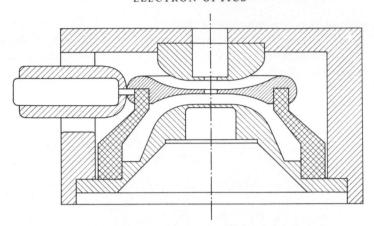

FIG. 75b. The cross-section of a three-electrode lens (scale: $^3/_5$).

FIG. 75c. Some typical electrodes.

this case than in the case of the immersion objective which is discussed later; an error of a few per cent is of little importance, as the electron velocity never drops below some 20 per cent of the maximum velocity, whereas near a cathode, the corresponding velocity might be less than one-thousandth of its value at the anode.

In a unipotential lens with all three openings the same size, and with a thick central electrode, there is a different method of approximating to the real situation which gives a better expression for $\varphi(z)$ in the neighbourhood of the centre of the lens, particularly if the holes are large in diameter with respect to the inter-electrode distance (a situation in which Regenstreif's formulae can no longer be applied).

The general solution of the Laplace equation can be put into the form

$$\Phi(z, r) = \int_0^\infty [A(k) \sin kz + B(k) \cos kz] I_0(k\,r)\,\mathrm{d}k. \qquad (8.18)$$

If the potential is known over the surface of a cylinder of radius R, $A(k)$ and $B(k)$ can be determined and the axial potential $\varphi(z)$ deduced. We suppose, therefore, that the potential over the cylinder on which the peripheries of the three apertures lie varies in some simple fashion; we suppose, for example, that the potential, which is constant over the electrode surfaces, changes linearly between them, an assumption fully justified by measurements in the electrolytic tank.

With a potential Φ_2 applied to the outer electrodes and Φ_1 to the central one, the formula which is finally obtained for $\varphi(z)$ (for $z \geq 0$) is

$$\varphi(z) = \Phi_1 - \frac{\Phi_2 - \Phi_1}{2\omega(z_2 - z_1)} \log \left[\frac{\cosh \omega(z + z_2) \cosh \omega(z - z_2)}{\cosh \omega(z + z_1) \cosh \omega(z - z_1)} \right] \qquad (8.19)$$

(see Septier (1953), for example), in which $\omega = 1\cdot318$, z_1 is the abscissa of the face of the central electrode, and z_2 that of the inner face of the outer electrode. The origin $z = 0$ is at the centre of the lens, and the radius of the openings is taken as the unit of length.

The case $z_1 = z_2$ corresponds to the limiting situation of a three-cylinder lens, with an infinitely small gap between each pair of cylinders. The potential reduces to

$$\varphi(z) = \Phi_1 - \tfrac{1}{2}(\Phi_2 - \Phi_1) [\tanh \omega(z + z_2) - \tanh \omega(z - z_2)]. \qquad (8.20)$$

The situation in which the diaphragms have different radii has been studied by Ehinger and Bernard (1954).

8.2.2 The Cardinal Elements

The formulae of the preceding section are too complicated to be substituted into the ray equation with any hope of successfully solving it. Regenstreif, following an idea of Rüdenberg (1948), represented the curve $\varphi(z)$ by three smoothly joined parabolic arcs. The Gaussian or transgaussian equation of motion is soluble in terms of hyperbolic functions (with arguments which are circular functions) or of circular functions (with hyperbolic functions as arguments). Each arc is joined continuously and with the same gradient at the frontier between each parabolic region; although the calculation is long, it is straightforward and possesses the advantage that the resulting formulae are particularly tractable in that

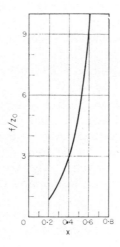

FIG. 76. The focal length of a weak lens as a function of x.

for a whole range of microscope lenses, the physical data only appear in the form of a single, and thus most convenient, parameter x, given by

$$x = \frac{\varphi(0)}{\varphi(z_0)} = \frac{\Phi_1 + \dfrac{\Phi_2 - \Phi_1}{1 + \dfrac{z_0}{R_1} \tan^{-1}\left(\dfrac{z_0}{R_1}\right)}}{\Phi_1 + (\Phi_2 - \Phi_1)\left[1 - \dfrac{\dfrac{R_2}{2R_1}}{1 + \dfrac{z_0}{R_1} \tan^{-1}\left(\dfrac{z_0}{R_1}\right)}\right]}; \quad (8.21)$$

x is simply the ratio of the axial potential at the centre, $\varphi(0)$, to the potential on the axis in the plane of one of the outer electrodes. From among Regenstreif's numerous results on this series of lenses, we list five important examples:

(i) The focal length of a lens of *weak or medium* convergence is given by

$$\frac{f}{z_0} = \frac{8}{3} \frac{x}{(1 - x)^2} \quad (8.22)$$

(see Fig. 76); the principal planes then coincide with the central electrode.

(ii) The focal length f of strong lenses and the position z_F of their foci are given by

$$\frac{f}{z_0} = \frac{0 \cdot 72}{\sin(0 \cdot 707 \log_e x + 0 \cdot 355)},$$

$$\frac{z_F}{z_0} = 1 + 0 \cdot 764 \frac{\sin(0 \cdot 707 \log_e x - 0 \cdot 887)}{\sin(0 \cdot 707 \log_e x + 0 \cdot 355)} \qquad (8.23)$$

(see Fig. 77).

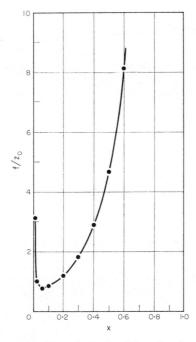

FIG. 77. The behaviour of the focal length of a strong lens.

(iii) The critical potential Φ_c of the central electrode at which the lens is transformed into a mirror is given by

$$\frac{\Phi_c}{\Phi_2} = - \frac{1}{\dfrac{z_0}{R_1} \tan^{-1}\left(\dfrac{z_0}{R_1}\right)} \qquad (8.24)$$

(see Fig. 78).

(iv) The focal length of convergent or divergent mirrors is given by

$$\frac{f}{z_0} = \frac{0 \cdot 72}{\sin[0 \cdot 707 \log_e(-x) + 0 \cdot 355]} \qquad (8.25)$$

(see Fig. 79).

(v) The cardinal elements of lenses in which the central electrode is positive and accelerating in its action are given by Regenstreif, and the behaviour of the focal length is illustrated in Fig. 80.

The formulae which we have at our disposal, therefore, are adequate to describe very fully indeed the behaviour of this family of lenses. These theoretical predictions are in satisfactory agreement with the measurements of Heise and Rang (1949).

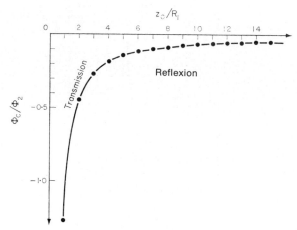

FIG. 78. The repulsive potential Φ_c/Φ_2 as a function of z_0/R_1.

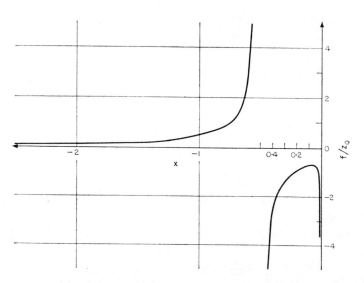

FIG. 79. The focal lengths of mirrors.

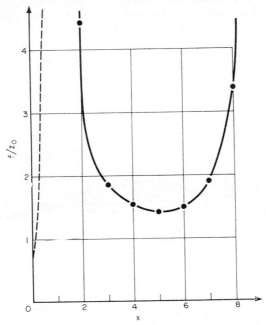

FIG. 80. The focal lengths of lenses with a positive central electrode.

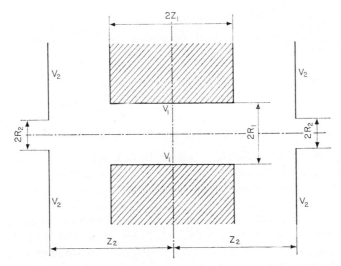

FIG. 81. The appearance of a lens with a thick central electrode.

Finally, we mention the case in which the central electrode is very thick, for which the formulae are more complicated but no more difficult—the potential is represented by a wide "plateau" in the centre of the lens, and to the parabolic arcs, we must add a fourth section, a horizontal straight line; the results are again in good agreement with experiment.

In Fig. 81, the lens is shown schematically; in Fig. 82, the values of the focal length are plotted. The parameter v is defined by $v = z_1/z_2$.

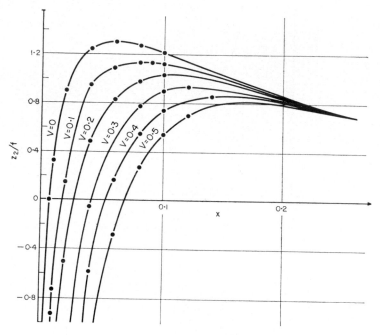

FIG. 82. The focal length of a lens with a thick central electrode as a function of x, for various thicknesses.

8.2.3 Aberrations; Ellipticity Astigmatism

The results of the preceding section have a special relevance when we consider aberrations, since the focal length, distortion, and spherical aberration all pass through their minima simultaneously (Bruck and Romani, 1944; Heise, 1949). With the aid of Regenstreif's formulae, therefore, we can easily establish the conditions for these minima which correspond to the value $x_m = 5\cdot8 \times 10^{-2}$ ($f_m = 0\cdot7730$) when the central electrode is thin. Only with difficulty can the aberration coefficients C_s and C_c of the lens be determined using this approximate method; we need to represent $\Phi(z)$ by a single, well-chosen, analytic function, which would enable us to integrate the equation of the paraxial rays. Glaser (1952) and Glaser and Schiske (1954, 1955) have suggested the following model:

$$\Phi_0(z) = \Phi_0 - \frac{\Phi(O)}{1+(z/d)^2} = \Phi_0\left(1 - \frac{k^2}{1+(z/d)^2}\right),$$

with

$$k^2 = \frac{\Phi(O)}{\Phi_0}.$$

Φ_0 denotes the accelerating voltage and d is the half-width of the bell-shaped curve representing $\Phi(z)$; $\Phi_0(O) = \Phi_0 - \Phi(O)$ is the potential at $z = 0$ (Fig. 82a).

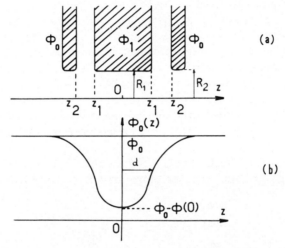

FIG. 82a. The potential distribution in an einzel (unipotential) lens, the Glaser–Schiske model; d denotes the half-width at half the maximum height.

The factor k^2 can be calculated by Regenstreif's method, recently improved by Kanaya *et al.* (1966), for each type of lens and each value of Φ_1. The following table gives the values of k^2 and d for various lenses, when the central electrode is held at cathode potential ($\Phi_1 = 0$).

R_1/R_2	z_1/R_1	z_2/R_1	k^2	d/z_2
1	0	1	0·550	0·875
		2	0·740	0·677
		4	0·860	0·590
2	0	2	0·720	0·615
		4	0·855	0·581
1	1	3	0·950	0·700
	1·5	3·5	0·980	0·705
	1·5	4·5	0·988	0·659
	2	4	0·996	0·750

Integration of the paraxial equations of motion leads to well-tabulated elliptic integrals; the cardinal elements and the aberration constants were then calculated by Glaser and Schiske. Kanaya *et al.* have cast these into a simplified form. Throughout the range $0 < k^2 < 1$, the following formulae are accurate to within a few per cent:

$$\frac{f}{d} = \frac{64}{3\pi}\left(\frac{1-k^2}{k^4}\right), \qquad \frac{C_c}{d} = 2\left(\frac{f}{d}\right)\left(\frac{1+k^2/4}{1-3k^2/4}\right), \qquad \frac{C_s}{d} = \frac{1}{2k^2}\left(\frac{f}{d}\right)^3.$$

The values of f/d (the "immersion" focal length), f'/d (the asymptotic focal length), z_H/d (the absolute value of the position of the principal plane), C_s/f and C_s/d are shown in Fig. 82b.

In theory, we should be able to achieve very small values of the aberration coefficient C_s for k^2 close to 1 ($C_s < d$). In fact, however, the object must be placed at the focus in object space in an electron microscope objective, outside the field (or, at least, in a region where $\Phi(z) \simeq \Phi_0$); under these circumstances, we must have $f \simeq f'$ and hence $k^2 < 0\cdot8$ which implies $C_s/f > 20$ ($C_s/d > 40$) as Fig. 82b shows. These values agree well with those measured on real objectives.

The stronger excitations will only be usable when the object is virtual, therefore (projective lenses).

The astigmatism which results from an ellipticity of the aperture also passes through a minimum at the same point as the focal length. Intuitively,

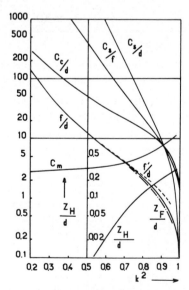

FIG. 82b. The cardinal elements (f/d and z_F/d) of the symmetric three-electrode einzel lens (Kanaya *et al.*, 1966), together with the aberration coefficients C_s, C_c and C_m (C_m denotes the chromatic aberration of the magnification), as a function of k^2.

this is reasonable, as the focal lengths of two lenses with exactly circular openings, one of radius equal to the major axis of the actual lens, the other of radius equal to the length of the minor axis, are in fact the same. To prove rigorously that the minima occur simultaneously is less easy. Regenstreif has succeeded in demonstrating that it would be so in a lens of which the central opening is only slightly elliptical (1951b), a restriction which allowed him to make certain approximations.

Regenstreif (1951c) has even gone so far as to consider strongly elliptical openings which might be very useful, as we have already seen, in the correction of spherical and chromatic aberration, where we need lenses with a pronounced ellipticity.

8.2.4 Application to the Electrostatic Electron Microscope

The results of the previous sections have proved to be very valuable as guides towards possible improvements of the electrostatic electron microscope.

The objective

Our formulae suggest that we can construct a remarkable type of objective, in which the focus is within a few tenths of a millimetre from the face of the lens at which the electrons enter, and the "working distance" is thus a minimum; further, the focal length corresponds to the minimum of the curve which represents f as a function of x—this, in turn, means that the objective is very stable and completely insensitive to small variations of the potential such as those which can only with difficulty be suppressed since they originate in stray currents; finally, by choosing this value of the focal length, the spherical aberration and ellipticity astigmatism are automatically minimized.

This kind of objective can be built as a unipotential lens in which the potential of the central electrode and that of the cathode are the same; the central electrode, therefore, will be quite thick. There is, however, advantage to be gained by taking into account the insensitivity to variations in the potential, and polarizing the central electrode negative with respect to the cathode (by a few kV, provided by an autobias resistance). Under these conditions, for the same focal length, the ellipticity astigmatism is reduced for machining of the same precision, since the central opening can be made perceptibly larger.

Projective lenses

With the aid of the formulae quoted above, a whole range of projective lenses free of distortion can be designed—it proves always to be valuable to use a thick central electrode, but in this case, we are far less restricted in our choice of the various dimensions.

8.2.5 Other Studies

A new and simple calculation procedure has been suggested by Bernard (1951a). Useful information is also to be found in Mahl (1940), Dosse (1941c), Chanson (1947), Plass (1942) and Ramberg (1942), Lippert and Pohlit (1952), see § 17.3.2. Mirrors, which have been studied in particular by Regenstreif (1951 a), have been used by Mahl and Pendzich (1943) in a microscope at magnification 5000. More recent work, which has appeared since the first edition of this volume, and which contains copious theoretical and experimental results on the symmetrical three-electrode einzel lens, has been collected together into a special section (17.3).

8.3 "MESH" OR "GRID" LENSES

(cf. § 4.1.3)

Lenses of this kind were used first—and very brilliantly—to improve the clarity of the mass spectrograph of J. J. Thompson (Cartan, 1937); nowadays, their principal application is in linear ion accelerators, and Bernard (1951a, 1953a) has recently established the full exact theory of their mode of operation.

The only difference between this sort of lens and the familiar three-electrode lens lies in the fact that the central electrode is replaced by a network of fine mesh, held in a suitable support (Fig. 83a). This grid is held at a potential Φ_G, and the outer electrodes at Φ_0; we define σ to be the quantity $\sigma = \dfrac{\Phi_G - \Phi_0}{\Phi_0}$. The axial potential can be calculated in the form of a Fourier integral, which results in a complicated expression which is, however, well represented by the function

$$\varphi(z) = \Phi_0 \left[1 + \sigma - \frac{\sigma}{2\omega d} \log \frac{\cosh \omega \dfrac{z + d}{R}}{\cosh \omega \dfrac{z - d}{R}} \right], \qquad (8.26)$$

$$\omega = 1 \cdot 318 \, .$$

The distribution of potential on the axis is symmetrical with respect to the plane of the grid, and is shown in Fig. 83b; the reason why this particular design of lens is of interest is evident, for the potential curve is always concave in the same sense. The sign of $\varphi''(z)$ and the sense of the radial electric field, therefore, remain unchanged throughout the whole length of the lens. Lenses of this kind are thus *either uniformly divergent or uniformly convergent. The differential effect from which the usual type of lens suffers has been eliminated.*

An approximate representation of the potential enables us to calculate the cardinal elements, just as in the case of the two-cylinder lens (§ 8.1.2). Picht's reduced form of the Gaussian equation is used, with a characteristic function $T = \varphi'/\varphi$ in the form of two of the bell-shaped fields of Glaser, one for each half of the lens.

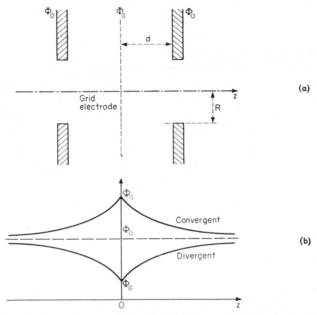

(a)

(b)

FIG. 83. The form of the electrodes and the distribution $\varphi(z)$ in a lens with a mesh electrode.

$$T(z) = \frac{T_0}{1 + \left(\dfrac{z - z_0}{a}\right)^2}, \qquad z < 0,$$

$$T(z) = \frac{T_0}{1 + \left(\dfrac{z + z_0}{a}\right)^2}, \qquad z > 0.$$

(8.27)

The parameters T_0, z_0 and a are chosen in such a way as to minimize the systematic error which is introduced by this approximate representation. The following results are obtained:

$$\frac{1}{f_i} = -\frac{1}{f_0} = -\frac{2\sin^2(K\alpha)}{K^2 a}\left[K\cot(K\alpha) - \cot\alpha + \frac{1}{4}\frac{T_1\alpha}{\sin^2\alpha}\right],$$

$$z_{f_i} = -z_{f_0} = a K \cot(K\alpha) + f_i,$$

in which

$$K = \sqrt{1 + \frac{3}{16} T_0^2 a^2}\,, \tag{8.28}$$

and

$$\alpha = -\cot^{-1}\left(\frac{z_0}{a}\right), \tag{8.29}$$

and T_1 is the value of the characteristic function at $z = 0$.

The numerical values which have been calculated from this formula prove to be in good agreement with the experimental values obtained by Knoll and Weichart (1938). The comparison between the convergence of a grid lens and the corresponding three-electrode lens in which the grid is replaced by an electrode with an opening of radius $\frac{1}{2}R$ is shown, on the same scale, in Fig. 84. At the same potential, the grid lens is always far more convergent than a normal lens, above all for small values of σ. Grid lenses are likely to be useful, in consequence, for focusing fast particles (the velocity of which corresponds to a high value of the potential Φ_0) when one can apply only $\Phi_G - \Phi_0$, small compared with Φ_0, as focusing potential; this is just the situation in linear accelerators.

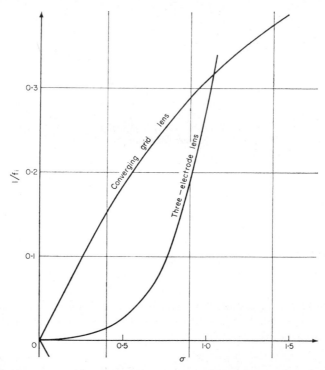

FIG. 84. A comparison between the convergence of a lens with a grid and the corresponding three-electrode lens.

It has been assumed, in the preceding discussion, that the grid behaves likes a continuous membrane. In fact, it is only rarely possible to construct metallic membranes which combine mechanical robustness with the extreme transparency which is necessary if the majority of the electrons are to pass through unretarded. In practice, a fine-mesh network is used, the holes of which produce a slight perturbation of the potential in their vicinity— the ideal trajectories are slightly modified, and the Gaussian image, whose appearance can be calculated from the cardinal elements, is slightly blurred; it suffers, we might say, from "grid aberration". By regarding each hole in the grid as a weak lens, Bertein (1951) has established a simple expression, with the aid of the parametric representation of the Gaussian trajectories, for the radius r of the small aberration disc by which the ideal point is replaced:

$$r = \varrho \, \frac{\sigma}{(1 + \sigma)^{3/4}} \cdot \frac{f_i}{2d} \cdot \frac{\sin K\alpha}{K\sin\alpha} \cdot \tanh\left(\omega \, \frac{d}{R}\right); \qquad (8.30)$$

ϱ is the radius of the holes in the mesh. Further calculation shows that the coefficient of ϱ is always less than unity, and thus that the radius of the spot into which the electrons are spread is less than that of the hole itself.

The most finely meshed grid at present commercially available seems to be that which is known as "electroformed" in nickel, with holes of some $15 \, \mu$ across. The construction of this type of gauze is described in detail in an article by Wilder (1949). The aberration is not, therefore, a sufficiently important defect to prohibit the use of this type of lens.

The geometrical aberrations inherent in all lenses are no less apparent in grid lenses; although spherical aberration is still present, Scherzer's theorem according to which the spherical aberration is always positive (§ 7.7.2) can no longer be proven, as a result of the discontinuity in E_z. Rather, as the full calculation by Bernard (1952b) shows, grid lenses may possess either positive or negative spherical aberration. We can conceive, therefore, of an ensemble of stigmatic systems in which a grid lens cancels the spherical aberration of the other lenses (Gianola, 1950; Seman, 1952).

That the grid always produces its own aberration must not, however, be forgotten—such a combination of mesh lens and ordinary lenses, therefore, may well be capable of a full correction of spherical aberration only when it is possible to replace the grid by a continuous membrane, virtually transparent to the particles which are being used. For ions, such a goal seems indeed far off, but for electrons, it is already possible to construct beryllium membranes about $10 \, m\mu$ thick (Hast, 1948) which are quite transparent to electrons, as the following results of Möllenstedt (1950) show

Kilovolts   30   40   50;   Transparency   0·53   0·3   0·15.

Transparency is defined to be the ratio of the number of electrons which pass through the thin foil and are slowed down (losing a few eV) to the number which are scattered elastically.

8.4 THE IMMERSION OBJECTIVE

8.4.1 The Optical Properties of an Accelerating Region of Constant Field

The simplest type of immersion objective is the two-electrode version—
it consists of a plane cathode, K; and a thin plane anode, A, in which a
hole of small diameter has been punched (Fig. 85a). The electrons are
emitted by K with a mean velocity fixed by Φ_0; they then cross the space
between K and A, of length a, in which the electric field is assumed to be
constant and given by

$$E = -\frac{\Phi_a - \Phi_0}{a}. \tag{8.31}$$

The electrons finally enter the field-free region beyond A ($E_2 = 0$), after
being refracted as they pass through the anode.

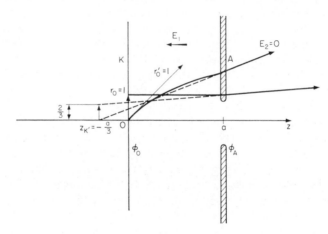

FIG. 85a. The two-electrode immersion objective (K = cathode; A = anode) with
two particular trajectories.

(i) *The trajectories in the inter-electrode region*

We have

$$\varphi(z) = \Phi_0 + E_z = (\Phi_a - \Phi_0)\frac{z}{a} + \Phi_0,$$

and hence

$$\varphi'(z) = \frac{\Phi_a - \Phi_0}{a},$$

and

$$\varphi''(z) = 0.$$

The general equation of motion becomes

$$r'' + r' \frac{\varphi'}{2\varphi} = 0 \qquad (8.32)$$

in the region $0 \leq z \leq a$ so that on integrating,

$$r' \sqrt{\varphi} = r'_0 \sqrt{\Phi_0} = \text{const.}, \qquad (8.33)$$

whence

$$r(z) = r_0 + \int_0^z \frac{r'_0 \sqrt{\Phi_0}}{\sqrt{\varphi(z)}} \, dz,$$

and finally

$$r(z) = r_0 + 2r'_0 \frac{\sqrt{\Phi_0 \, \varphi(z)}}{(\varphi(z) - \Phi_0)} z. \qquad (8.34)$$

At $z = a$ (where the electrons arrive at the anode),

$$r(a) \simeq r_0 + 2r'_0 \sqrt{\frac{\Phi_0}{\Phi_a}} \, a,$$

since

$$\frac{\Phi_0}{\Phi_a} \ll 1,$$

and

$$r'(a) = r'_0 \sqrt{\frac{\Phi_0}{\Phi_a}}.$$

(ii) *The passage through the anode*

We assume that the anode is thin enough for the electric field to vary suddenly from E_1 on the left to E_2 on the right over an infinitely short interval between $z = a^-$ and $z = a^+$. The value of φ'' becomes infinite, but φ, φ', r and r' remain finite. We obtain, on integrating,

$$\int_{a^-}^{a^+} r'' \, dz = - \int_a^{a^+} \left(\frac{\varphi'}{2\varphi} r' + \frac{\varphi''}{4\varphi} r \right) dz,$$

or

$$r'_{a+} - r'_{a-} = \frac{\Phi'_{a-} - \Phi'_{a+}}{4\Phi_a} = \frac{E_2 - E_1}{4\Phi_a} r_a. \qquad (8.35)$$

The anode is clearly equivalent to a thin lens with convergence

$$\frac{1}{f} = - \frac{\triangle r'}{r_a} = - \frac{E_1 - E_2}{4\Phi_a}. \qquad (8.36)$$

When $E_2 = 0$, we have simply

$$- \frac{1}{f} = + \frac{E_1}{4\Phi_a} \simeq \frac{1}{4a}. \qquad (8.36')$$

or $f = -4a$.

The lens is divergent; on passing through such a lens, the slope of a trajectory is altered by

$$\triangle r'_a = \frac{r_a}{4a}.$$

(iii) *The position of the image of K; the magnification*

By calculating a trajectory such that $r_0 = 0$ and $r'_0 = 1$, we can determine the position, $z_{K'}$, of the image of K.

When the electrons emerge from the anode,

$$r_a = 2a \sqrt{\frac{\Phi_0}{\Phi_a}},$$

$$r'_{a+} = \frac{3}{2} \sqrt{\frac{\Phi_0}{\Phi_a}}, \tag{8.37}$$

so that

$$z_{K'} = a - \frac{r_a}{r'_{a+}} = -\frac{a}{3}. \tag{8.38}$$

A second trajectory, parallel to the axis, such that $r_0 = 1$ and $r'_0 = 0$ produces $r_a = 1$ and $r'_{a+} = \frac{1}{4a}$. This trajectory intersects the plane $z = -\frac{a}{3}$ at $r\left(-\frac{a}{3}\right) = \frac{2}{3}$. The image, K', of K is thus virtual, upright, lies at $z = -\frac{a}{3}$ and is linearly magnified, $G = +2/3$.

The simple two-electrode objective cannot, therefore, give by itself a real magnified image of K. It must be used in conjunction with a lens L placed beyond A, which produces a real image of K' on the final screen; this combination has been employed effectively in emission microscopes (L may be either electrostatic or magnetic).

8.4.2 The Three-electrode Objective: the Complexity of the Problem

This type of lens is used for two quite distinct purposes. In the electron guns of microscopes and cathode ray oscillographs, it serves to concentrate as intense a beam of electrons as possible into the smallest possible spot, or more generally, it produces an electron beam of some given shape and intensity. In emission microscopy and image conversion, on the other hand, the immersion objective produces a magnified image of the emissive cathode on the fluorescent screen (Fig. 85b).

These two uses are totally dissimilar; in the first, the beam of electrons is intense, and space charge effects have often to be considered—it is for the sake of the very fine crossover that this kind of system is studied. In the other application, space charge is quite negligible, as the intensities are very feeble—a few μA for high tensions of the order of tens of kilovolts.

Further, it is not the crossover that is of interest, but the *real* image of the emitting surface. Only a first approximation to the real situation in electron guns is provided by results obtained with, for example, the emission microscope (see § 8.4.3).

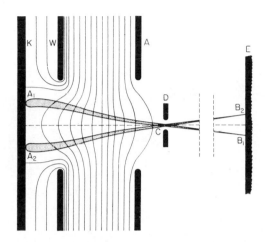

FIG. 85 b. The electron trajectories in an immersion objective.

8.4.3 Properties of the Three-electrode Immersion Objective

Two complications appear in this situation: firstly, the curve $\varphi(z)$ has a most complex shape, as is shown by the electrolytic tank measurements which have been made on objectives with plane electrodes by Septier (1954b) —see Fig. 86a—and Duchesne (1953)—see Fig. 86b, c; the notation is explained by Fig. 87. Septier's lens was designed for a microscope, and hence was to provide strongly magnified images; Duchesne's, on the other hand, had only to produce the unit magnification required in an "electron telescope". The result is that the ray equation cannot be handled directly, and the reduced equation is no advantage, as $T(z) = \dfrac{\varphi'(z)}{\varphi(z)}$ goes to infinity at the origin.

The other reason why this lens is not easy to study arises from the fact that the ray paths are peculiarly susceptible to small variations in the lens power near the cathode, where the electrons are still moving rather slowly; such variations can occur as a result of small variations in the potential applied to the control electrode or Wehnelt (the electrode adjoining the cathode), or of slight changes in the shape of the control electrode, or even the cathode itself. Trifling alterations in the potential distribution produce considerable repercussions on the focusing and magnification of the image.

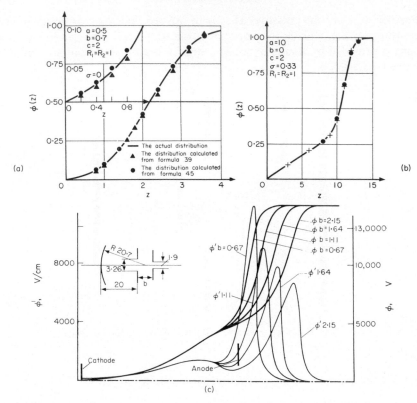

FIG. 86. The form of the distribution $\varphi(z)$. (a) In a very convergent objective
(Septier). (b) In an objective with a weak magnification. (c) In an objective with
a large aperture (Duchesne).

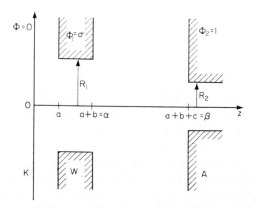

FIG. 87. The appearance of an objective with thick electrodes.

Only with extreme care, therefore, can the optical properties of this type of lens be calculated with methods similar to those we have been using for other kinds of lens, namely straightforward calculation using a convenient but approximate model for $\varphi(z)$ or for $T(z)$. In particular, such models must be chosen with the utmost care in the region around the cathode.

A recent study which Septier (1954 b) has made of an immersion objective with thick plane electrodes shows that it is possible to calculate the axial potential in such a lens quite accurately; a step-by-step calculation then gives the cardinal elements (focal length, focus, and principal plane, on the image side) to within one or two per cent.

There are two formulae with which the potential $\varphi(z)$ in an objective with thick plane electrodes (see Fig. 87) can be calculated quite accurately; they overestimate and underestimate the potential near the cathode (which is chosen to be the origin of potential) respectively. The first, which can only be applied to lenses with all openings of the same size, has the form

$$\varphi(z) = \frac{\sigma}{2\omega\,a} \log \frac{\cosh \omega(z + a)}{\cosh \omega(z - a)}$$

$$+ \frac{1 - \sigma}{2\omega\,c} \log \frac{\cosh \omega[z + (a+b+c)] \cosh \omega[z - (a+b)]}{\cosh \omega[z - (a+b+c)] \cosh \omega[z + (a+b)]}, \quad (8.39)$$

in which $R_1 = R_2 = 1$, $\Phi_1/\Phi_2 = \sigma$, and $\omega = 1\cdot318$.

The second, valid for any situation, is given by

$$\varphi(z) = \frac{\sigma}{a\,\pi} \left[(z + a) \tan^{-1}\left(\frac{z + a}{R_1}\right) - (z - a) \tan^{-1}\left(\frac{z - a}{R_1}\right) \right]$$

$$+ \frac{1 - \sigma}{c\,\pi} \left[(z + \beta) \tan^{-1}\left(\frac{z + \beta}{R_2}\right) - (z - \beta) \tan^{-1}\left(\frac{z - \beta}{R_2}\right) \right. \quad (8.40)$$

$$\left. - (z + \alpha) \tan^{-1}\left(\frac{z - \alpha}{R_1}\right) + (z - \alpha) \tan^{-1}\left(\frac{z - \alpha}{R_1}\right) \right],$$

in which $\alpha = a + b$, and $\beta = a + b + c$.

A comparison between the results of these two formulae and the experimental results for a particular lens is shown in Fig. 86a.

A great deal of theoretical work has been devoted to the calculation of the resolving power of this kind of system, which can always be written in the form

$$\delta = k\,\frac{\Phi_c}{E_0}, \quad (8.41)$$

in which Φ_c is the most probable energy with which the electrons are emitted (measured in volts), and E_0 is the electric field at the cathode; k is a constant, to which different authors ascribe values which vary between 4 and 1·2—the latter seems the most likely. (For further details, see Langmuir, 1937; Recknagel, 1941; Artcimovitch, 1944; and Septier, 1954a.) If, however, instead of calculating the diameter of the electron beam which has originated at an object point in or close to the Gaussian image plane, we determine the electron intensity distribution in the same cross-section, and if we define the resolving power as the diameter of the circle which contains about 90 per cent of the electrons of the beam, it becomes apparent that in the neighbourhood of the point of the caustic of the beam which lies in front of the Gaussian image plane, the value of k can be lowered to about 0.2 or 0.3, and theoretical values of d can be obtained which are far lower than those which are normally admitted (Septier, 1955). We can also explain the fact that the resolving powers which have been measured by certain workers have been smaller than the theoretical resolving power which is given by the formulae quoted above, with $k = 1.2$.

If we now stop down the elementary beams which emerge from different points on the object, the angular aperture of these beams can be reduced and the resolving power improved. The actual mechanism of this improvement will be examined in § 13.2.2.

The first experimental exploration of the objective with very thin plane electrodes was performed by Johannson (1933, 1934), while more recently, Septier (1952, 1953a, b, 1954b) has examined the same type of objective with much thicker electrodes. We shall now list the main features of the Gaussian optics of such a lens: in the work of Duchesne (1953), the reader will find a study of a similar objective used at low magnification as an image converter, where rays steeply inclined to the axis, originating in a large cathode, are considered. In addition Soa (1959) has recently studied several three-electrode immersion objectives; his results, which show the way in which the optical characteristics vary with the various geometrical parameters, complement the experimental data outlined below.

The focal curve and the focal length at high magnification

The parameters upon which the optical properties depend are a, σ, R_1, R_2, b and c (see Fig. 87); σ is the relative polarization of the control electrode.

Once the parameters which describe the actual construction of the lens have been fixed (R_1, R_2 and c), the image is focused onto some fixed point D for an unlimited family of pairs of values (a, σ). The curve $a = f(\sigma)$ is known as the focal curve, and is characteristic of the objective. To obtain this curve, a is set at a number of known values, and at each the image is focused by varying σ. For each (a, σ) pair, the magnification can be measured either photographically, or by measuring the lateral displacement of the image which corresponds to a known displacement of the object. As the objective is far from the screen ($D \gg a + b + c$), the focal length at the focus can be calculated from the formula $f = \dfrac{D}{G}$. In Fig. 88a, the

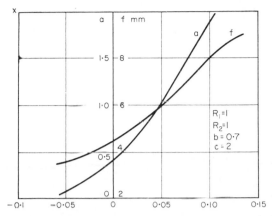

FIG. 88a. Curves showing the distance, a, and the focal length as functions of the polarization, σ, the image being at a fixed distance $D \gg a + b + c$.

values of a and f are shown (as functions of σ) for an objective defined by $R_1 = R_2 = 1$, $c = 2$ and $b = 0.7$. For all the objectives which were studied, f proved to be an increasing function of σ, which could be regarded as a hyperbola in the usual region of operation $(-0.05 < \sigma < 0.05)$, $f = \dfrac{A}{\sigma - B}$. G, correspondingly, varies linearly with σ; for $\sigma = 0$, the immersion objective is strongly convergent $(3 \text{ mm} < f < 4.5 \text{ mm})$.

Figure 88b (Soa, 1959) shows how the positions of the focus, the principal plane and the image vary when σ is varied but a is held constant together with the variation of f as a function of σ for the same objective. It can be seen that the condition $f = f_{\min}$ corresponds to an image located very close to the anode; if we are to obtain a high magnification image, we shall automatically have $f > f_{\min}$. This phenomenon is common to all lenses of this type.

The influence of the geometrical parameters

The control electrode–anode distance. This distance, c, has to be kept within a rather narrow band of values around $c = 2$ mm. Below this, there is a danger of electrical breakdown with $\Phi_2 = 25$ or 30 kV; above, the electric field is inadequate. An increase in c is equivalent to a decrease in Φ_2. The focal curve can be deduced from the curve corresponding to $c = 2$ mm, through their similarity parallel to the σ-axis.

The radius of the anode. No new effect is produced by varying R_2, as an increase in this radius corresponds to a decrease in Φ_2. To transfer from the focal curve for an objective with $R_2 = 1$ to an objective with $R_2 = 2.5$, for example, we simply multiply the values of the abscissa σ by 0.8. The focal length increases with R_2, visible evidence of the progressively stronger

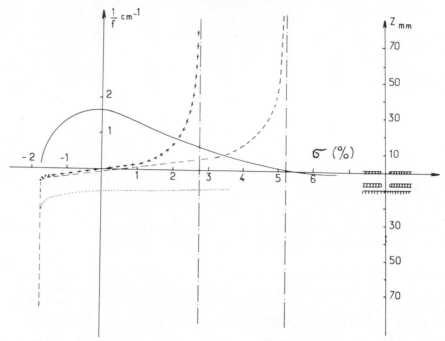

FIG. 88b. The convergence (—) and the positions of the focus (———), principal plane (· · ·) and image (+++) as functions of the polarization σ.

divergent action of the opening in A. For negative polarizations, however, this effect is slight, a few per cent only between apertures with radii in the ratio $2\cdot5 : 1$.

Control electrode thickness. An increase $\triangle b$ in b shifts the focal curve along the ordinate axis *a* towards the origin. Defining the angular aperture α of the

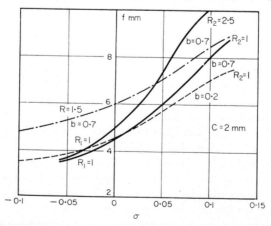

FIG. 89. The variation of f with the various geometrical parameters.

control electrode by the expression $\tan\alpha = \dfrac{R_1}{a+b}$, α is effectively constant near $\sigma = 0$ when b alone is varied. For $\sigma = 0$, the focal length f remains practically constant; for $\sigma < 0$, f increases slightly as b decreases, although never more than 3 per cent for variations in b between 0·2 and 0·4 mm. The change is in the opposite sense if σ is positive.

Control electrode radius. The focal curves are again shifted parallel to the Oa axis, but the effect of a change in R_1 is more pronounced than a change in b, and is in the opposite sense, as increasing R_1 corresponds to increasing a. *The aperture of W decreases* when R_1 increases, irrespective of the polarization. R_1 proves to be the vital parameter of the objective, for the focal length varies rapidly with R_1—for example, with $R_2 = 1$, $b = 0·7$ and $c = 2$, $f_{\sigma-0} = 0$ changes from 4 to 6 mm when R_1 increases from 1 to 1·5 mm.

Figure 89 summarizes the effects upon f of the various parameters, in a few numerical examples.

The field E_0 at the surface of the cathode

The resolving power in the central region is restricted only by the spherical aberration of the component pencils, which in turn is a decreasing function of the field E_0 which prevails in the neighbourhood of K if the objective is unstopped. It is most important, therefore, to know the magnitude of this field. Knowing the focal curves for various combinations of the geometrical parameters, the values and variations of E_0 for various values of the objective parameters σ, R_1, R_2 and b can be studied in the electrolytic tank. E_0 will be measured in volts per centimetre if we regard the anode potential Φ_2 as one volt.

For any given geometrical arrangement, E_0 is unaffected by wide variations in a (see Fig. 90), and begins slowly to decrease only for large positive values of σ (and hence large values of a). (See Septier, 1954b; Soa, 1959.)

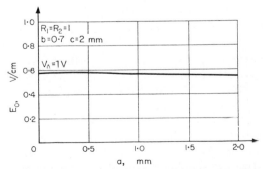

FIG. 90. The field E_0 as a function of the distance between the cathode and the grid.

Increasing the anode radius R_2 produces a decrease in E_0, and is thus equivalent to reducing Φ_2. The effect of decreasing the control electrode thickness b becomes progressively more pronounced the larger the opening in the control electrode, and is a slight increase in E_0; E_0 falls rapidly, however, if the radius of the hole in the control electrode is increased. In every case, therefore, this latter proves to be the most important of the parameters. In the following table, a few numerical examples of these changes are given ($c = 2$ mm throughout).

R_1 (mm)	R_2 (mm)	b (mm)	E_0 (V cm^{-1})
1	0·25	0·7	0·63
1	2·5	0·7	0·48
1	1	0·7	0·60
1	1	0·5	0·62
1	1	0·2	0·64
0·75	1	0·7	0·62
1·2	1	0·7	0·54
1·5	1	0·7	0·45

There would be an advantage, therefore, in using a very thin control electrode with an extremely small hole, but in fact, it is not feasible to work other than with $a \geq 0·2$ mm, $b \geq 0·2$ mm, and $R_1 \geq 0·5$ mm (the opening in the control grid inevitably becomes dirty, and if this hole is too small, the ensuing perturbations are excessive). For $R_1 = 0·5$ mm, $R_2 = 0·25$ mm, $b = 0·2$ mm, $E_{0\,\mathrm{max}}$ is of the order of 1 V cm^{-1}; if, therefore, we take

$$\Phi_c = 0·1 \text{ V}, \quad \Phi_2 = 30 \text{ kV}, \quad \text{and} \quad \delta_{\min} = 1·2\frac{\Phi_c}{E_0},$$

the resolving power can never exceed

$$\delta_{\min} = 40 \text{ m}\mu.$$

Field curvature

Off the axis, the resolving power deteriorates rapidly as a result of field curvature. For given a, the effect of increasing the polarization σ from the value at which the centre of the image is in focus is to produce an annular region of increasing mean diameter d_m in which the image is in focus.

Just how important is the curvature of the image surface is shown by plotting the curve $d_m = f(\triangle \sigma)$ for a few values of σ; the limiting case of a plane image surface corresponds to $d_m = \infty$ for $\triangle \sigma = 0$.

For a given geometrical design, this curvature is smallest, though not zero, near $\sigma = 0$, increases slowly for $\sigma < 0$, but tends rapidly towards infinity for $\sigma > +0·1$. If we select some mean value δ_M for the resolving power, we can measure the diameter d of the sharply imaged zone, on the edges of which $\delta = \delta_M$; d, therefore, is greatest for $\sigma \simeq 0$, and if the point of observation is at a distance $L = 66$ cm, $d_{max} \simeq 5$ cm. Under these operating conditions, d_{max} is substantially the same for all the objectives; decreasing R_2 has a slightly adverse effect upon the image quality; b has no appreciable influence. If R_1 is very small, the image can be focused only for positive values of σ, and below $R_1 = 0.5$ mm, the field curvature, which decreases with R_1 (as the "aperture" of W increases), again becomes troublesome.

The optimum electrical and geometrical parameters for an immersion objective with plane electrodes suitable for electron microscopy will be, therefore, $\sigma = 0$, $R_1 = 0.5$ to 0.6 mm, $b = 0.2$ to 0.3 mm and $c = 2$ mm (this means that 0.2 mm $< a < 0.4$ mm); R_2 can be taken equal to 0.25 mm.

A new type of objective

In order to increase E_0, and hence the resolving power, we might construct a system consisting of an accelerating unit and a weak three-electrode lens L. A thin diaphragm D with a large opening has, however, to be inserted between the cathode and the accelerating anode (Fig. 91) to reduce the aberration. It is tied to the cathode, and although this lowers the field E_0, this latter is still considerably larger than in the objectives described above (Septier, 1952, 1954b). If the electrode A is very thin, and contains an extremely small opening, the diaphragm D is unnecessary, and we can then obtain $E_0 \simeq 150$ kV cm^{-1} (Düker and Illenberger, 1962).

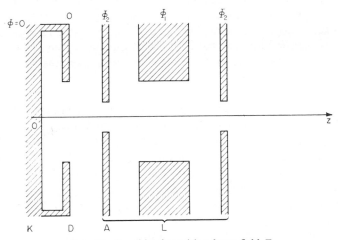

FIG. 91. An objective with a large field E_0.

Theoretical work on the immersion objective

In an article by Hahn (1958) is to be found a full study of the trajectories in an immersion objective, setting out from a series expansion of the potential $\varphi(z)$ and the first order equation of motion for the Gaussian optical properties of an axially symmetrical system; the trajectories are also obtained in series form. The cardinal elements and the third order aberrations are then studied, and in the original paper is to be found a table of the various coefficients (transverse and longitudinal chromatic aberration, spherical aberration, coma, astigmatism, distortion and field curvature); these are primarily of theoretical interest.

8.5 THE TRIODE ELECTRON GUN

8.5.1 Introduction

The three-electrode immersion objective is very often used as an electron source, under the name of *triode gun*. It consists of an emissive cathode, a focusing electrode or "grid" (or Wehnelt in the German literature) and an accelerating electrode (or anode).

If we characterize the gun by the quantity $P = \dfrac{I}{V^{3/2}}$, which is measured in $AV^{-3/2}$ and is known as the *perveance*, guns can be classified into two broad categories (I represents the total intensity and V the potential of the anode).

(i) The first category comprises guns with a very low perveance ($P < 10^{-8}$), in which the role of the space charge of the beam is negligible (these are the guns used in electron microscopes and instruments of a similar nature) or of little importance (as in oscilloscope guns where the only effect of the space charge is a slight widening of the spot—see Chapter 12). The electron trajectories are identical to those in the immersion objective which we have already discussed, and in particular, they cross the axis to form a zone of minimum diameter known as the "crossover".

(ii) The second category contains all guns with a large perveance ($P > 10^{-7}$), in which we are trying to extract the maximum useful current while applying a potential which is sometimes very small. This is the case with numerous modern ultra-high-frequency generators (klystrons and travelling-wave tubes, for example) and devices which make use of the energy in an electron beam for melting, cutting and welding the refractory metals *in vacuo*. The best electrostatic guns at present lie in the range $10^{-6} < P < 5 \cdot 10^{-6}$ (on this topic, the survey article by Süsskind, 1956, should be consulted).

Guns with a very high perveance are distinguished from those in the categories above by two very special characteristics: the cathode emission is limited by the space charge (there is no electric field at the cathode) and the trajectories, subjected both to the converging action of the three-electrode system and to the diverging action of the space charge, no longer cross the axis. The beam has only a zone where its cross-section shrinks to a minimum (or "throat") which lies at the exit of the gun; subsequently it diverges very rapidly. We make a special effort, therefore, to intensify the convergence of the triode system to the uttermost, by choosing special shapes for the electrodes.

We shall first of all review the different types of gun with very low perveance, essentially those to be found in electron microscopes and other such devices; after this, we shall describe succinctly the methods of calculation which are used in connection with very high perveance guns of simple construction.

8.5.2 Electron Microscope Guns

For any given electrode system, the electrolytic tank provides the potential distribution $\varphi(z)$ and we can thus calculate the optical properties of the gun—the emissive surface area of the cathode, the position of the crossover, and the aperture of the beam where it leaves the anode. Ellis (1947) has made a simplified analysis in which $\varphi(z)$ is regarded as being composed of two straight lines, and integration of the equations of motion is thus straightforward. Jacob (1948, 1950a, 1950b) too has performed calculations for such a system in which $\varphi(z)$ is described by a function of the form

$$\varphi_1(z) = A_1 \sinh(kz)$$

near the cathode, and by another function

$$\varphi_2(z) = A_2 e^{kz}$$

far from the cathode; again, the equations of motion can be integrated exactly.

A more realistic but more complicated theory has been worked out by Maloff and Epstein (1938) and by Plocke (1951, 1952). In these articles, the possible modifications which space charge could make to $\varphi(z)$ are taken into account. The effect of space charge is not only to reduce the emissive zone of the cathode and of the electric field at the cathode, and to move the crossover further away from the cathode, but also to decrease the angular aperture of the beam. All these theoretical predictions agree well with the observations which have been made upon oscillographs. Review articles on cathode ray tube guns have been published by Morton (1946), Moss (1945), and Knoll and Thiele (1941).

Just as in an ordinary immersion objective, the potential distribution in the region near the cathode plays an overwhelming role; the following three parameters are therefore of great importance:

the distance between cathode and control electrode, or grid, h;

the radius of the opening in the grid, R, and its thickness;

the polarization of the grid $(-\sigma)$.

The emissive area of the cathode is restricted to the interior of the equi-

potential surface $\Phi = 0$. The trajectories which originate in this zone intersect the axis at a distance from the cathode which lengthens as the convergence is reduced; it is this "crossover" which is regarded as the effective
electron source. For high voltage guns, the crossover lies between the grid
and the anode. As the polarization σ is reduced, the crossover approaches
the anode, then leaves the gun and moves farther and farther away (the
same effects occur if the cathode is brought nearer to the grid). This axial
displacement is accompanied by a variation in the diameter—the further the
crossover penetrates into the interior of the gun, the smaller is its diameter
d_c. (If the real crossover lies between the grid and the anode, rays which
emerge from the anode will seem to have come from a virtual crossover,
namely, the image of the real crossover formed by the optical system which
consists of the region between the real crossover and the anode. The
values of d_c which are measured refer to this virtual crossover.)

We can distinguish between two kinds of gun, according to the way in
which this system is used:

"*Short-focus*" *guns* in which the crossover is formed internally, and is
of very small diameter (3 to 5×10^{-2} mm); these guns are used together
with an additional (condenser) lens, which forms the image of this crossover on the preparation.

"*Long-focus*" *guns* which form the crossover externally, and are used
without a condenser. By varying σ, the position of the crossover is varied
in the neighbourhood of the specimen (placed at a distance from the
anode which may vary between 5 and 15 cm). Even in the best conditions,
however, $0\cdot3 < d_c < 0\cdot5$ mm; nevertheless, the aperture of the beam α is
smaller than in the preceding case, as the Lagrange–Helmholtz relation
shows.

At first sight, it may seem somewhat artificial to make this distinction
between the two "kinds" of gun, as the same gun can be used in both
cases. In reality, however, every electrode system has an optimum mode
of operation, which gives the highest brightness, and if this optimum is
to be achieved, the gun must be operated in the right conditions.

The vital feature of these guns is neither the intensity which is emitted,
nor the current density, but the *brightness*, measured in some cross-section
of the beam, the crossover in particular. The brightness is defined as the
current per unit area per unit solid angle (see § 4.3.5). In theory, this quantity depends only upon the current density j_0 emitted at the cathode, the
working temperature $T°K$ and the accelerating potential Φ_0, thus:

$$B_{th} = j_0 \frac{e \Phi_0}{\pi k T} \qquad \text{A cm}^{-2} \text{ sterad}^{-1}. \qquad (8.42)$$

The current density at the crossover is given by

$$j_c(r) = j_c(0) \exp \left\{ - \left(\frac{r}{r_c} \right)^2 \right\}, \qquad (8.43)$$

in which $j_c(0) = B \pi \alpha^2$ (α being the natural semi-aperture of the beam), and r_c is the "natural half-width" of the bell-shaped curve above, which is related to the radius r_0 of the emissive surface of the cathode by the expression

$$r_c = \frac{r_0}{\alpha} \sqrt{\frac{kT}{e\Phi_0}} . \tag{8.44}$$

To increase B_{th} when Φ_0 is fixed, the only possibility is to search for cathodes with a high specific emission j_0, working at a low temperature. From this point of view, oxide-coated cathodes would be superior to tungsten ones, but their lifetime is too short in the vacuum which obtains in a microscope, which is relatively poor (10^{-4} to 10^{-5} mm Hg) and often dirty (containing oil or grease molecules, and water vapour).

It is better to use a tungsten filament bent into a "V", of which only the point, which is placed on the axis, emits in normal conditions. This point is then equivalent to a spherical cap of very small radius (~ 0.1 mm) and the gun produces an axially symmetric beam. This is no longer true if the grid potential is too small; the two branches of the filament or the part behind the point emit electrons which are then "accepted" by the gun but which follow different trajectories. This gives the illusion that the main beam is broken up into several parts.

For a filament of about $1/10$ mm diameter, the lifetime t is given effectively by

$$t\,(\text{hours}) \simeq \frac{32}{j_0} \quad (j_0 \text{ in A cm}^{-2}) \tag{8.45}$$

(Bloomer, 1957). For $j_0 = 1$ A cm^{-2}, the working temperature is of the order of 2650°K and $t \sim 32$ hours. Under these conditions, the value of B_{th} is of the order of 5×10^4 A cm^{-2} sterad^{-1}. A number of authors have suggested that the filament should be replaced by a genuine point, with a radius of curvature of the order of a fraction of a micron, to which an electric field of the order of 10^7 V cm^{-1} is applied. The current density may reach 10^5 A cm^{-2}, but this field emission is unstable and the lifetime of the source is very short even if an extremely high vacuum is attained (10^{-7} mm Hg). Recent work has shown that it is possible to obtain sources of this type with a useful lifetime by employing mixed or "thermal-field" emission (thermionic emission, $T \sim 2000$°K, reinforced by the Schottky effect. Maruse and Sakaki, 1958; Sakaki and Maruse, 1960; Swift, 1960; Drechsler, Cosslett and Nixon, 1958; and Pilod and Sonier, 1961). The brightness which is obtained is 1000 or 1500 times better than the brightness which would have been produced by simple thermionic emission (at $T = 2000$°K, j_0 would be of the order of $2 \cdot 10^{-3}$ A cm^{-2}) and about ten times better than we should obtain if a filament of the ordinary kind were used; $B \sim 3 \cdot 10^5$ A cm^{-2} sterad^{-1}. The stability of the emission from a heated point has been studied by several workers (Elinson, 1958; Trolan and Dyke, 1958). The total current emitted by the point is only of the order of 1 to 10 μA, as opposed to the 50 to 200 μA emitted by other types of filament.

When the crossover is real, and lies beyond the anode, we can obtain a source having practically the theoretical value of the brightness by stopping down the peripheral part of the crossover; if, on the contrary, we use the whole crossover as the source, the mean measured brightness falls to about 10 or 20 per cent of the theoretical brightness. To measure B, the following method may be followed. A magnified image of the crossover is formed on a screen containing a very small hole, with the aid of a lens L of known magnification; a deflection system moves the image across the hole. By measuring the current which is collected in a Faraday cage, we can plot the curve $j_i(r)$ across a diameter of the image and hence calculate the distribution $j_c(r)$ over the crossover. The lens L is then removed and the natural divergence of the beam α measured; with these data, B can be calculated.

If B is to be compared with B_{th}, the temperature T must be known very accurately as must the emissivity j_0; the former is measured pyrometrically, while the latter is either taken from the tables or measured using experimental diodes. Experiment shows (Haine and Einstein, 1952) that in a triode gun with plane electrodes and a tungsten filament, designed for use with an internal crossover, the two important parameters in the adjustment are the cathode–grid distance h and the (negative) grid potential $(-\sigma)$. When this potential is reduced, starting at cut-off, the brightness increases rapidly as a result of the increase in the emissive area of the cathode, passes through a maximum, and finally diminishes very gradually, doubtless influenced by the aberrations of the system which spread out the crossover (Fig. 92 A 2); h is meanwhile held constant. For each value of h, there is thus an optimum potential σ_{opt} which becomes greater as h is reduced (see Fig. 92 A 1 where the cathode is behind the grid).

If we always select the potential σ_{opt}, the brightness remains practically constant for a given temperature T when h is varied; there is even so a rather ill-defined optimum value of h, h_{opt}.

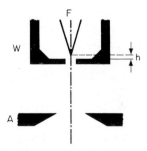

FIG. 92 A 1. Short-focus gun with plane electrodes (Haine and Einstein).

When σ is reduced from the cut-off value, and h is kept fixed, α increases (since the emissive surface of the cathode increases); this increase is very rapid when h is large, less so when h is small. The diameter of the cross-over, on the other hand, remains virtually constant (see Fig. 92 A2).

If h is now fixed and T increased, the value of σ_{opt} decreases slightly; B increases very rapidly with T, since j_0 varies exponentially with T. In this way, values of B of the order of 10^4 A cm^{-2} sterad^{-1} have been measured on a gun with plane electrodes, at a temperature T of 2700°K and accelerating potential $\Phi_0 = 50$ kV; this represents a value which is about 95 per cent of B_{th}. The aperture α is then of the order of 6 milliradians and $d_c \simeq 0.04$ mm. The total current I_c which leaves the cathode when $\sigma = \sigma_{opt}$ decreases steadily when σ_{opt} is increased, when h is increased and when the diameter of the opening in the grid is diminished. These are all factors which we have already seen to be important in the preceding section—they all have the effect of reducing the emissive area of the cathode. The theoretical brightness B_{th} is independent of the geometry of the gun, but in practice the maximum brightness attainable, B_{max} (which is close to B_{th}), is obtained for different values of the aperture α and the total current I_c in each different type of gun. The best gun will be the one which yields B_{max} with the smallest possible values of α and I_c.

Another gun with a conical grid has also been studied by Haine and

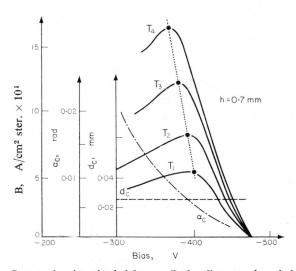

FIG. 92 A2. Curves showing the brightness B, the diameter d_c and the angular aperture α_c of the crossover as functions of the polarization of the grid (W) for a given distance between the filament (F) and W, for the case of the gun illustrated in Fig. 92 A1 (after Haine, Einstein and Borcherds, 1958). B varies with temperature, T ($T_1 = 2,585$°K; $T_2 = 2,650$°K; $T_3 = 2,720$°K and $T_4 = 2,800$°K), but α_c and d_c are virtually independent of T.

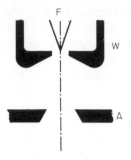

FIG. 92 A 3. Short-focus gun with a conical grid.

Einstein (1958; see Fig. 92 A 3). The results obtained resemble those outlined above, but the values of h_{opt} and σ_{opt} are clearly different. We should also mention the work of Hillier and Baker (1945), Dosse (1940), Ellis (1947), von Borries (1948) and more recently of Boersch and Born (1958) whose results confirm those of Haine.

A great deal of effort has gone into the experimental determination of the forms which triode guns should take if the crossover is to be formed outside ("long-focus guns") and the brightness is to reach its maximum value. Bricka and Bruck (1948) were the first to find a satisfactory solution to this problem; their gun, shown schematically in Fig. 92 A 4, has spherical electrodes and the crossover always lies beneath the hole in the anode. Bricka and Bruck quote a maximum measured current density of 25 mA cm^{-2} at the centre of the crossover which lies 150 mm away from the anode, as opposed to the 3 mA cm^{-2} which would be obtained with a gun with plane electrodes used in the same conditions. The aperture was 8×10^{-3} and hence the brightness B of the order of 1 or 2×10^4 A cm^{-2} sterad^{-1} which is about 20 per cent of the theoretical brightness.

Castaing (1951, 1960) has shown that it is possible to obtain a brightness close to the theoretical brightness with this gun, working under the same conditions as a near-focus gun, by stopping down the crossover ($B = 5\cdot8 \times 10^4$ A cm^{-2} sterad^{-1} for $j_0 = 2$ A cm^{-2}, $T = 2700°$K and $\Phi_0 = 30$ kV or $B \sim 0\cdot7\,B_{th}$). The crossover is, however, very close to the anode.

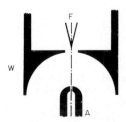

FIG. 92 A 4. The Bruck and Bricka long-focus gun (with spherical electrodes).

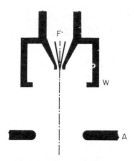

FIG. 92 A5. The Steigerwald *Fernfokus.*

More recently, another solution has been proposed by Steigerwald (1949); this is the *Fernfokus,* shown schematically in Fig. 92 A5. In normal operation, the region near the point of the grid is divergent; the polarization is low and the current emitted is in consequence intense. The beam then converges in the grid–anode region, but the overall magnification is small; at distances between 15 and 50 cm from the anode, the crossover is of the order of 0·3 to 0·5 mm in diameter.

A thorough study of the gun has been made by Braucks (1958, 1959) in which the influence of all the geometrical parameters and the polarization upon the performance of the gun (spot diameter, aperture, and brightness) is analysed; Braucks also examines the effect of placing an additional electrode within the ordinary grid. With the ordinary three-electrode gun, very high current densities (~ 50 or 60 mA cm^{-2}) can be obtained in a crossover 23 cm from the anode with an aperture α of the order of 3×10^{-3} radians. B is then of the order of 4×10^{4} A cm^{-2} sterad^{-1} for $\Phi_0 = 30$ kV, and the total current leaving the gun is $100\,\mu$A.

A long-focus gun based on the *Fernfokus* has been suggested by Bas (1954) and Bas and Gaydou (1959) which is designed to be used in an X-ray tube with a fine focus. The cathode is now the flat end of a tungsten cylinder about 0·5 mm in diameter, heated by lateral electron bombardment (Fig. 92 A6). The current density measured at the centre of the cross-

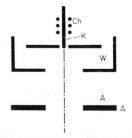

FIG. 92 A6. Bas' gun with a double grid.

over, which lies 5 mm beneath the anode, leads to a maximum value of the brightness of the order of 60 per cent of B_{th}. The difference between measured and theoretical brightness is due to a broadening of the cross-over; this in turn is a consequence of the crater which is produced at the centre of the anode by the impacts of ions created in the residual gas and accelerated in the sense opposite to that of the electrons. With this gun, a crossover 0·2 mm in diameter can be obtained at a great distance (150 to 300 mm); further, this gun is capable of giving very large currents if the diameter of the cathode is increased.

Other long-focus guns have been studied by Ehrenberg and Spear (1951) and Spear (1952).

In conclusion, it is worth considering the effect of sending electrons through a microscope with a high magnification in the opposite direction—the linear magnification will be the inverse of its usual value, and we shall obtain a spot of very small diameter, the minimum being simply the limit of resolution in the same instrument when used in the normal way. The production of these ultrafine spots, which are usually known as "electron probes", has been studied in detail by Castaing (1951). They have such important applications as shadow microscopy, scanning microscopy and microanalysis. Castaing's probe instruments are described in § 21.5.

8.5.3 Electrostatic Guns with High Perveance

When the beams of electrons (or ions) are very intense but have fairly low energies, the role of the space charge becomes dominant within the gun. In the beam itself, the particles are subjected to a transverse defocusing force which tends to make them diverge rapidly. To obviate this defocusing, a transverse field must be created between the cathode and the anode which opposes the defocusing force; this can be done if the shape of the electrodes which are placed around the beam is suitably chosen.

This shape can only be calculated rigorously in a very limited number of cases, and even then, certain simplifying hypotheses must be made. We shall outline the principle of this calculation rapidly.

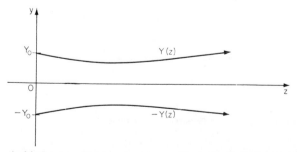

FIG. 92 S 1. A thin beam, with plane symmetry. The cathode lies in the plane $z = 0$.

Within the beam, the potential satisfies Poisson's equation:

$$\nabla^2 \Phi = \frac{\varrho}{\varepsilon_0} \quad \text{(R.M.K.S. units)}, \qquad (8.46)$$

in which ϱ represents the charge density, which is assumed to be constant in the transverse cross-section; it is related to the total current I, the longitudinal component v_z of the velocity and the cross-section $S(z)$ by

$$\varrho = \frac{I}{S(z)v_z}. \qquad (8.47)$$

We consider first of all a system with translation symmetry, infinite in the Ox direction. The width of the cathode which is plane is $2\,Y_0$; the cathode lies in $z = 0$ and is symmetrical with respect to the yOz plane. We denote the marginal trajectory of the beam by $Y(z)$; this is the trajectory which emerges from the edge of the cathode, $y = Y_0$, $z = 0$, on the side where the values of y are positive (see Fig. 92 S 1).

The potential $V(z, y)$ in the neighbourhood of the axis can, by virtue of the symmetry, be expanded in the form:

$$V(z, y) = V_0(z) + \frac{1}{2} y^2 \left(\frac{\partial^2 V(z, y)}{\partial y^2} \right)_0, \qquad (8.48)$$

in which $V_0(z)$ is the potential on Oz.

To a first approximation, the motion of the particles is described by the two equations

$$\frac{d^2 z}{dt^2} = \frac{e}{m} \frac{\partial V(z, y)}{\partial z} = \frac{e}{m} \frac{dV_0(z)}{dz}, \qquad (8.49\,a)$$

$$\frac{d^2 y}{dt^2} = \frac{e}{m} \frac{\partial V(z, y)}{\partial y} = \frac{e}{m} y \left(\frac{\partial^2 V(z, y)}{\partial y^2} \right)_0. \qquad (8.49\,b)$$

Equation (8.49 a) is readily integrable, and we obtain

$$v_z^2 = \frac{2e}{m} V_0(z). \qquad (8.50)$$

In the transverse direction, the equation is linear in y: if $y = Y(z)$ is a trajectory, therefore, $y = c\,Y(z)$ is also a possible trajectory, with $0 < c < 1$; the flow is laminar (this is true only on the assumption that the transverse emission velocities of the particles can be neglected).

On eliminating the time between equations (8.50) and (8.49 b), we obtain

$$\sqrt{V_0(z)} \frac{d}{dz} \left[\sqrt{V_0(z)} \frac{dy}{dz} \right] = \frac{1}{2} y \left(\frac{\partial^2 V(z, y)}{\partial y^2} \right)_0. \qquad (8.51)$$

The right-hand side of equation (8.51) can be evaluated with the aid of equation (8.46). If I is the current emitted per unit length of the cathode in the Ox direction, we have

$$\varrho = \frac{I}{2Y(z)\sqrt{\dfrac{2e}{m}\,V_0(z)}},$$

(8.52)

and

$$\left(\frac{\partial^2 V(z,y)}{\partial y^2}\right)_0 = \frac{\varrho}{\varepsilon_0} - \frac{d^2 V_0(z)}{dz^2}.$$

(8.53)

Finally, therefore, on substituting from equation (8.53) into equation (8.51), we obtain

$$\sqrt{V_0(z)}\,\frac{d}{dz}\left[\sqrt{V_0(z)}\,\frac{dy}{dz}\right] + \frac{1}{2}\,V_0''(z)\,y$$

$$= \frac{I}{4\varepsilon_0\sqrt{\dfrac{2e}{m}\,V_0(z)}}\,\frac{y}{Y(z)}.$$

(8.54)

By hypothesis, however, $Y(z)$ is a particular trajectory, a solution of (8.54). If y is replaced by $Y(z)$, we obtain

$$Y(z)\,V_0''(z) + Y'(z)\,V_0'(z) + 2\,Y''(z)\,V_0(z) = \frac{I}{2\varepsilon_0\sqrt{\dfrac{2e}{m}\,V_0(z)}}.$$

(8.55)

First special case

Here, we try to obtain a beam of constant width parallel to the axis, $Y(z) = Y_0$. When the emission at the cathode is space charge limited, the condition

$$V_0'(z) = 0$$

(8.56)

is obtained in $z = 0$. If we take the zero of potential at the cathode, equation (8.55) simplifies to

$$V_0(z) = \left(\frac{9}{4}\,k_1\,I\right)^{2/3} z^{4/3} = A_1\,z^{4/3}.$$

(8.58)

On substituting from equations (8.58), (8.53), and (8.52) into (8.48) it becomes apparent that at the edge of the beam $y = \pm\,Y_0$, the distribution $V(z,y)$ is such that

$$\sqrt{V_0(z)}\,V_0''(z) = \frac{I}{2\varepsilon_0\,Y_0\sqrt{\dfrac{2e}{m}}} = k_1\,I,$$

(8.57)

in which k_1 is a constant. The solution of this equation gives the axial potential in the form

$$V(z,\,Y_0) = V_0(z),$$

which entails

$$\frac{\partial V(z, Y_0)}{\partial y} = 0, \qquad (8.59\,\text{a})$$

and

$$\frac{\partial^2 V(z, Y_0)}{\partial y^2} = 0. \qquad (8.59\,\text{b})$$

Along an arbitrary trajectory within the beam, $y(z) = c\,Y_0$, we find the same law:

$$V(z, c\,Y_0) = V_0(z).$$

The equipotential surfaces are therefore planes perpendicular to Oz.

At the boundaries, the potential within the beam and the external potential V_{ext} must coincide; V_{ext} satisfies the Laplace equation

$$\frac{\partial^2 V_{\text{ext}}}{\partial z^2} + \frac{\partial^2 V_{\text{ext}}}{\partial y^2} = 0, \qquad (8.60)$$

and must satisfy the condition:

$$\frac{\partial^2 V_{\text{ext}}}{\partial y^2} = 0 \quad \text{at} \quad y = Y_0,$$

to conform with equation (8.59 a). In the upper half-plane, we select a new axis $O_1 z$ which coincides with the edge of the beam $y = Y_0$ (see Fig. 92 S2). From the general theory of potential (Durand, 1952), we know that both the real and the imaginary parts of any analytic function of a

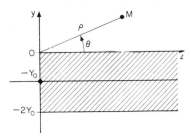

Fig. 92 S2. A flat beam with constant cross-section: the change of coordinates.

complex variable, $u = z + j\,y$, are possible solutions of the Laplace equation. If, in equation (8.58), z is replaced by $z + j\,y$, a solution of the Laplace equation for which $V = f(z)$ for $y = 0$ is $V = \text{Re}[f(u)]$. We can thus obtain an expression for $V(z, y)$ which satisfies equation (8.60) [see Pierce, 1949, in which we find $V(z, y) = \text{Re}(A_1 u^{4/3})$]; in polar coordinates (r, θ), we should have

$$V_{\text{ext}}(r, \theta) = \left(\frac{9}{4} k_1 I\right)^{2/3} r^{4/3} \cos\frac{4}{3}\theta \quad \text{for} \quad 0 \le \theta \le \frac{3\pi}{2}. \qquad (8.61)$$

By writing $V_{ext} = $ const, we can trace the equipotentials outside the beam. If in particular $V = 0$, we shall obtain a *straight line*, defined by $\theta_0 = \dfrac{3\pi}{8} = 67° \; 30'$.

The appearance of the equipotentials which are obtained in this way is shown in Fig. 92 S3, in which the accelerating potential has been put arbitrarily equal to unity. In a real system, for an accelerating potential V_0, we should have

$$V_0(z) = V_0 \quad \text{at} \quad z_0 = \left(\frac{V_0}{A_1}\right)^{3/4}.$$

Alternatively by fixing V_0 and z_0, we can calculate the maximum current I which leaves the cathode, and the perveance $P = I/V_0^{3/2}$ of the gun.

Certain of the equipotential surfaces are materialized in the form of metal electrodes. The cathode K lies in the plane $z = 0$, and is $2 Y_0$ in width. In general, the grid W follows the equipotential $V = 0$ and is connected electrically to the cathode, but we may use a negative electrode which will be placed behind K. The anode and the equipotential $V = V_0$ coincide.

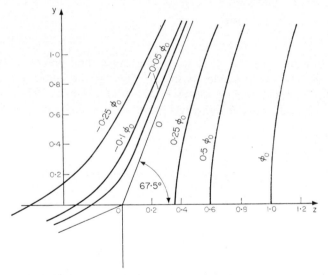

FIG. 92 S3. Equipotentials outside a flat beam of constant cross-section, when the cathode emission is limited by space charge. The divergence which occurs during the passage through the anode is neglected.

After passing through the anode, the beam enters a region free of electric field; the opening in the anode behaves like a divergent lens, the focal length of which we can calculate (see § 8.3.1), and further, the beam diverges under the action of its own space charge.

If we are to obtain a convergent beam at the exit of the gun, the shape of the electrodes must be altered.

Second special case: the convergent beam with a plane of symmetry

By using the solution of Langmuir and Blodgett (1923) for the linear flow of charge in the space between two concentric cylindrical electrodes when the conditions are space charge limited, the shapes of the electrodes of the gun can still be calculated by Pierce's method (1949), but only in an approximate fashion. Alternatively, we can set off from equation (8.54), in which we write

$$Y(z) \simeq Y_0(1 - z/a) \tag{8.62}$$

(the trajectory produced backwards intersects the axis again behind the anode at $z = a$).

Equation (8.54) then becomes

$$\frac{d}{dz}\left[\left(1 - \frac{z}{a}\right)\frac{dV}{dz}\right] = k\frac{I}{Y_0\sqrt{V}} \tag{8.63}$$

(Wakerling and Helmholz, 1949), in which I represents the total intensity emitted per unit length of the cathode in the Oz direction.

We try to find a solution of the form

$$V(z) = A z^{4/3} F(z), \tag{8.64}$$

which satisfies the two conditions imposed by the space charge limited

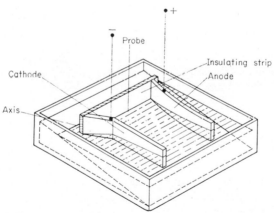

Fig. 92 S4. Diagram of an electrolytic tank to be used to determine the shape to be given to the electrodes of the gun. The beam is represented schematically by an insulating strip.

emission in the plane $z = 0$, namely

$$V(z) = 0 \quad \text{and} \quad V'(z) = 0.$$

The function $F(z)$ is obtained in an approximate manner in series form; in the plane $z = 0$, $F(0) = 1$. If we revert to the coordinate system shown in Fig. 92 S2, where Oz is the edge of the beam, we find:

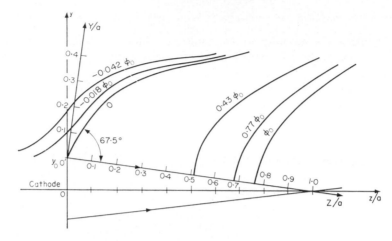

FIG. 92 S5. The case of a convergent beam with plane symmetry and straight
edges. The anode corresponds to the equipotential φ_0; the real beam diverges
after passing through this electrode.

$$V(r, \theta) = A \left(r^{4/3} \cos \frac{4}{3}\theta + \frac{8}{15} r^{7/3} \cos \frac{7}{3}\theta + \cdots \right). \qquad (8.65)$$

The shape of the grid is obtained by putting $V = 0$. For $r = 0$, we see
that as before, the electrode leaves the cathode at an angle of $\dfrac{3\pi}{8}$ with the
edge of the beam, but that this electrode is no longer plane. An approximate
calculation is normally made using the two terms above, and a suitable
form for the electrodes which will lead to the distribution of equation
(8.64) is then sought with the aid of an electrolytic tank. The method
used is as follows: the electrodes are built of thin metal sheets, and placed
in a flat-bottomed tank. The edges of the beam are simulated by two insulating strips, so that the condition $\partial V/\partial n = 0$ at the edge of the beam
is imposed (that is, the normal field is zero). The potential distribution
along a strip is measured, outside the beam, and the grid and anode are
then bent and shaped until the desired distribution is obtained.

The general arrangement of the tank in the case of an axially symmetric
beam is illustrated in Fig. 92 S4.

The form of the equipotentials obtained in this way is illustrated in
Fig. 92 S5.

Third special case: the thin beam with arbitrary cross-section

$Y(z)$ is now an arbitrary function; it might be part of a hyperbola, for
example (Wakerling and Helmholz, 1949), of the form

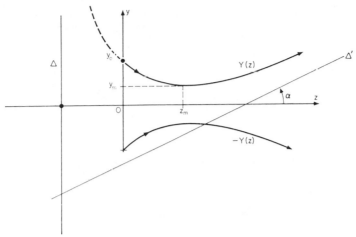

FIG. 92 S6. A beam bounded by two hyperbolic arcs, with asymptotes $\varLambda$ and $\varLambda'$.

$$Y(z) = (Y_0 - Y_m)\frac{\left(\dfrac{x - x_m}{x_m}\right)}{1 + \dfrac{Y_0 - Y_m}{\cdot_m}\dfrac{x}{x_m}\cot\alpha}. \tag{8.66}$$

The meanings of Y_0, Y_m, x_m and α are shown in Fig. 92 S6.

The case $\alpha = \pi/2$ corresponds to a parabolic trajectory. Equation (8.54) is solved with the aid of an electronic computer, which furnishes the function $V_0(z)$. We can then determine the potential *outside the beam* (which is assumed to be thin enough for a series expansion of the potential to be adequately represented by terms up to those in y^2) by using the equation

$$V(z, y) = V_0(z) - \frac{1}{2}y^2\frac{d^2 V_0(z)}{dz^2} + \frac{1}{2}y\frac{k I}{\sqrt{V_0(z)}}, \tag{8.67}$$

in which we write $V(z, y) =$ constant. Again using an electronic computer, we solve for y and thus obtain the equation $y(z)$ for each of the different equipotentials.

Figure 92 S7 is an example of such a distribution, in a particular case.

We see that it is possible to obtain the desired beam with the aid of electrodes which are wholly outside the beam, and that it is even possible to slow the beam down after it has been accelerated; this can be extremely useful in certain research in which intense beams of electrons or ions are in use.

Fourth special case: the axially symmetrical beam

The problem now becomes even more complicated, as the equation which gives $V_0(z)$ is still less tractable than equation (8.54), save when the

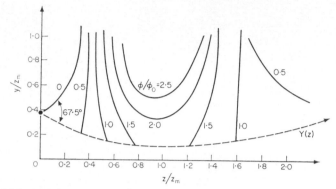

FIG. 92 S7. A chart of the potential when the edges of the beam are hyperbolic; φ_0 represents the maximum energy of the particles on the axis.

cross-section of the beam remains constant. We find, in fact:

$$R(z)\, V_0''(z) + 2R'(z)\, V_0'(z) + 4R''(z)\, V_0(z)$$

$$= \frac{I}{2\pi\,\varepsilon_0\,\sqrt{\dfrac{2e}{m}\,V_0(z)}} \cdot \frac{1}{R(z)}. \qquad (8.68)$$

If $R(z) = R_0$, we can again find a solution of the form

$$V(z) = A\,z^{4/3}. \qquad (8.69)$$

To a first approximation, and when the transverse dimensions of the beam are very slight, the distribution on the edge of the beam resembles that of equation (8.69) and the shapes of the electrodes which are necessary can be determined with the aid of an electrolytic tank with a sloping bottom. The family of equipotentials which is obtained is represented in Fig. 92 S8.

The case of a convergent beam with a rectilinear border (emitted by a cathode in the shape of a spherical cup) has also been solved approximately, with the aid of the results of Langmuir and Blodgett (1924) on rectilinear particle flow between two concentric spheres. The shape of the electrodes is shown in Fig. 92 S9.

Equation (8.68) could obviously be solved with the aid of a computer by means of a method resembling that described in the preceding sections. In practice, however, axially symmetric beams are far more common than plane beams and a great deal of attention has thus been devoted to the former; both the approximate calculation of the electrode shape, and the experimental determination of this shape have been carried out, when a beam with given characteristics is to be produced.

Requirements which are frequently encountered are that the beam should converge as it leaves the anode, and pass through a point of minimum cross-section at a distance z_c from the anode as far from the latter as possible

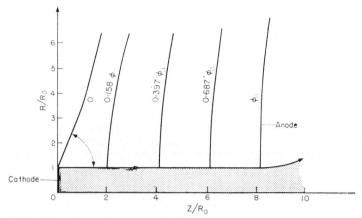

FIG. 92 S8. Theoretical chart of the equipotentials for a beam of circular cross-section, parallel to the axis.

without sacrificing the highest possible intensity. In this connection, we mention the calculations of Guénard (1945) and Huber (1949) and the experimental work of Helm, Spangenberg and Field (1947), Samuel (1945), Heil and Ebers (1950) and Müller (1956); the articles by Süsskind (1956), devoted to guns descended from the "Pierce gun", and by Ivey (1954) which is concerned with the focusing of intense beams, may also be consulted with profit.

We shall mention yet one further method, suggested by Charles and Septier (1948), which depends only upon the chart of the potential $\Phi(r, z)$ which is obtained in the electrolytic tank. It allows us to calculate the shape of the beam and the maximum intensity which the cathode can emit at a given potential when the gun is of arbitrary shape and is working in space

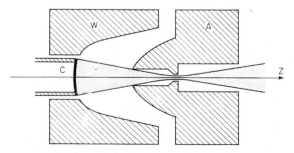

FIG. 92 S9. An example of the electrodes with which we can produce a convergent axially symmetric beam which has straight edges between C and A. The electrodes C and W are connected electrically to one another. The beam cross-section passes through a minimum in the opening in the anode, and subsequently diverges rapidly.

charge limited conditions $\left(\dfrac{\partial V}{\partial z} = 0 \text{ on } K\right)$. This method has in particular been used for klystron guns.

In conclusion, it should be stressed that calculation can provide only approximate solutions, which represent a first approximation to the shape of the electrodes. Further, none of the preceding calculations is valid unless the influence of the opening in the anode and above all, the transverse components of the particle velocities at the cathode, are neglected. Finally, experiment shows that electrode systems with forms as simple as those illustrated in Figs. 92 S5 and 92 S9, while always ensuring that the potential at the edges of the beam is distributed in a manner closely resembling equation (8.69), give very good results.

CHAPTER 9

MAGNETIC LENS MODELS

9.1 THE LONG SOLENOID

For a detailed study of the uniform magnetic field which is produced by a long solenoid, the reader is referred to the standard textbooks; a full description of the application of the solenoid in β-ray spectrographs is to be found in an article by Grivet (1950b, 1951a), together with an extensive bibliography. Here, we shall attempt no more than to summarize the basic formulae. In a rotating meridian plane (cf. § 5.1) the trajectory is of the form

$$r = 2\varrho \sin \alpha \sin \left(\frac{z}{2\varrho \cos \alpha} \right), \tag{9.1}$$

in which α is the angle between the initial velocity at the object point (which lies within the field) and the direction of the axis. Oz is an axis which is parallel to the field at the object point through which it passes. The distance between the object A and the image A' is given by

$$A A' = 2\pi \varrho. \tag{9.2}$$

Expanding r in powers of α about A', the coefficients of spherical aberration C_s and chromatic aberration C_c prove to be equal, thus

$$\begin{aligned} C_s &= \pi \varrho, \\ C_c &= \pi \varrho. \end{aligned} \tag{9.3}$$

The magnification G in the rotating meridian plane is -1, but since this plane rotates through just $180°$ between the object and its image, the actual magnification is $+1$: image and object are equal. The number of ampère-turns per unit length which corresponds to some given value of ϱ is given by

$$n I = \frac{5 \cdot 366 \sqrt{\Phi(1 + 0 \cdot 979 \times 10^{-6} \Phi)}}{2\varrho} = \frac{5 \cdot 366 \sqrt{\Phi^*}}{2\varrho} \quad \text{(A, V, m)} \tag{9.4}$$

for electrons of energy Φ eV; the relativistically corrected value of the accelerating potential for electrons is Φ^*, but for ions $\Phi^* \simeq \Phi$.

9.2 MODELS REPRESENTING SHORT SYSTEMS

In the majority of magnetic lenses, the axial field $B(z)$ is bell-shaped in distribution. Step-by-step integration is feasible, but it is more attractive to try to find a simple model to represent this bell-shaped field, of such a kind that the trajectory equation is integrable in terms of tabulated functions.

9.2.1 The Glaser Model (1941b)

In a symmetric lens, the bell-shaped curve $B(z)$ can be closely described by the function

$$B(z) = \frac{B_c}{1 + (z/a)^2}, \tag{9.5}$$

in which B_c is the field at the centre of the lens, and a is the distance between the centre of the lens and the points at which the longitudinal component of the field has fallen to $B_c/2$.

If we write

$$x = z/a, \quad y = r/a, \quad k^2 = \frac{eB_c^2 a^2}{8m_0 \Phi^*}, \quad p = \sqrt{1+k^2},$$

the trajectories can be expressed in terms of elementary functions. (For electrons, $k = 1 \cdot 48 \times 10^5 \, aB_c/\sqrt{\Phi^*}$ in m.k.s. units.) This can be seen as follows. The equation of the Gaussian trajectories is of the form

$$\frac{d^2y}{dx^2} + \frac{k^2 y}{(1+x^2)^2} = 0 \tag{9.6a}$$

and after a new change of variable,

$$x = \cot \phi \quad (\pi > \phi > 0),$$

which gives

$$dx = -\frac{d\phi}{\sin^2 \phi} \quad \text{and} \quad 1 + x^2 = \frac{1}{\sin^2 \phi},$$

we obtain

$$\frac{d^2y}{d\phi^2} + 2 \cot \phi \frac{dy}{d\phi} + k^2 y = 0. \tag{9.6b}$$

Figure 93a shows how the variables z and ϕ can be visualized simply. We write

$$y(\phi) = \frac{v(\phi)}{\sin \phi} = v(\phi) \sqrt{1+x^2}$$

and equation (9.6b) becomes

$$v''(\phi) + p^2 v(\phi) = 0, \tag{9.6c}$$

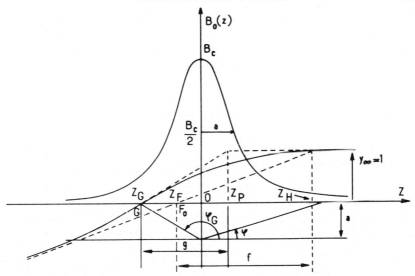

FIG. 93a. The Glaser model: definition of the variable Φ; the classical cardinal elements (F, H, f) and the immersion elements $(G, P$ and $g)$.

which gives the general solution

$$\frac{r}{a} = y = \frac{1}{\sin\phi}\,(C_1 \sin p\phi + C_2 \cos p\phi) = C_1\,\frac{\sin p\,(\phi + C_2)}{\sin\phi}, \quad (9.6\text{d})$$

in which

$$\phi = \cot^{-1}(z/a).$$

Let us now consider a trajectory $y_1(\phi)$, parallel to the axis for $z = +\infty$ ($\phi = 0$) and traversing the lens from right to left (Fig. 93a). It remains at a finite distance from the axis ($y = 1$ for $\phi = 0$), so that

$$C_2 = 0 \quad \text{and} \quad y = C_1\,\frac{\sin p\phi}{\sin\phi} \quad \text{with} \quad C_1 = 1/p. \quad (9.7)$$

The slope of this trajectory at any point is given by

$$y' = \frac{dy}{d\phi}\frac{d\phi}{dx} = \frac{1}{p}\,(\sin p\phi \cos\phi - p \sin\phi \cos p\phi),$$

which gives $y' = 0$ for $\phi = 0$, as required. At $x = -\infty$ ($\phi = \pi$), we have

$$y'(-\infty) = \frac{\sin p\pi}{p}.$$

This trajectory intersects the axis at points where $\sin p\phi = 0$, that is where $\phi_n = n\pi/p$, in the range $0 < \phi < \pi$. For high values of p, and hence of the excitation parameter k^2, we can thus have several real foci. For $p \le 2$ (or $k^2 \le 3$), there is only a single focus; the lens is more commonly used in this range of excitations.

It is to be noticed that the focus defined in this way is "immersed" in the field (see §4.3.4); it is not the "asymptotic" focus that we have used hitherto to ascertain the behaviour of the rays in a lens.

Glaser's bell-shaped field was the first example known of "Newtonian" fields, fields which can lead to two sets of cardinal elements—"classical" cardinal elements and "immersion" elements—for both of which the relations for conjugacy between object and image remain valid.

The classical cardinal points (Fig. 93a). The asymptote to the trajectory of equation (9.7) is

$$y = -\cos p\pi + x\frac{\sin p\pi}{p}, \tag{9.8a}$$

and the usual lens constants (focal length, position of the focus and principal plane) for a symmetrical lens are, on the image side,

$$\frac{f}{a} = \frac{p}{\sin p\pi}, \quad \frac{z_F}{a} = p\cot(p\pi), \quad \frac{z_H}{a} = p\cot\left(\frac{p\pi}{2}\right). \tag{9.8b}$$

The image rotation between the object point at infinity and the image z_F is given by

$$\theta_{oi} = \pi k.$$

These "asymptotic" elements are used whenever we wish to form a real image at very high magnification ($z_i \to \infty$) with a virtual object; this is the situation at the projective lens in an optical system containing several lenses. In object space, the rays converge towards a point near the

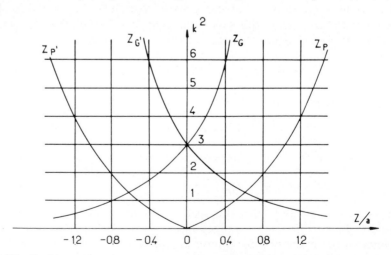

FIG. 93b. Position of the immersion foci and principal planes (object and image) as a function of k^2.

classical object focus F_0 (Fig. 93a); the final magnification G is given by

$$G = \left| \frac{A'B'}{AB} \right| \simeq \frac{z_i}{f_0}.$$

It can be shown (see Glaser, 1952) that, under these conditions, the magnification passes through a maximum for $k^2 = 1 \cdot 05$; the minimum focal length $f_{0\min} \simeq 1 \cdot 5a$ and $z_{FO} \simeq 0 \cdot 8a$, so that the virtual object is then immersed a relatively short distance. This is the arrangement that is generally used in electron microscopy with real objects.

The immersion cardinal elements (Fig. 93b). Using equation (9.7) directly, with an immersed object, we find that on the image side,

$$\frac{g}{a} = -\frac{1}{\sin \pi/p}, \qquad \frac{z_G}{a} = \cot \frac{\pi}{p},$$

$$\frac{z_P}{a} = -\cot \frac{\pi}{2p}, \qquad (9.9a)$$

and the rotation is given by

$$\theta_{oi} = \pi k/p.$$

If we consider two particular trajectories,

$$u(\phi) = \frac{\sin p\phi}{\sin \phi} \quad \text{and} \quad v(\phi) = \frac{\cos p\phi}{\sin \phi}$$

and an object plane defined by $\phi = \phi_o$ with the corresponding image plane $\phi = \phi_i$, the magnification G will be given by

$$G = \frac{u(\phi_i)}{u(\phi_o)} = \frac{v(\phi_i)}{v(\phi_o)}.$$

This implies

$$\sin p(\phi_i - \phi_o) = 0$$

or, for $n = 1$

$$\phi_i - \phi_o = \frac{\pi}{p}. \qquad (9.9b)$$

(This is a property common to all Newtonian fields.)

The image will again be rotated through an angle $\theta = k\pi/p$ with respect to the object.

In Fig. 93b, the variation of the positions of the foci and the principal planes with k^2 is shown, and in Fig. 93c the behaviour of the reduced convergence $C = a/g$ is illustrated.

For $k^2 = 3 (p = 2)$, the immersion foci coincide at the centre of the lens and we have $g = g_{\min} = a$. In this situation, we have (for electrons)

$$g_{\min} = 1 \cdot 17 \times 10^{-5} \frac{\sqrt{\Phi^*}}{B_c}. \qquad (9.10a)$$

(g in metres, Φ^* in volts and B_c in tesla).

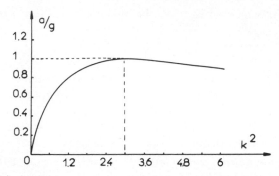

FIG. 93c. Dependence of the convergence $C = a/g$ as a function of k^2 ($g/a = 1$ for $k = 3$).

For a given value of the ratio $\sqrt{\Phi^*}/B_c$, we can derive the value of a that yields the absolute minimum value of g. We regard a as a function of k:

$$a = \sqrt{\frac{8m_0}{e}} \cdot \frac{\sqrt{\Phi^*}}{B_c} \cdot k,$$

and seek the minimum of

$$g(k) = \frac{a(k)}{\sin \dfrac{\pi}{\sqrt{1+k^2}}}.$$

We find that $g = g_{min}$ for $k^2 = 0{\cdot}8$, which yields

$$a_{min} = 0{\cdot}54 \frac{\sqrt{\Phi^*}}{B_c} \times 10^{-5},$$

$$g_{min} = 0{\cdot}84 \frac{\sqrt{\Phi^*}}{B_c} \times 10^{-5}$$

(9.10b)

(in the same units as (9.10a)).

A magnetic lens differs from an electrostatic lens in that a real (non-magnetic) object can be placed near the immersion focus and hence a highly magnified image will be obtained. In electron microscopy, however, the object is "illuminated" with an electron beam from a gun, accelerated through a potential difference Φ; these electrons will therefore experience the action of a large part of the field $B(z)$ before they reach the object. It will thus be difficult to ascertain the exact conditions of illumination at the object and this is why an external (or only slightly immersed) focus has, for so long, always been used. (In reality, all the real foci will be immersed with Glaser's mathematical model, since the field extends from $-\infty$ to

$+\infty$, but we can assume that when $B < 10^{-2}B_c$, say, we are "outside" the field.) Very recently, however (see Chapter 17), a microscope objective in which the object is placed at $z = 0$ has been developed by Riecke (1962) and Ruska (1962, 1966).

Asymmetrical distributions. It is always possible to represent an asymmetrical bell-shaped curve by two functions of the foregoing type (Dosse, 1941):

$$B(z) = \frac{B_c}{1 + (z/a_1)^2} \quad \text{for} \quad z \leqslant 0; \; z = a_1 \cot \phi; \; p_1 = \sqrt{1 + k_1^2},$$

$$B(z) = \frac{B_c}{1 + (z/a_2)^2} \quad \text{for} \quad z \geqslant 0; \; z = a_2 \cot \phi; \; p_2 = \sqrt{1 + k_2^2}.$$

A trajectory is calculated in one of the regions ($z > 0$, for example); from its characteristics (y and y') at $z = 0$, we can determine the constants C_1 and C_2 that appear in the general expression for the same trajectory in the second field region ($z < 0$). Object and image are related thus (Glaser, 1952):

$$\tan\left\{p_2\left(\frac{\pi}{2} - \phi_i\right)\right\} = \frac{a_1 p_2}{a_2 p_1} \tan\left\{p_1\left(\frac{\pi}{2} - \phi_0\right)\right\}.$$

If we characterize the degree of asymmetry by the ratio $q = a_1/a_2$, we can show that the cardinal elements do not change much as q is varied from 0 to infinity. We notice that the focal lengths on the object and image sides remain equal, but the foci are no longer symmetrical about $z = 0$.

Aberration coefficients. Glaser and Lammel (1943) succeeded in putting the aberration coefficients of their lens in a form that could be calculated directly.

For a lens with very high magnification (object situated near the object focus), the spherical aberration coefficient (referred to the object plane) is given by the general expression

$$\frac{C_s}{a} = \frac{1}{\sin^4 \phi_0}\left\{\frac{\pi k^2}{4p^3} + \frac{1}{8}\frac{4k^2 - 3}{4k^2 + 3}(\sin 2\phi_i - \sin 2\phi_0)\right\}, \qquad (9.11)$$

which becomes

$$\left(\frac{C_s}{a}\right)_{\phi_i = 0} = \left(\frac{\pi k^2}{4p^3} - \frac{1}{8}\frac{4k^2 - 3}{4k^2 + 3}\sin\frac{2\pi}{p}\right)\frac{1}{\sin^4 (\pi/p)}. \qquad (9.12)$$

When k^2 is increased, C_s initially decreases steadily (see Fig. 93d); for $k^2 = 3$, in particular, we have $C_s \simeq 0\cdot3a$. Between $k^2 = 3\cdot5$ and $k^2 = 7$ it remains virtually constant and then begins to increase very slowly. The minimum ($C_s = 0\cdot25a$) occurs in a range of excitations that are too high to be used in practice.

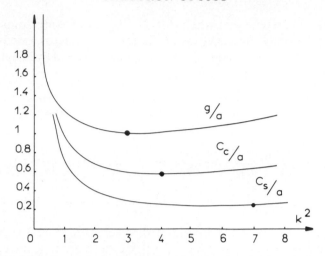

FIG. 93d. Variation of the immersion focal length, g/a, of C_s/a and of C_c/a. The dots indicate the minima.

If Φ^* and B_c are fixed, we can vary a to give C_s its minimum value. We find that now

$$k^2_{\min} = 2\cdot8, \qquad a_{\min} = 113\frac{\sqrt{\Phi^*}}{B_c}$$

and

$$(C_s)_{\min} = 0\cdot30\,a_{\min}$$

(a in mm, Φ^* in volts and B_c in gauss).

We see, therefore, that for an objective lens with $k^2 = 3$ and the minimum focal length ($f_{\min} = a$), we shall have the focus at $z = 0$ and very nearly minimum aberration if we choose a lens for which a is given by the foregoing formula.

The chromatic aberration constant is given by

$$\frac{C_c}{a} = \frac{\pi k^2}{2p^3}\frac{1}{\sin^2 \phi_0},$$

or, for high magnification (object at the focus),

$$\frac{C_c}{a} = \frac{\pi k^2}{2p^3}\frac{1}{\sin^2 (\pi/p)}.$$

C_c passes through a minimum near $k^2 = 4$ and we have

$$\left(\frac{C_c}{a}\right)_{\min} \simeq 0\cdot58.$$

If we try to extract the maximum magnification from the objective, we have for $k^2 = 3$ ($p = 2$)

$$\frac{C_c}{a} \simeq 1 \cdot 8 \simeq 3 \left(\frac{C_c}{a} \right)_{min},$$

$$\frac{C_s}{a} \simeq 0 \cdot 3.$$

The dependence of C_s and C_c upon k^2 is illustrated in Fig. 93d.

Correction

This Glaser model is capable of describing to within a few per cent the properties of magnetic lenses with a wide variety of pole-pieces, and under very diverse conditions of excitation; it is very well adapted to the case of iron-free coils, which may well be used in electron microscopes or diffraction devices working at very high voltages, now that superconductors with high critical field have become available (§9.5). There do exist, however, cases in which the experimental $B(z)$ curve differs markedly from the Glaser model, in particular when the iron is far from saturation. In such cases, however, the precision with which the model describes the actual situation can be considerably improved by choosing the half-width a of the "bell" suitably. Normally, the model curve is given the same half-width and height as the actual curve (measured, generally speaking) for the lens, and the agreement elsewhere is found to be sufficiently good. In this situation, on the other hand, a satisfactory representation of the real characteristic is not possible; we no longer attempt, therefore, to match the two curves, other than by making them of the same height, but rather choose a half-width, $a_1 = ma$, for the model curve (a being the real half-width) in such a way that the error in one of the cardinal elements—the focal length f_0 or f_i is usually selected—is minimized. Glaser (1950a) has given an exact solution of this problem, but often a less rigorous procedure, suggested by Grivet (1951b, 1952a) proves to be both considerably more simple and adequately precise. If T_1 and T_2 are the equations of the two bell-shaped curves, we simply write

$$\int_{-\infty}^{\infty} (T_1^2 - T_2^2) \, dz = 0. \tag{9.13}$$

9.2.2 The Grivet–Lenz Model

Glaser's model curve is very satisfactory at the centre of the lens, less so at its two edges—in fact, the field falls off very rapidly, following, in general, an exponential law. This exponential law seems to be characteristic of the way in which the field decreases whenever this decrease is produced as a result of deliberate "screening". Glaser's curve, however, decreases very slowly, much too slowly, and it was this fact that led Grivet (1951 b, 1952a, b) and Lenz (1951) to consider a model less unsatisfactory in this respect. The new model which was suggested takes the form:

$$B(z) = B_0 \operatorname{sech}(z/b), \tag{9.14}$$

in which

$$b = 0 \cdot 7593a, \tag{9.15}$$

where a is the half-width of the "bell" at one-half of its maximum height.

The Gaussian rays

If b is chosen as the unit of length, and $x = z/b$, $R = r/b$, the Gaussian ray equation is of the form

$$R'' \cosh^2 x + h^2 R = 0, \tag{9.16}$$

in which $h = 0 \cdot 7593\, k$, and k is the usual parameter of magnetic lenses defined by

$$k^2 = \frac{e\, B^2}{8m\, \Phi_0}\, a^2.$$

A change of variable, $u = \tanh x$, transforms this equation into the Legendre equation

$$(1 - u^2)\frac{d^2 R}{du^2} - 2u\frac{dR}{du} + \nu(\nu + 1)\, R = 0, \tag{9.17}$$

so that a general ray can be written in the form

$$R = A\, P_\nu(\tanh x) + B\, P_\nu(-\tanh x), \tag{9.18}$$

in which $P_\nu(u)$ is the Legendre *function* of order ν, with $\nu(\nu + 1) = h^2$. Some values of this function, which has been tabulated by Schelkunoff, are reproduced in Table 4 (see pp. xx–xxi). The cases for which ν is an integer, n, are physically significant and mathematically special—the *function* $P_\nu(u)$ goes over into the *polynomial* $P_n(u)$, while $P_n(-u)$ has to be replaced by $Q_n(n)$, a Legendre *function* of the second kind.

The particular values $\nu = n$ (integral)

Whether or not ν be an integer, the ray $r = P_\nu(\tanh x)$ is always a principal ray, parallel to the axis and unit distance away from it as $x \to \infty$, since $P_\nu(1) = 1$ is always true, and $P'_\nu(1)$ is finite; the relation $u = \tanh x$ ensures that the asymptotic behaviour is acceptable. If, however, ν takes some integral value n, this principal ray emerges parallel to the axis for $x \to -\infty$ as well, since $P_n(-1) = 1$ if n is an integer; in the classical sense of the word, the lens is afocal. Inside the lens, the ray intersects the axis n times, giving n immersion foci, G_n, symmetrically placed about the centre of the lens which itself is a focus if n is odd—if $n = 1$, it is the only focus, and in this case

$$r = A \tanh x + B x \tanh x.$$

The values $v = n$ for which the system is afocal separate the values of v which correspond to convergent systems ($0 < v < 1$; $3 < v < 4$; $5 < v < 6$;...) from those which correspond to divergent systems ($2 < v < 3$; $4 < v < 5$;...). In actual instruments, multiple foci are always avoided, and we need only study the region $0 < v \leqq 1$, that is $0 < h^2 \leqq 2$.

The classical cardinal elements

These elements define the homographic correspondence between the two asymptotes of each ray; the expansion of $P_v(u)$ in the neighbourhood of $u = -1$ is known, and after a certain amount of transformation, we obtain

$$P_v(\tanh x) \rightarrow \frac{\sin v\pi}{\pi} [2x + 2\{\psi(v) + C\} + \pi \cot \pi v],$$

in which $C = 0.5772$ is Euler's constant, and $\psi(v)$ is the derivative of $\log(v!)$. For the position of the focus z_F, the focal length f, and the image rotation φ we eventually obtain

$$\frac{z_F}{b} = \psi(v) + C + \frac{\pi}{2} \cot(\pi v), \tag{9.19}$$

$$\frac{f}{b} = \frac{\pi}{2 \sin(v\pi)}, \tag{9.20}$$

$$\varphi = \pi v.$$

When the lens is weak, the disparity between the results obtained with this model and those obtained with Glaser's model is slight; it becomes important, however, when the convergence is strong, and the most powerfully convergent projective lens proves to be defined not by $k = 1$, $f = 1.5a$ but by $v = 1/2$, $k = 1.14$, $f = 1.19a$.

The function $\psi(v)$ is well tabulated, in the tables of Jahnke and Emde in particular; it varies only between 0 and 1 when v itself varies from 0 to 1. The quantities z_F and f are plotted as functions of k in Fig. 94a.

Immersion

When either the object or the image (or both) lies inside the field, there is no longer a rigorous homographic correspondence between the two as there is in Glaser's model field; cardinal elements valid for every object and image no longer exist, as can easily be verified for the simple case $v = 1$. The cardinal elements defined by Glaser—which we shall simply call "immersion"—are, however, useful to describe the connexion between the object and its image in the neighbourhood of one particular pair of conjugate points, namely a focus and infinity. The focus G lies at the point at which

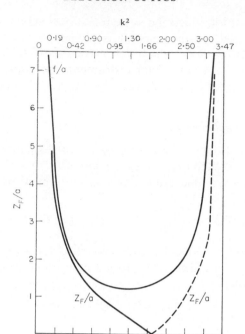

FIG. 94a. The normal cardinal elements, calculated from the Grivet–Lenz model.

$$P_\nu(u_\nu) = 0, \qquad u_\nu = \tanh x_G, \qquad z_G = b\, x_G, \tag{9.21}$$

in which u is one of the n real roots of $P_\nu(u) = 0$, and n is the integer nearest to and larger than ν; the root is unique if $0 < \nu \leq 1$. The immersion focal length, g, is given by

$$\frac{b}{g} = \frac{P'_\nu(u_\nu)}{\cosh^2 x_G} = \frac{2\sin(\nu\pi)}{\pi\, P_\nu(-u_\nu)}, \tag{9.22}$$

the second form being convenient when ν is small. The more $P_\nu(-u_\nu)$ differs from unity, the greater is the difference between g and its classical counterpart f; for $\nu \leq 0.3$ (so that $P_\nu \geq 0.98$), the immersion is negligible, and for small ν, the immersion focus G tends towards the familiar focus $F(x_F = x_G = 1/2\nu)$; see Fig. 94b. The image rotation varies with the position of the object, but between infinity and a focus it is given by

$$\varphi = \left\{ \frac{\pi}{2} - \tan^{-1}(\sinh x_G) \right\} h. \tag{9.23}$$

The value g_0 of g for which the focus G is situated at the centre of the lens characterizes an important type of microscope objective. The numerical value proves to be $g_0 = 0.7593a$ for $\nu = 1$ and $k^2 = 2$, which is appreciably

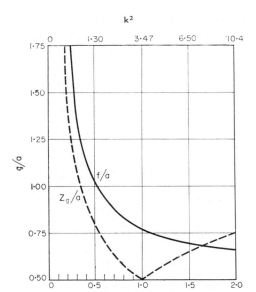

FIG. 94b. The immersion cardinal elements, obtained from the Grivet–Lenz formulae.

different from the value furnished by the Glaser model, namely, $g_0 = a$, $k^2 = 3$, and which is in fact closer to reality than the latter; in particular, the relation between ampère-turns $n\,I$, half-width a (millimetres), and the field at the centre of the lens B_0 (kilogauss) is

$$B_0 a = 5\cdot27\,n\,I,\tag{9.24}$$

for the Grivet–Lenz model, while for Glaser's field the numerical coefficient is 4. Experimentally, the value of the coefficient ranges between 4·4 and 5·8.

Another point of agreement with reality is that g_0 is not the minimum value of g; the cardinal elements of this model are tabulated numerically below.

9.2.3 Generalization of Glaser's Formulae

A number of authors have considered the properties of the more general type of bell-shaped field:

$$B(z) = \frac{B_0}{[1 + (z/a)^2]^\nu},\tag{9.25}$$

for various values of ν using numerical integration, (Glaser and Lenz, 1951; Lenz, 1950). Sturrock (1951b), on the other hand, has used a method due to Regenstreif to extend Glaser's results to fields of the form

THE CARDINAL ELEMENTS IN THE GRIVET–LENZ MODEL

v	k^2	$\dfrac{z_G}{a}$	$\dfrac{g}{a}$	$\dfrac{f}{a}$	$\dfrac{z_F}{a}$
0·1	0·1908	3·787	3·86	3·850	3·787
0·2	0·4163	1·861	2·025	2·029	1·860
0·3	0·6765	1·190	1·450	1·474	1·176
0·4	0·9539	0·8302	1·180	1·254	0·779
0·5	1·301	0·5914	1·0276	1·193	0·466
0·6	1·659	0·4200	0·932	1·254	0·1464
0·7	2·065	0·2847	0·864	1·474	−0·2700
0·8	2·496	0·1745	0·820	2·029	−0·987
0·9	2·965	0·0810	0·786	3·860	−2·962
0·95	3·211	—	—	7·525	−9·796
1	3·469	0	0·7593		
1·1	4·007	0·0717	0·739		
1·2	4·579	0·1359	0·722		
1·3	5·186	0·1933	0·708		
1·4	5·828	0·2477	0·697		
1·5	6·504	0·2968	0·688		
2	10·407	0·500	0·658		

$$B(z) = \frac{B_0}{1 - (z/a)^2}, \tag{9.26}$$

$$B(z) = \frac{B_0}{(z/a)^2 - 1},$$

or, by writing $a^* = j\,a$, to

$$B(z) = \frac{B_0}{1 + (z/a^*)^2}. \tag{9.26'}$$

B is clearly infinite at $z = \pm a$.

For $|z| < a$, we replace the quantities u, a, k, p of § 9.2.1 by

$$\frac{\pi}{2} + j\varphi, \quad j\,a, \quad j\,k, \quad \omega\sqrt{k^2 - 1},$$

and finally obtain

$$y = \operatorname{sech}(C_1 \sin\omega\varphi + C_2 \cos\omega\varphi),$$
$$z = a\tanh\omega\varphi.$$

The image is defined by

$$\varphi_i - \varphi_0 = \frac{\pi}{\omega}, \tag{9.27}$$

so that

$$-z_{f_0} = z_{f_i} = a \coth \frac{\pi}{\omega},$$

$$-f_0 = f_i = a \operatorname{cosech} \frac{\pi}{\omega},$$

$$-z_{H_0} = z_{H_i} = a \coth \frac{\pi}{2\omega},$$

and the principal planes, which are not crossed, lie outside the lens. The spherical and chromatic aberration coefficients are given by

$$C_s = a \left[\frac{\pi k^2}{4\omega^3} + \frac{1}{8} \frac{4k^2+3}{4k^2-3} (\sinh 2\,\varphi_i - \sinh 2\,\varphi_0) \right] \operatorname{sech}^4 \varphi_0, \qquad (9.28)$$

$$C_c = a \frac{\pi k^2}{2\omega^3} \operatorname{sech}^2 \varphi_0. \qquad (9.28')$$

The β-ray spectrograph of Siegbahn and Slätis (1949) is an example of a case where ω is small.

If $|z| > a$, however, the Glaser symbols u, a, B_0, k and μ are replaced by $j\varphi$, ja, $-B_0$, $-jk$ and $\omega = \sqrt{1-k^2}$ respectively, and we find

$$y = \operatorname{cosech} \varphi [C_1 \sin \omega \varphi + C_2 \cos \omega \varphi], \qquad (9.29)$$

$$z = a \coth \varphi.$$

Since the image is defined by

$$\varphi_0 - \varphi_i = \pi/\omega,$$

the cardinal elements are

$$z_{f_0} = -z_{f_i} = a \coth \pi/\omega, \qquad (9.30)$$

$$z_{H_0} = -z_{H_i} = a \coth \pi/2\,\omega,$$

$$f_0 = -f_i = a \operatorname{cosech} \pi/\omega,$$

and

$$C_s = a \left[-\frac{\pi k^2}{4\,\omega^3} + \frac{1}{8} \frac{4k^2+3}{4k^2-3} (\sinh 2\,\varphi_0 - \sinh 2\,\varphi_i) \right] \operatorname{cosech}^2 \varphi_0, \quad (9.31)$$

$$C_c = a \frac{\pi k^2}{2\omega^3} \operatorname{cosech}^2 \varphi_0. \qquad (9.32)$$

The case for which ω is small corresponds to a microscope objective.

9.3   THE UNSHIELDED SHORT COIL

9.3.1 The Field

Lyle (1902) made a detailed study of the fields around short toroidal coils, and gave very simple formulae for the fields along their axes.

Torus of square cross-section

The distribution of the longitudinal component B_z of the magnetic field along the z-axis is only very slightly different from that which is produced by a single turn carrying the same current i, lying coaxially in the symmetry

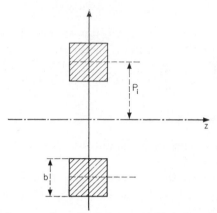

FIG. 95. Cross-section of a short, unshielded electron lens.

plane of the coil. Its radius ϱ_2 is related to the mean radius ϱ_1 of the coil by the expression

$$\varrho_2 = \varrho_1 \left(1 + \frac{b^2}{24\,\varrho_1^2}\right), \tag{9.33}$$

in which b is the length of a side of the square cross-section (Fig. 95); ϱ_2 will be referred to as the mean corrected radius.

Torus of rectangular cross-section

We shall assume that the radial thickness of the coil b is greater than its axial extent c; the field component B_z is then the same as that which would be produced by two circular coplanar turns each carrying a current $\frac{1}{2}i$, of radii $\varrho_2 + e$, $\varrho_2 - e$, where

$$\varrho_2 = \varrho_1 \left[1 + \frac{c^2}{24\varrho_1^2}\right] \quad \text{and} \quad e^2 = \frac{b^2 - c^2}{12}. \tag{9.34}$$

The formulae applicable to a number of other possible cases are to be found in the article by Lyle already mentioned; there seems little point in listing them here, as the results given above are the only ones as yet to have found practical application. Readers who are interested in more rigorous methods of calculating the induction, $B(z)$, produced by coils of rectangular cross-section should consult the article by Garrett (1951).

9.3.2 Adjustment of Glaser's Model

The field on the axis of a circular current-carrying loop of radius ϱ_2 is given by

$$B(z) = \frac{\dfrac{\mu_0 i}{2\varrho_2}}{[1 + (z/\varrho_2)^2]^{3/2}}, \tag{9.35}$$

which is not in a very convenient form for calculating the optical properties. We replace this exact expression, therefore, by a Glaser bell-shaped field, and optimize the value of a by Grivet's procedure.

(i) *Square cross-section*

Starting from

$$B_c = \frac{\mu_0 i}{2\varrho_2} \qquad \text{(weber m}^{-2}\text{, At, m),}$$

a is adjusted to satisfy

$$\rho_2 = 1\cdot 333a.$$

(ii) *Rectangular cross-section*

With

$$B_c = \frac{\mu_0 i}{2\varrho_2}\left[1 - \frac{e^2}{\varrho_2^2}\right]^{-1},$$

a is given by

$$\rho_2 = 1\cdot 33a\left[1 + 3\cdot 84\left(\frac{e}{\varrho_2}\right)^2\right].$$

This expression is partially empirical, as after arranging that the bell-shaped fields are equally wide, the coefficient $1\cdot 3$ is replaced by $1\cdot 33$, so that the expression reduces to the value chosen in the previous paragraph when $e = 0$.

An old formula of Ruska (1933) is still useful to give a fist approximation:

$$\frac{d}{f} = \frac{1}{(220)^2}\frac{(n\,I)^2}{\Phi_0}, \qquad n\,I = 220\sqrt{\Phi_0\frac{d}{f}} \tag{9.36}$$

(nI is the number of ampère-turns on the coil, $\Phi_0\text{kV}$ is the accelerating potential, f is the focal length, and d is the mean diameter of the coil). Experimental data are to be found in profusion in studies of the cathode ray tube.

9.4 SHIELDED COILS

When a coil is shielded save for a narrow slit s around the bore, the field on the axis is sharply concentrated, and much higher for any given number of ampère-turns than its unshielded counterpart. Iron-clad coils are, in this sense, more economical than unshielded coils.

Ruska's formula still gives an order of magnitude estimate of the focal length, when the gap between the iron is narrow, provided d is regarded as the interior diameter of the opening.

9.4.1 The Axial Field

A precise theory can be established by assimilating the magnetic circuit into the simple model shown in Fig. 96 studied experimentally by van Ments and Le Poole (1947). We shall consider only the case in which the permeability of the iron is sufficiently large throughout the pole pieces to be legitimately regarded as infinite. Ampère's circuital theorem expresses the difference between the scalar magnetic potential at the two poles, $\Psi_2 - \Psi_1$, in terms of the ampère-turns $n I$, thus:

$$\Psi_2 - \Psi_1 = n I = \frac{1}{\mu_0} \int B_z \, \mathrm{d} r.$$

The formula established by Bertram (1940, 1942)

$$B = \frac{\mu_0 n I}{2 R} \cdot \frac{\sinh \omega \varepsilon}{\varepsilon \cosh \omega\left(z + \dfrac{\varepsilon}{2}\right) \cosh \omega\left(z - \dfrac{\varepsilon}{2}\right)}, \tag{9.37}$$

can now be applied with $\varepsilon = s/R$, s being the size of the gap and $\omega = 1 \cdot 318$.

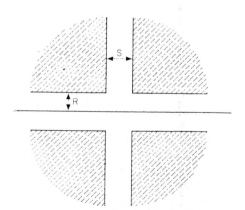

FIG. 96. Cross-section of the pole-pieces of a shielded magnetic lens.

Introducing the maximum value B_m of the axial field, which obtains at the centre of the lens, we find

$$B = B_m \frac{\cosh^2 \dfrac{\omega \varepsilon}{2}}{\cosh \omega \left(z + \dfrac{\varepsilon}{2}\right) \cosh \omega \left(z - \dfrac{\varepsilon}{2}\right)}, \tag{9.38}$$

where B_m is given by

$$B_m = \frac{\mu_0 n I}{R \varepsilon} \tanh \frac{\omega \varepsilon}{2}. \tag{9.39}$$

Comparing B_m with the value of the field in the gap, B_0, which is given by

$$B_0 s = \mu_0 n I,$$

we find

$$\frac{B_m}{B_0} = \tanh \frac{\omega \varepsilon}{2}. \tag{9.40}$$

For the half-width a of the real curve, at the point at which it has fallen to half its maximum height, we obtain

$$\cosh \left(2\omega \frac{a}{R}\right) = 2 + \cosh \left(\omega \frac{s}{R}\right). \tag{9.41}$$

The quantities a/R and B_m/B_0 are plotted in Fig. 97, and prove to be in excellent agreement with the experimental curve obtained by Le Poole.

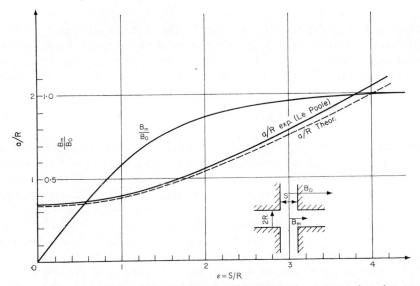

FIG. 97. The magnetic characteristics of a shielded lens—a comparison between theory and experiment.

The agreement is very good, too, with the curves calculated precisely by Liebmann (1951), Liebmann and Grad (1951) and Lenz (1950), using the relaxation method.

9.4.2 Adapting the Model

Adapting Glaser's model by Grivet's procedure, to calculate the cardinal elements, we find

$$\frac{a'}{R} = \frac{1}{2} \left(\coth \frac{\omega \varepsilon}{2} \right)^2 \left(\frac{\omega \varepsilon}{\tanh \omega \varepsilon} - 1 \right). \tag{9.42}$$

That a' and a prove, on evaluation, to be almost identical shows that the model is a good one (for $s = 0$, for example, $a_0 = 0\cdot667$ while $a_0' = 0\cdot669$). The formulae of § 9.2.1 yield values which approximate closely to the values of f, z_F, g, and z_G which are observed in practice.

In particular, if we follow Le Poole and write

$$K = \frac{(n\,I)^2}{\Phi}$$

(in which $n\,I$ is the number of ampère-turns, and Φ is the accelerating potential of the electrons), we find that

$$K = 28\cdot796 \left(\frac{\varepsilon}{\tanh \dfrac{\omega \varepsilon}{2}} \right)^2 \frac{R^2}{a^2} k^2, \tag{9.43}$$

which agrees very well with experiment. This formula also gives a very satisfactory value $K_{\min}$ of K, at which a projective lens has its minimum focal length, namely $K_{\min} = 1\cdot3$; more precision can be attained by distinguishing between a' of the model and a of the actual lens, in which case we find

$$K_{\min} = 115\cdot184\, \varepsilon^2 \tanh^2 \frac{\omega \varepsilon}{2} \frac{1\cdot3}{(\omega \varepsilon \coth \omega \varepsilon - 1)^2}. \tag{9.44}$$

The predicted value $f_{\min}$ of f is also accurate, and if we make the parallel calculation for the immersion focus, determining the value K_0 of K for which G lies at the centre of the lens and the associated focal length g_0, the agreement with experiment is again close; a comparison of the following table with Le Poole's curves shows this and suggests that the greatest disparity appears when the induction is largest.

In the following table, a few numerical values, calculated with the aid of the preceding formulae, are listed; together with K/k^2, the values K_0 of K for which the immersion focus G lies at the centre of the lens are tabulated, as are the corresponding focal lengths g_0; likewise, the values $K_{\min}$ of K for which the ordinary focal length f is a minimum, $f_{\min}$.

$\dfrac{s}{R}$	$\dfrac{a}{R}$	$\dfrac{K}{k^2}$	K_{min}	$\dfrac{f_{min}}{R}$	K_0	$\dfrac{g_0}{R}$
0	0·669	148·5	193·2	0·798	515·1	0·508
0·5	0·698	146·3	190·35	0·833	507·6	0·530
1	0·783	141	183·4	0·934	489	0·596
2	1·104	126	164	1·317	437	0·838
3	1·515	122	158·7	1·807	423	1·151
4	2·008	116·7	152	2·396	404·8	1·526
5	2·5	115·9	151	2·982	402	1·9

9.4.3 Magnetic Microscope Lenses

These lenses have been extensively studied; the basic problem is to design a lens with a very short focal length, of the order of millimetres, so that the microscope is as compact as possible. The objective is the most difficult, as it must also satisfy the further conditions of minimum spherical aberration

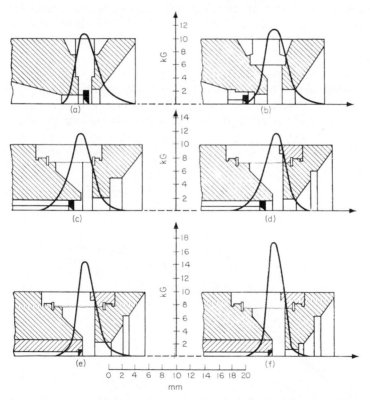

FIG. 98. Cross-sections of the pole-pieces of the various magnetic lenses built by Ruska, together with the field distributions on the axes of the systems.

and ellipticity astigmatism. Another difference between the design of objective and projective lenses stems from the fact that the former must leave enough room for the specimen-holder, which means that the focus must not lie too near the centre of the objective if the advantage of working at the minimum focal length is not to be sacrificed. The focal length f, therefore, must be 8–10 mm rather than 3 mm, and it is often useful to add another lens to the microscope in order to keep its overall length within acceptable bounds. In practice, the specimen-holder is often so cumbersome that the focal length of the objective has to be longer than that of the projective (3 mm instead of 1·5, for example). At the projective lens, however, only distortion need be considered, as the beam is very fine.

The pole-pieces in both kinds of lens may be made of a cobalt alloy with a very high saturation induction (26,000 gauss), in order to produce a very high field density in a restricted volume.

With the aid of Glaser's model, the aberrations can be calculated (Dosse, 1941 c, d; Glaser and Lammel, 1943)—the results have been verified experimentally for the cases of spherical aberration and distortion. This latter aberration, though, can be eliminated as Hillier (1946) has demonstrated. Mechanical defects and imperfections in the magnetic homogeneity of the objective can also be corrected (Hillier and Ramberg, 1947), and in consequence, a resolution of better than 1 mμ is now attainable.

Saturation of the field implies that the field distribution along the axis is flatter than a Glaser bell-shaped curve, and this effect becomes the more pronounced as the opening in the pole-pieces is reduced or, equivalently,

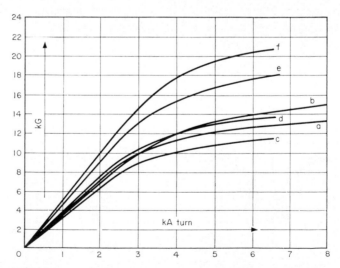

FIG. 99. The maximum field obtained in Ruska's objectives, as a function of the number of ampère-turns.

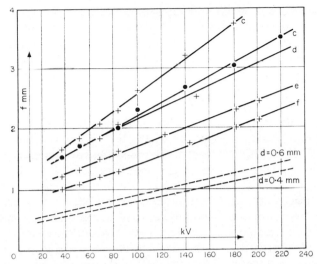

FIG. 100. The focal length as a function of the accelerating potential of the electrons, in the various objectives studied by Ruska.

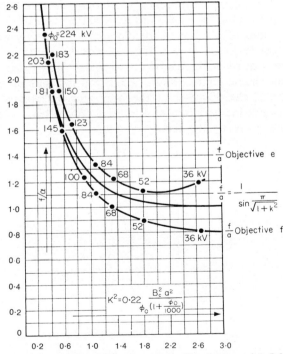

FIG. 101. The focal length as a function of the maximum axial field, for two of Ruska's objectives.

as f is shortened. The effect of saturation may become important when the magnetizing current is high, but it does not alter the general behaviour of the lens; only the numerical values of the lens constants are modified, as the simple theory outlined above is no longer adequate. In particular, if we study the variation of the focal lengths f and g with the parameter k^2, defined by $k^2 = \dfrac{e\,a^2\,B^2}{8\,m\,\Phi_0}$, we find that both f and g reach a minimum. In this region, the lens is insensitive to small variations of k^2 produced by the magnetizing current, perhaps, or by the high tension at the electron gun; it is, in other words, achromatic. Although the effect is more marked for f (the projective lens) than for g (the objective), it is perceptible and can be made use of even in the latter case. For the corresponding value of k^2, the pole-pieces are nearly saturated, and the effects of magnetic hetero-geneities less pronounced. For these reasons, these are the operating con-ditions under which objectives are designed to work. In Fig. 98, the various stages in the construction of the magnetic objective designed by Ruska (1944) are shown; the values of the maximum field are plotted as a function of the number of magnetizing ampère-turns in Fig. 99, while Figs. 100 and 101 summarize the optical properties.

Only two parameters are required to characterize the behaviour of the symmetrical projective lens shown in Fig. 102: the angle of the cone and the ratio of the width of the gap to the diameter of the part which is bored away. The cone angle, always of the order of 70° or 75°, is of very little use as a means of reducing f, in order to make the instrument less bulky. The optimum choice for the width of the gap depends upon the velocity of the electrons, the diameter of the bore and the magnetizing current. Figure 102 shows the variation of $B(z)$ in a typical lens, while the optical properties are given by the theory of § 9.2.

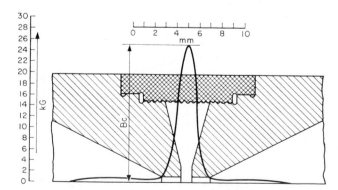

FIG. 102a. Cross-section of the pole-pieces of a magnetic projective lens, and the axial field distribution.

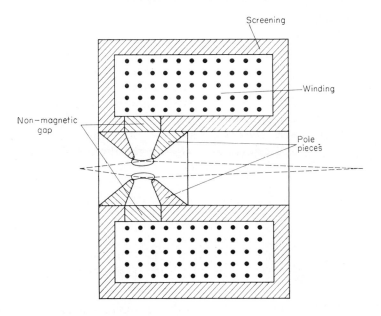

Fɪɢ. 102b. Cross-section of the whole lens.

The cardinal elements prove to be described very satisfactorily by the formulae which we have obtained earlier, when compared with the experimental results obtained by Liebmann and Le Poole.

Figure 103, in conclusion, shows a cross-section of a complete magnetic lens.

The very thorough studies which have been made by Durandeau and Fert (1957) are to be found summarized in § 17.4. They comprise a complete account of the properties—magnetic as well as optical—of these lenses; with their aid, any desired type of objective, projective or condenser lens can easily be constructed.

9.5 SUPERCONDUCTING COILS

For some years now, superconducting metal compounds have been known that retain their superconducting property in the presence of very high magnetic fields. The value of the magnetic field which, at a given temperature, destroys this zero-resistance state ($\rho = 0$) and introduces an ohmic resistance is known as the "second critical field", H_{c2}.

The "first critical field", H_{c1}, is the value of the field at which the lines of force begin to penetrate the superconductor; for $H < H_{c1}$, there is a "total Meissner effect" and no lines of force penetrate the superconductor, which behaves as a perfectly diamagnetic material ($B = 0$ inside the metal). For $H_{c1} < H < H_{c2}$, the magnetic field does penetrate the superconductor, but this does not destroy the "superconductivity" and the resistance of the metal remains zero. For $H_{c2} < H$, the resistance reappears, the superconductivity is destroyed and the metal (or alloy) behaves like an ordinary conductor with a normal Joule effect. It ceases to be suitable for the application we are considering.

The following table gives the values of the critical temperature T_c ($\rho = 0$ if $T < T_c$) and critical field H_{c2} for the three materials that are at present manufactured commercially.

		T_c (°K)	B_c (kG)
Alloys	Nb Zr	12	60–70
	Nb Ti	13	100–150
Compound	Nb_3 Sn	18	200–250

Figure 103 shows the variations of the maximum current density that can be obtained in the superconducting wire or ribbon. (In fact, if the coil is to behave in a stable fashion, the superconducting material must be covered with a thick layer of copper or aluminium and the current densities shown in Fig. 103 must then be roughly halved for small coils and divided by about 10 or 20 for very big coils (Donadieu, 1965).

We see that by using these materials, it will be possible to reduce the cross-section of the magnetizing coil of ordinary microscope lenses considerably, but their main virtue arises from the possibility of obtaining very high fields with iron-free coils, unlimited by saturation phenomena. We can always reduce the longitudinal extent of the curve $B(z)$, (i) by using an outer screen of cobalt steel (which will saturate only close to the coil) or (ii) by introducing cylindrical superconducting tubes close to the axis. (A superconducting substance provides a perfect magnetic screen so long as the magnetic field does not exceed H_{c1}.) Finally (iii) we might use the new materials that saturate only at high fields and which are ferromagnetic at liquid helium temperature (these are the rare earth metals).

The use of these lenses involves us in delicate technological problems: the coils are permanently immersed in a liquid helium bath ($T_c = 4\cdot2$°K), but offers new opportunities for focusing beams of very high energy particles (electrons between 1 and 5 MeV, for example, see Chapter 17).

With the aid of formulae (9.33) or (9.34), or the formula below, we can calculate the field B_c at the centre of a coil of length $2b$, inner radius a_1 and outer radius a_2, in which a mean current density j (A/m²) is flowing:

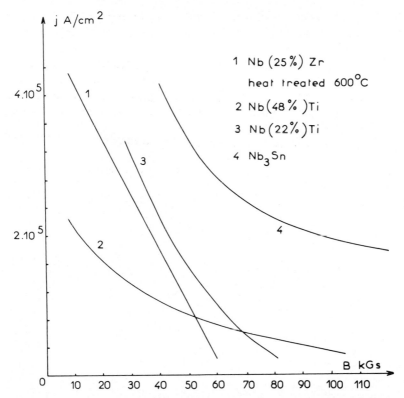

Fig. 103. Curves showing the maximum current density as a function of the induction B for various superconductors.

$$B_c = \mu_0 jb \log \frac{\dfrac{a_2}{a_1} + \sqrt{\left(\dfrac{a_2}{a_1}\right)^2 + \left(\dfrac{b}{a_1}\right)^2}}{1 + \sqrt{1 + (b/a_1)^2}}. \qquad (9.45)$$

With a coil made of niobium–tin ribbon 12 mm wide and 0·03 mm thick carrying 500 ampères, a field of 50 kG could be created without difficulty, with a half-width a of the bell-shaped field equal to 6 mm; treating the distribution as a Glaser bell-shaped curve, we could then obtain the following values of the various parameters:

Φ (MV)	Φ^* (MV)	k^2	g (mm)	C_s (mm)
2	6	5·8	6·3	1·5
3	12	2·9	6·5	1·8
5	30	1·15	7·5	3·6

With particle energies of a few hundred keV, and $a < 2$ mm, we could achieve short focal lengths and aberration constants C_s of a few tenths of a millimetre.

Nevertheless, great precautions will have to be taken in constructing the coils, to ensure that the field $B(z)$ has perfect rotational symmetry. The wires will need to be positioned with the utmost care, and deformation of the conductors during cooling and under the action of the intense field will have to be avoided.†

†Work on microscope lenses (by Fernández-Morán, among others) will be described in Chapter 17.

CHAPTER 10

STRONG FOCUSING LENSES

10.1 THE STRONG FOCUSING PRINCIPLE

10.1.1 The Inability of Ordinary Lenses to Focus High Energy Particles

During the last ten years, the range of applications of corpuscular optics has widened considerably, and with the construction of large particle accelerators a number of fresh problems have appeared; the basic problem is to focus high energy ions or electrons, or even just roughly to channel them over large distances.

In an electrostatic lens, the convergence diminishes rapidly when the accelerating potential Φ_0 is increased, then passes through a minimum, and subsequently increases regularly, as we have seen in § 4.4; for $\Phi_0 > 5 \times 10^6$ V, the convergence is comparable with its value at low energies. This, however, requires the ratio $k = \dfrac{\Phi_1}{\Phi_0}$ to be held constant (Φ_1 represents the potential applied to the lens), and although it may be easy to speak of holding k constant in theory, in practice such an operation is not possible; breakdown phenomena, which affect the optical properties adversely, restrict Φ_1 absolutely to about 100 kV. Further, for a given value of Φ_1, the convergence of a lens varies effectively as $(\Phi_1/\Phi_0)^2$. Electrostatic lenses formed either from tubes or from diaphragms are therefore out of the question where the focusing of particles with energies higher than 1 MeV is concerned ($\Phi_0 = 10^6$ V).

The domain in which electrostatic lenses can be used is somewhat extended by using grid lenses (§ 8.4), the convergence of which varies as (Φ_1/Φ_0); a certain number of particles are intercepted by the grid, however, and the worst drawback is not the loss of current but the secondary emission which is provoked by the impacts of the primary particles at the wires of the grid and which can be extremely troublesome.

Magnetic lenses too are inadequate at high energies. Their convergence falls steadily when Φ_0 is increased, and further, it is not at present possible to increase the induction on the axis, B_z, beyond the limiting value of 2·5 or 2·6 tesla which is imposed by saturation of the iron in the magnetic circuit. Another consideration is that even for fairly low energies of a few tens or hundreds of keV, these lenses focus beams of *heavy* ions inefficiently.

These difficulties have been surmounted in a few special cases by using lenses without iron; very short current pulses, several hundreds or even thousands of ampères in amplitude, are fed into specially constructed coils and an induction which may reach several hundred kilogauss is created. A pulsed supply of this kind is absolutely vital if excessive Joule heating is to be avoided. In this way, it has been possible to focus intense beams of protons with energies of a few hundred keV; here, the principal cause of the divergence is the mutual repulsion of the ions due to space charge.

We may conclude, therefore, that the normal types of lens which have been described earlier can no longer be used with electrons when their energy is more than a few hundred MeV, nor with protons beyond a few hundred keV. The reason for this is to be found in an analysis of the convergence mechanism itself; both in electrostatic and magnetic lenses, all the energy which is expended goes into creating a principal field (either electric or magnetic) parallel to the direction of motion of the particles, and therefore without *direct* focusing action.

We consider, first of all, the case of the unipotential three-electrode lens—the particles are slowed down in the first half of the lens, and accelerated in the second. The variation of E_z in the Oz direction gives rise to a radial field E_r which produces defocusing at the two extremities and focusing in the central zone where the particles move most slowly. The final convergence is thus a *differential* effect.

In a magnetic lens, the radial field B_r which exists at the point where the particles enter the lens endows the particles, which had been travelling parallel to the axis, with a transverse velocity v_θ, and it is only when this deviation takes effect that the particles are usefully affected by the principal field B_z. The action of B_r is of the opposite sign at the exit, however, so that here again the overall effect is a secondary effect, proportional to B^2.

A field which is perpendicular to the direction of motion (the field in the deflector of an oscillograph or in a magnetic prism, for example) will, on the contrary, act *directly* upon the convergence, as the force exerted upon the particles is proportional to the field strength B. If the system is to have the properties of a lens—if, namely, it is to be capable of concentrating the particles which are emitted by an axial point source into another axial (image) point—this force must everywhere be proportional to the distance from the axis, and hence the field itself must be proportional to this distance. The lenses which we are about to describe, which were proposed by Courant, Livingston and Snyder (1952), only partially satisfy this requirement that the field should be transversal, and proportional in strength to the distance from the axis. Here, too, we have a differential phenomenon, but far more favourable than in ordinary lenses. These are "strong convergence" lenses, which are also known as quadrupole lenses on account of the particular way in which the electrodes or pole-pieces of which they are composed are disposed around the axis.

10.1.2 "Strong Focusing" or "Quadrupole" Lenses†

These lenses consist of four cylindrical electrodes (or pole-pieces), parallel to the Oz axis; the lenses have four symmetry planes which intersect along this axis, and which are separated by angles of $\pi/4$. They produce an electric field E or a magnetic field H, the intensity of either of which at any point M is proportional to the distance between M and the axis Oz; the gradient

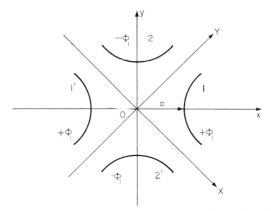

FIG. 104. Transverse cross-section of a system of electrodes with quadrupole symmetry.

of the field in the radial direction is constant throughout the whole useful region of space. The theoretical interest of these field distributions was pointed out by Melkich in 1944 in a general study of cylindrical systems, but their practical application as lenses dates only from 1952. Since then, several theoretical and experimental studies have been devoted to elucidating their properties; a very complete bibliography relating to this question is to be found in the articles by Grivet and Septier (1960) and Septier (1961a).

We consider a transverse section of such a lens (Fig. 104) and we suppose first of all that the electrodes are infinitely long in the Oz direction. One pair of diametrically opposed electrodes (or pole-pieces), 1 and 1′, is held at a potential (which may be either electrostatic or scalar magnetic) $+\Phi_1$, with respect to the earth and the other pair, 2 and 2′, at a potential $-\Phi_1$. The aperture of the lens is defined by the radius a of the circular channel, which is tangential to the four electrodes. The two symmetry planes which do not intersect the electrodes we designate zOX and zOY; the other two symmetry planes which do pass through the electrodes we label zOx and zOy.

First, let us consider the electrostatic case. If a positive particle is incident parallel to the axis in the plane zOx, it will experience a repulsion due

†Quadrupole optics is reviewed in Hawkes (1966).

to the electrode 1 (or 1′), but will not be affected by the presence of 2 and 2′ as a result of the symmetry. The particle will remain in the plane zOx, and will converge towards the axis. In the plane zOy, the trajectory will also be planar, but the particle will be attracted by 2 (or 2′) and will diverge away from the axis. Particles which are incident at the lens other than in the planes zOx and zOy will follow skew trajectories, approaching Oz in the Ox direction, but moving away from Oz in the Oy direction. In one direction the effect is of convergence, therefore, and in the other, of divergence.

In the magnetic case, the Biot–Savart law shows that the force which acts is perpendicular to the field, so that the privileged planes in which the trajectories remain planar are zOX and zOY.

If we want finally to obtain convergence in every meridian plane, two of these lenses, Q_1 and Q_2, of length l, must be placed end to end, differently oriented about the axis; the second lens is identical to the first save for a rotation about the axis of 90°. The divergent plane of Q_2 corresponds to the convergent plane of Q_1, and vice versa. The two lenses should also be as close as possible to one another to prevent the particles which diverge away from Q_1 getting too far from the axis before being subjected to the focusing action of Q_2. If f_{1x} is the focal length in the convergent plane of Q_1, f_{2y} the focal length in the divergent plane of Q_2, and e the width of the gap between the two lenses, the overall effect can be calculated roughly by assuming that each lens is a thin lens. From the standard formula for combinations of lenses, we can show immediately that even for $f_{2y} = -f_{1x}$, the whole ensemble is convergent; the power is given by

$$\frac{1}{f} = \frac{1}{f_{1x}} + \frac{1}{f_{2y}} - \frac{e}{f_{1x}f_{2y}},$$

and if $f_{2y} = -f_{1x}$,

$$\frac{1}{f} = + \frac{e}{f_{1x}^2}.$$

The effect of such a doublet is thus again a differential one, but the immense progress with respect to ordinary lenses resides in the fact that by lengthening the lenses, we can increase the convergence as much as we like without altering the electric or magnetic excitations. It is physically impossible to produce the same effect—convergence (or divergence) in every radial plane—with a single quadrupole lens; the potential is obliged to satisfy the Laplace equation within the useful region, and we can demonstrate that in consequence the lens must behave in the manner just described. This is proved in the course of the analysis of the field distribution which is given in the following paragraph; the higher terms of the general expansion are of a different symmetry from that of the basic quadrupole component, and cannot counteract the effect of the latter.

10.2 THE FIELD DISTRIBUTION

10.2.1 The Potential

We shall discuss the scalar electrostatic or magnetic potential, first of all making the assumption that the electrodes are infinitely long in the Oz direction; Φ is then a function of the two variables x and y only (or X and Y according to the coordinate system chosen: one is appropriate in the electrostatic case, the other in the magnetic case). If we write

$$s = X + iY, \quad W(s) = V(X, Y) + i\Phi(X, Y),$$

and take the symmetry into account, we obtain the following expression for $W(s)$:

$$W(s) = \frac{1}{2} h_2 s^2 + \frac{1}{6} h_6 s^6 + \frac{1}{10} h_{10} s^{10} + \cdots.$$

The equipotential surfaces can be extracted thus:

$$\Phi(X, Y) = \text{Im}\{W(s)\}$$

$$= h_2 XY + h_6 XY\left(X^4 - \frac{10}{3} X^2 Y^2 + Y^4\right) + \cdots.$$

With respect to the axes Ox, Oy we find
$$\Phi(x, y) = \text{Re}\{W(s)\}$$

$$= \frac{1}{2} h_2(x^2 - y^2) + \frac{1}{6} h_6\{x^6 - 15x^2 y^2(x^2 - y^2) - y^6\} + \cdots.$$

On the electrodes, at a distance a from Oz, the potential is $\pm\Phi_1$, so that for *magnetic lenses*:

$$\frac{\Phi(X, Y)}{\Phi_1} = \frac{2K_2}{a^2} XY + \frac{2K_6}{a^6} XY\left(X^6 - \frac{10}{3} X^2 Y^2 + Y^6\right) + \cdots,$$

and for *electrostatic lenses*:

$$\frac{\Phi(x, y)}{\Phi_1} = K_2 \frac{x^2 - y^2}{a^2} + K_6 \frac{x^6 - 15x^2 y^2(x^2 - y^2) - y^6}{a^6} + \cdots.$$

The coefficients K_n are constants which depend only upon the form of the cross-section of the electrodes.

The field components E_x and E_y, B_X and B_Y are given by:

$$E_x = -\frac{\partial \Phi(x, y)}{\partial x}, \qquad E_y = -\frac{\partial \Phi(x, y)}{\partial y},$$

$$B_X = -\frac{\partial \Phi(X, Y)}{\partial X}, \qquad B_Y = -\frac{\partial \Phi(X, Y)}{\partial Y}.$$

If the gradient of the field is to be constant, the coefficients K_{2n} must be zero for $n > 1$, so that

$$\Phi(x, y) = K_2 \frac{\Phi_1}{a^2} (x^2 - y^2).$$

The electrodes must then coincide with the equipotential surfaces $\pm\Phi_1$, the equation of which has the form

$$\frac{x^2 - y^2}{a^2} = 1 \quad \text{or} \quad XY = \frac{1}{2} a^2,$$

which implies $K_2 = 1$.

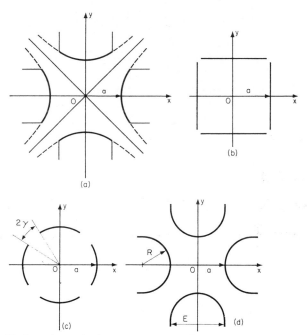

FIG. 105. Various electrode shapes. (a) Hyperbolic segments. (b) Plane electrodes. (c) Concave circular electrodes. (d) Convex circular electrodes.

The ideal distribution for which we are searching would be produced, therefore, by four electrodes the cross-sections of which are of the form of a pair of rectangular hyperbolae with asymptotes OX and OY. In practice, however, it is not possible to build electrodes which are infinitely long in these two directions and even when the electrodes consist of portions of hyperbolic cylinders (Fig. 105a), higher order terms reappear in the expression for the potential; these latter are fortunately of little importance as we shall see later. The coefficients $K_2, K_6, \ldots$ can be calculated in a few simple cases (plane electrodes—see Fig. 105 b; right cylinders formed from a segment of a circle with an angular separation of 2γ—see Fig. 105c). The terms K_2 and K_6 in the expansion $\dfrac{\Phi(X, Y)}{\Phi_1}$ then take the following values:

	K_2	K_6
Infinite hyperbolae	1	0
Plane electrodes	1·037	0·009
Concave circular electrodes.	$1·273\,\dfrac{\sin 2\gamma}{2\gamma}$	$0·042\,\dfrac{\sin 6\gamma}{6\gamma}$

In the final case, we can clearly reduce K_6 to zero by choosing $\gamma = \pi/6 = 30°$; this means that over an extended region in the neighbourhood of the axis, the gradient is accurately constant.

The sharp transition from hyperbolae to planes modifies the ideal distribution relatively little; we might consider taking advantage of this insensitivity by constructing electrodes which are circular in section (Fig. 105 d) rather than hyperbolic, as the former are considerably simpler to machine. Some measurements of the coefficients in magnetic lenses have produced the following results:

Hyperbolic pole-pieces

Width $E = 2a$ $\qquad\qquad$ $K_6 = 0·5 \times 10^{-2}$

Circular pole-pieces

Width $E = 2·5\,a$

Radius $R = 1·25\,a$ $\qquad$ $K_6 = 2 \times 10^{-2}$ $\qquad$ $K_{10} = -1·24 \times 10^{-2}$

$R = 1·125\,a$ $\qquad\qquad$ $K_6 = -0·5 \times 10^{-2}$ $\qquad$ $K_{10} = -1·25 \times 10^{-2}$

With circular pole-pieces of width $E = 2·5\,a$ and radius $R = 1·15\,a$, K_6 is effectively zero.

The potential distribution established above is only valid in a real lens far from the extremities of the electrodes and in particular, at the centre of the lens (where $z = 0$) provided the mechanical length l is long in comparison with a. Near the longitudinal extremities, the potential and field become functions of all three variables—this is the region of *leakage* or *overlap fields*. There are still symmetry planes, however, and we can express $\Phi(x, y, z)$ in a form in which the simplifications which result from the symmetry have been made; the general expression in which only the symmetry conditions have been inserted can be written

$$\frac{\Phi(x, y, z)}{\Phi_1} = \alpha\,\frac{x^2 - y^2}{a^2} + \beta\,\frac{x^4 - y^4}{a^4}$$

$$+ \gamma\,\frac{x^6 - y^6}{a^6} + \delta\,\frac{x^2 y^2(x^2 - y^2)}{a^6}.$$

Φ has also to satisfy the Laplace equation

$$\nabla^2 \Phi(x, y, z) = 0,$$

which implies

$$\beta = - \frac{1}{12} \frac{d^2\alpha}{dz^2} a^2,$$

$$\delta = - 15\gamma + \frac{1}{24} \frac{d^4\alpha}{dz^4} a^4,$$

so that finally

$$\frac{\Phi(x, y, z)}{\Phi_1} = \frac{k_2(z)}{a^2} (x^2 - y^2) - \frac{1}{12} \frac{d^2 k_2(z)}{dz^2} \frac{x^4 - y^4}{a^4} + k_6(z) \frac{x^6 - y^6}{a^6}$$

$$- \left(15 k_6 - \frac{1}{24} \frac{d^4 k_2}{dz^4} a^4\right) \frac{x^2 y^2 (x^2 - y^2)}{a^6}.$$

At the centre $(z = 0)$:

$$k_2(z) = K_2 \quad \text{and} \quad k_6(z) = K_6.$$

If we consider only the fundamental term, the function $k_2(z)$ which is equal to unity at the centre falls away to zero far from the lens as shown in Fig. 106. We shall write

$$k_2(z) = k(z).$$

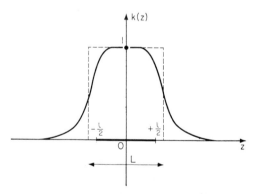

FIG. 106. The appearance of the characteristic function $k(z)$ for a lens of mechanical length l; the "equivalent length" L is greater than l.

The behaviour of $k(z)$, which is the "characteristic function" of the lens, can only be obtained experimentally; at present, the function can only be measured easily in the magnetic case.

We have performed the various calculations relative to the coordinate system xOy only; in the various special cases considered later, we change to the axes XOY by a rotation through $\pi/4$.

10.2.2 The Transverse Gradient

To obtain the first order optical properties, we assume that the potential distribution in the lens is given by the terms of lowest degree in the expansion of Φ, namely

$$\frac{\Phi(x, y, z)}{\Phi_1} = \frac{k(z)(x^2 - y^2)}{a^2},$$

or

$$\frac{\Phi(X, Y, z)}{\Phi_1} = 2\frac{k(z) X Y}{a^2}.$$

In the magnetic case, we use the second expression, and we obtain

$$B_X = -\frac{2\Phi_1}{a^2} Y k(z),$$

$$B_Y = -\frac{2\Phi_1}{a^2} X K(z),$$

$$B_z = -\frac{2\Phi_1}{a^2} X Y \frac{dk(z)}{dz}.$$

If we assume that the surface of each pole is a magnetic equipotential (and experiment shows that this is always true, since the effects of saturation always begin at a point of the yoke far from the poles) and that the consumption of ampère-turns in the iron is negligible, we can write:

$$\Phi_1 = \mu_0 n I,$$

where $n I$ represents the number of ampère-turns per pole.

The gradient, therefore, is given by

$$K(z) = \left| \frac{\partial B_X}{\partial Y} \right| = \left| \frac{\partial B_Y}{\partial X} \right| = \frac{2\mu_0 n I}{a^2} k(z).$$

At $z = 0$, $\quad K(0) = \dfrac{2\mu_0 n I}{a^2}$, $\quad k(0) = 1$.

If the permeability of the iron is not infinite, we can always measure the potential difference, $2\Phi_1$, between two adjacent poles, and we shall then have

$$K(0) = \frac{2\Phi_1}{a^2}.$$

In the electrostatic case, we find in the same way

$$K(0) = \frac{2\Phi_1}{a^2} \quad \text{and} \quad K(z) = K(0) k(z).$$

10.2.3 The Equivalent Length

As in all optical systems with a transverse field, the lens action can be characterized by the integral

$$I = \int_{-\infty}^{\infty} B_r \, dz,$$

calculated along a line parallel to Oz; B_r is the amplitude of the transverse component of $\boldsymbol{B}$, a distance r from Oz. Along these lines parallel to the z axis, the amplitude of B_r is given to a first approximation by

$$B_r(z) = B_r(0) \, k(z) = r \, K(0) \, k(z).$$

We can, therefore, define a hypothetical lens which is equivalent to the real lens by supposing that a field $B_r(0)$ acts over a distance L, and that for the two lenses the integral I takes the same value. This is equivalent to replacing the $k(z)$ curve by a rectangle of unit height and length L in the direction of the lines parallel to Oz. It is in this way that the "equivalent length" L is defined:

$$L = \frac{1}{B_r(0)} \int_{-\infty}^{\infty} B_r(z) \, dz.$$

This length L is greater than the mechanical length l of the electrodes (Fig. 106), and measurement shows that to all intents and purposes,

$$L \simeq l + 1 \cdot 1 \, a.$$

10.3 THE FIRST ORDER OPTICAL PROPERTIES

10.3.1 The Equations of Motion

The relativistically correct forms of the basic equations:

$$\frac{d\boldsymbol{p}}{dt} = \frac{d(m\boldsymbol{v})}{dt} = e\boldsymbol{E}, \qquad \text{(electrostatic case)}$$

or

$$\frac{d(m\boldsymbol{v})}{dt} = -e(\boldsymbol{v} \times \boldsymbol{B}), \qquad \text{(magnetic case)}$$

lead to simple expressions which have the same form in both cases provided we refer to the axes Ox, Oy for $\boldsymbol{E}$ and OX, OY for $\boldsymbol{B}$. If s is a unit vector, tangent to the trajectory, and ds is an element of the trajectory, we have

$$ds = dz \left\{ 1 + \left(\frac{dX}{dz}\right)^2 + \left(\frac{dY}{dz}\right)^2 \right\}^{1/2} = dz \left\{ 1 + \left(\frac{dx}{dz}\right)^2 + \left(\frac{dy}{dz}\right)^2 \right\}^{1/2}.$$

We can eliminate the time t from the equations above, by writing

$$v = \frac{ds}{dt} = v \frac{ds}{ds} \quad \text{or} \quad \frac{d}{dt} = v \frac{d}{ds}.$$

Further, we know that

$$p = \{2 e m_0 \Phi_0 (1 + \varepsilon \Phi_0)\}^{1/2},$$

and therefore

$$v = \left(\frac{2e \Phi_0}{m_0}\right)^{1/2} \frac{1 + \varepsilon \Phi_0}{1 + 2\varepsilon \Phi_0} \quad \text{with} \quad \varepsilon = \frac{e}{2m_0 c^2},$$

where Φ_0 represents the total accelerating potential of the particles.
Finally we obtain

$$\frac{d}{dz} \left\{ \left[\frac{\Phi_0 (1 + \varepsilon \Phi_0)}{1 + X'^2 + Y'^2}\right]^{1/2} \frac{ds}{dz} \right\} = \left(\frac{e}{2m_0}\right)^{1/2} \left(\frac{ds}{dz} \times B\right),$$

or

$$\frac{d}{dz} \left\{ \left[\frac{\Phi_0 (1 + \varepsilon \Phi_0)}{1 + x'^2 + y'^2}\right]^{1/2} \frac{ds}{dz} \right\} = \frac{1}{2}(1 + 2\varepsilon \Phi_0) \left[\frac{1 + x'^2 + y'^2}{\Phi_0 (1 + \varepsilon \Phi_0)}\right] E.$$

To a first approximation, for trajectories which are only slightly inclined to the axis (where $ds \simeq dz$), and neglecting the variations in the velocity due to the local potential $\Phi(x, y, z)$ which is always small in comparison with Φ_0 we arrive at the following equations if we consider an ideal lens in which the radial gradient is constant; these equations are valid for any velocity, as they are relativistically correct.

(i) *The magnetic case*

$$X'' + \left(\frac{e}{2m_0 \Phi_0 (1 + \varepsilon \Phi_0)}\right)^{1/2} K(z) X = 0,$$

$$Y'' - \left(\frac{e}{2m_0 \Phi_0 (1 + \varepsilon \Phi_0)}\right)^{1/2} K(z) Y = 0,$$

or, expressed in a different form

$$X'' + \frac{K(z)}{B\varrho} X = 0, \qquad Y'' - \frac{K(z)}{B\varrho} Y = 0.$$

$B\varrho$ is the "magnetic stiffness" or "rigidity" of the particle, defined by

$$B\varrho = \frac{mv}{e} = p/e.$$

If the particles are slow enough for Newtonian mechanics to be a valid approximation, which occurs when ε is small, we find

$$X'' + \sqrt{\frac{e}{2m_0 \Phi_0}} K(z) X = 0,$$

$$Y'' - \sqrt{\frac{e}{2m_0 \Phi_0}} K(z) Y = 0,$$

or
$$X'' + \beta^2 k(z) X = 0,$$
$$Y'' - \beta^2 k(z) Y = 0,$$

where

$$\beta^2 = K(0) \sqrt{\frac{e}{2m_0 \Phi_0}} = \frac{\mu_0 n I}{a^2} \sqrt{\frac{2e}{m_0 \Phi_0}}.$$

$K(0)$ is the gradient at $z = 0$.

(ii) *The general relativistic electrostatic case*

$$x'' + \frac{1 + 2\varepsilon \Phi_0}{2\Phi_0(1 + \varepsilon \Phi_0)} K(z) x = 0,$$

$$y'' - \frac{1 + 2\varepsilon \Phi_0}{2\Phi_0(1 + \varepsilon \Phi_0)} K(z) y = 0.$$

Where classical mechanics is adequate ($\varepsilon \ll 1$),

$$x'' + \beta^2 k(z) x = 0,$$
$$y'' - \beta^2 k(z) y = 0,$$

with

$$\beta^2 = \frac{K(0)}{2\Phi_0} = \frac{1}{a^2} \frac{\Phi_1}{\Phi_0}.$$

We have assumed implicitly that the convergent plane of the lens is OX (or Ox) and the divergent plane, OY (or Oy).

10.3.2 Integration of the Equations

If $k(z)$ is replaced by the equivalent rectangular distribution of width L, $k(z) = 1$ everywhere within the lens and $k(z) = 0$ outside; the equations above give the values of X and Y (or x and y). If, for example, we take the origin of the Oz axis in the entry plane of the lens

$$X = X_0 \cos\beta z + \frac{X_0'}{\beta} \sin\beta z,$$

$$Y = Y_0 \cosh \beta z + \frac{Y_0'}{\beta} \sinh \beta z.$$

X_0, Y_0, X_0' and Y_0' represent the initial conditions in the plane $z = 0$; these are the coordinates of the point at which the trajectory in question is incident, and the projections of its slope onto the XOz and YOz planes.

The equations of motion can equally well be integrated for other forms of the function $k(z)$, but the calculations become more complicated. The approximation of the rectangular model gives results which are satisfactory to within a few per cent; in the domain of quadrupole lenses, the accuracy and simplicity of this model make it a basic tool.

The relations which give X_s, Y_s and the slopes X_s' and Y_s' at the exit of a lens of length L are of the form

$$
\begin{cases}
X_s = X_0 \cos \beta L + X_0' \dfrac{\sin \beta L}{\beta}, \\[2mm]
X_s' = -\beta X_0 \sin \beta L + X_0' \cos \beta L,
\end{cases}
$$

$$
\begin{cases}
Y_s = Y_0 \cosh \beta L + Y_0' \dfrac{\sinh \beta L}{\beta}, \\[2mm]
Y_s' = \beta Y_0 \sinh \beta L + Y_0' \cosh \beta L.
\end{cases}
$$

These equations are in a form in which a matrix method can well be employed, and this method proves to be particularly fruitful when we are studying combinations of lenses, as we shall see later on (§ 10.4.3).

Any lens can be characterized by a pair of transfer matrices, one $\|T_C\|$ in the convergent plane XOz and the other $\|T_D\|$ in the divergent plane YOz; we can then write:

$$
\begin{pmatrix} X_s \\ X_s' \end{pmatrix} = \|T_C\| \begin{pmatrix} X_0 \\ X_0' \end{pmatrix},
$$

$$
\begin{pmatrix} Y_s \\ Y_s' \end{pmatrix} = \|T_D\| \begin{pmatrix} Y_0 \\ Y_0' \end{pmatrix},
$$

with

$$
\|T_C\| = \begin{pmatrix} \cos \beta L & \dfrac{1}{\beta} \sin \beta L \\[2mm] -\beta \sin \beta L & \cos \beta L \end{pmatrix},
$$

and

$$
\|T_D\| = \begin{pmatrix} \cosh \beta L & \dfrac{1}{\beta} \sinh \beta L \\[2mm] \beta \sinh \beta L & \cosh \beta L \end{pmatrix}.
$$

10.3.3 The Cardinal Elements

If we consider a particular incident ray which is parallel to the axis Oz, the conditions at the origin are defined by X_0, Y_0 and $X_0' = Y_0' = 0$. We find, therefore, that

$$
\begin{pmatrix} X_s \\ X_s' \end{pmatrix} = \|T_C\| \begin{pmatrix} X_0 \\ 0 \end{pmatrix}, \qquad \begin{pmatrix} Y_s \\ Y_s' \end{pmatrix} = \|T_D\| \begin{pmatrix} Y_0 \\ 0 \end{pmatrix}.
$$

The cardinal elements are defined by the emergent rays, and are given by the following relations:

$$
f_X = -\frac{X_0}{X_s'}, \qquad f_Y = -\frac{Y_0}{Y_s'},
$$

$$
u_X = \frac{X_s}{X_s'}, \qquad u_Y = -\frac{Y_s}{Y_s'},
$$

$$
v_X = u_X - f_X, \qquad v_Y = u_Y - f_Y :
$$

f_X and f_Y represent the image focal lengths in the two planes XOz and YOz. The abscissae of the foci, u_X and u_Y, and those of the principal planes v_X and v_Y, are measured from the exit face (see Figs. 107a and 107b). Finally,

$$f_X = \frac{1}{\beta \sin \beta L}, \qquad f_Y = -\frac{1}{\beta \sinh \beta L},$$

$$u_X = \frac{\cos \beta L}{\beta \sin \beta L}, \qquad u_Y = -\frac{\cosh \beta L}{\beta \sinh \beta L},$$

$$v_X = \frac{\cos \beta L - 1}{\beta \sin \beta L}, \qquad v_Y = \frac{1 - \cosh \beta L}{\beta \sinh \beta L}.$$

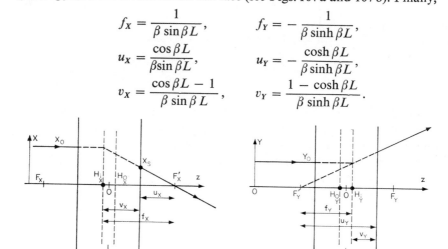

FIG. 107. The cardinal elements of a lens. (a) In the convergent plane XOz. (b) In the divergent plane YOz.

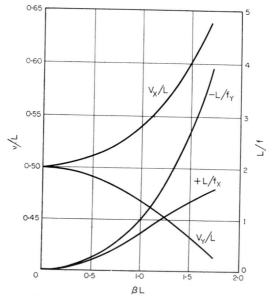

FIG. 108. The variation of the reduced convergence L/f and of the abscissa of the image principal plane v/L as functions of the excitation βL, in the two planes XOz and YOz for a single lens.

The "object" and "image" elements are symmetrically placed with respect to the centre of the lens. If the excitations are sufficiently slight, or if the particles are moving at very high energies, the factor β is small and we may write

$$\sin \beta L \simeq \sinh \beta L \simeq \beta L,$$

$$\cos \beta L \simeq 1 - \left(\frac{\beta L}{2}\right)^2, \qquad \cosh \beta L \simeq 1 + \left(\frac{\beta L}{2}\right)^2,$$

so that for weak lenses

$$f_X = -f_Y = \frac{1}{\beta^2 L},$$

and

$$v_X = v_Y = -\tfrac{1}{2} L.$$

The two focal lengths are equal in absolute magnitude, and the principal planes coincide at the centre of the lens. This is only a valid approximation when $\beta L < 0.2$, however; when $\beta L > 0.2$, the ratio f_X/f_Y increases rapidly and the principal planes move away from the centre (Fig. 108).

10.4 COMBINATIONS OF LENSES

10.4.1 The Simple Doublet

A doublet consists of two lenses Q_1 and Q_2, of lengths L_1 and L_2 and excitations β_1 and β_2 respectively, separated by a region of field-free space of length D. The convergent plane of Q_1 corresponds to the divergent plane of Q_2. The optical properties and in particular the cardinal elements can easily be calculated with the aid of the matrix formalism.

Two transfer matrices correspond to each lens—one $\|T_C\|$ for the convergent plane, the other $\|T_D\|$ for the divergent plane. A region of space of length D in which only the position of the current point is altered, and not the slope of the trajectory, is represented by the matrix

$$\|T_G\| = \begin{pmatrix} 1 & D \\ 0 & 1 \end{pmatrix}.$$

The transfer matrices of the doublet we shall denote by $\|M_X\|$ in the XOz plane (or the "convergent–divergent" plane, since Q_1 is convergent and Q_2 divergent in this plane) and by $\|M_Y\|$ in the YOz plane (the "divergent-convergent" plane); we obtain these matrices by forming the product of the individual matrices, written in the *order opposite* to that in which the particles encounter the corresponding elements. For example,

$$\|M_x\| = \|T_D\| \cdot \|T_G\| \cdot \|T_C\|.$$

or

$$\|M_X\| = \begin{pmatrix} \cosh \beta_2 L_2 & \dfrac{\sinh \beta_2 L_2}{\beta_2} \\ \beta_2 \sinh \beta_2 L_2 & \cosh \beta_2 L_2 \end{pmatrix} \begin{pmatrix} 1 & D \\ 0 & 1 \end{pmatrix} \begin{pmatrix} \cos \beta_1 L_1 & \sin \beta_1 L_1 \\ -\beta_1 \sin \beta_1 L_1 & \cos \beta_1 L_1 \end{pmatrix},$$

and similarly

$$\|M_X\| = \begin{pmatrix} \cos \beta_2 L_2 & \sin \beta_2 L_2 \\ -\beta_2 \sin \beta_2 L_2 & \cos \beta_2 L_2 \end{pmatrix} \begin{pmatrix} 1 & D \\ 0 & 1 \end{pmatrix} \begin{pmatrix} \cosh \beta_1 L_1 & \dfrac{\sinh \beta_1 L_1}{\beta_1} \\ \beta_1 \sinh \beta_1 L_1 & \cosh \beta_1 L_1 \end{pmatrix}.$$

And once again, in the exit plane of Q_2

$$\begin{pmatrix} X_s \\ X_s' \end{pmatrix} = \|M_X\| \begin{pmatrix} X_0 \\ X_0' \end{pmatrix},$$

$$\begin{pmatrix} Y_s \\ Y_s' \end{pmatrix} = \|M_Y\| \begin{pmatrix} Y_0 \\ Y_0' \end{pmatrix}.$$

For $X_0' = Y_0' = 0$, we obtain the focal lengths f_X and f_Y and the positions of the foci. The expressions for these quantities are very complicated in the general case, but are considerably simplified when the doublet is symmetrical ($\beta_1 = \beta_2$, $L_1 = L_2$) and for example

$$f_X = f_Y = [\beta (\sin \beta L \cosh \beta L - \cos \beta L \sinh \beta L + \beta D \sin \beta L \sinh \beta L)]^{-1}.$$

The relative positions of the cardinal elements are shown in Fig. 109. The relations which express the correspondence between the object and the image are calculated by the same method, but it must be admitted that the

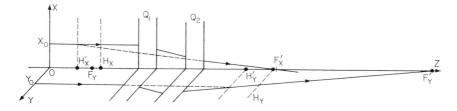

Fig. 109. The positions of the cardinal elements of a doublet. Q_1 and Q_2 have here the same excitation; the two focal lengths, f_X and f_Y, are equal.

full calculations for an optical system adapted to a particular problem are extremely complicated. In most cases, we are obliged to resort to an electronic computer. A great deal of data is to be found, however, in recent publications, both graphical (Enge, 1959) and in the form of tables of calculated values of the cardinal elements (Blewett, 1959).

10.4.2 The Pseudo-stigmatic Doublet and the Stigmatic Doublet

In general, a doublet produces two distinct images of an axial point source which lie in the planes zOX and zOY. Each image is a line focus so that the system has innate astigmatism. Nevertheless, by exciting Q_1 and Q_2 differently, the two focal lines can be brought into the same image plane; the combination is then called "pseudo-stigmatic", as the linear

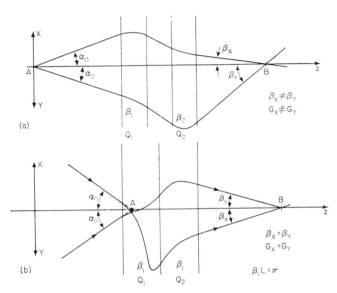

Fig. 110a. Pseudo-stigmatic operation; the beam is initially axially symmetric, and is highly astigmatic when it leaves the system. B is the image of A but $G_X \neq G_Y$. Q_1 and Q_2 have different excitations.

Fig. 110b. Stigmatic operation of a very strongly excited doublet ($\beta_1 L_1 = \beta_2 L_2 = \pi$); A is virtual and B is real. The beam is axially symmetric at both entry and exit.

magnifications along OX and OY are different (see Fig. 110a). This mode of operation is used very often to concentrate the maximum number of particles which emerge from a source into a region around a point.

At very high excitations, we can demonstrate theoretically—and confirm experimentally—that a symmetrical doublet can behave like an axially symmetrical lens; both the foci F_X and F_Y and the principal planes H_X and H_Y coincide. When $D = 0$, the doublet behaves in this way for $\beta L = \pi$. The lens can then be used in an electron microscope to form a faithful image (Septier, 1958); the asymptotic foci of the doublet lie, however, inside the lens so that the strongly convergent lens which is obtained in this way is only suitable as a projective lens.

10.4.3 The Triplet

The doublet discussed above possesses one marked drawback: the optical centre of the equivalent lens is not the same in the two planes zOX (or zOx) and zOY (or zOy), and moves when the excitations of the lenses are varied—this complicates the adjustment especially.

The present tendency is to replace the doublet by a symmetrical triplet, a diagram of which is given in Fig. 111. The two outer lenses Q_1 and Q_3 are identical and have the same excitation β_1; the central lens Q_2 is turned

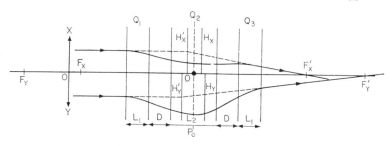

FIG. 111. The ray-paths in a symmetrical triplet.

through $90°$ with respect to Q_1 and Q_3, and has an independent excitation β_2. The principal planes are different in zOX and zOY, but remain symmetrical with respect to the centre of symmetry O. Moreover, for weak lenses, the triplet can be regarded as a pair of thin convergent cylindrical lenses situated at O.

The calculations associated with such a system are extremely complicated in the case of strongly convergent lenses. The cardinal elements of a few triplets are tabulated in Blewett (1959) and Enge (1961).

Under certain conditions of excitation, the symmetric triplet is stigmatic for a pair of conjugate planes, P and P', which are themselves symmetrically placed with respect to the centre, O. The magnification is then equal to unity. This particular system has been studied in detail by Septier and Dhuicq (1965), Septier (1966), and has recently been employed in electron microscopy by Bauer (1965–6) together with a symmetric doublet equivalent to a round lens. Images of good quality have been obtained.

10.4.4 The Quadruplet

Work by Dhuicq and Septier (1959) and Dhuicq (1960) has shown that it is possible to design an optical system with a very high convergence that is equivalent to a round lens for all pairs of conjugate points on the axis Oz, using four identical quadrupoles, symmetrical in pairs about the optical centre, $z = 0$ (Fig. 111a). The excitations will be symmetric when the scalar potentials at the electrodes or pole-pieces are the same in the two central lenses and are likewise the same in the two outermost lenses. The plane zOx is then "convergent–divergent–divergent–convergent" (the $CDDC$ arrangement).

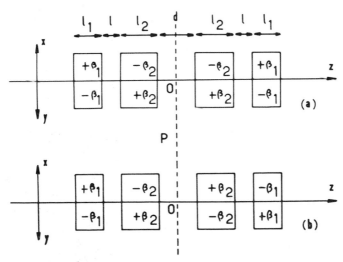

FIG. 111a. Quadruplets, electrically symmetrical (a) and antisymmetrical (b).

In this situation, the asymptotic foci will always be situated inside the outermost quadrupoles so that the lens can only be used as a projective lens (the real immersed foci do not coincide in the two symmetry planes and one cannot therefore place a real object at them).

In contrast, we can find working regions for which the foci lie beyond the outside quadrupoles and the focal length is very short by exciting the lens of such a system so that the zOx plane becomes $CDCD$ (Dymnikov and Yavor, 1963; Dymnikov et al., 1965a; Dhuicq and Septier, 1966a). These lenses are equivalent to round lenses, and could be used with high energy particles (electrons of several MeV, or even ions of a few hundred keV); the focal length would remain less than or equal to one centimetre.

For the case $l_2/l_1 = 2$, $l = 0$, the curves $D = d/l_1 = \text{const}$, $F = F_x/l_1 = F_y/l_1 = \text{const}$ and $Z_{Fx}/l_1 = Z_{Fy}/l_1 = \text{const}$ are shown in Fig. 111b as functions of the excitations, U and V, of the two pairs of lenses Q_1, Q_4 and Q_2, Q_3 respectively ($U = \beta_1^2 l_1^2$ and $V = \beta_2^2 l_2^2$).

Thus, for $U = 1.95$, $V = 4.5$ for example, the central gap is $d/l_1 = 0.155$ and the focal length is $f = 0.041 l_1$. The focus is located at $Z_F = 0.267 l_1$ from the terminal face. If $l_1 = 20$ mm, $l_2 = 40$ mm and $a = 4$ mm (a is the bore radius), we have $d = 3.1$ mm, $f = 0.8$ mm and $p = 6.4$ mm.

For 3 MeV electrons ($\Phi_0^* = 12$ MeV), we should need field gradients in the lenses of the order of $K_1 = 11{,}000$ G/cm and $k_2 = 15{,}000$ G/cm and these values are attainable. The superiority of this type of lens over ordinary lenses so far as convergence is concerned is clear. The aberration coefficients are much higher, however.

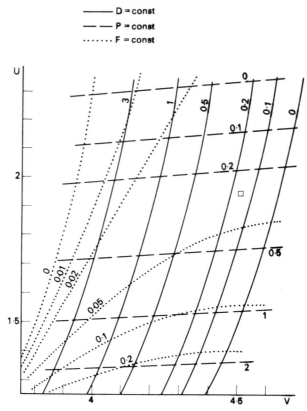

FIG. 111b. The optical elements of an antisymmetric quadruplet equivalent to a round lens, when $M = l_2/l_1 = 2$ and $L = l/l_1 = 0$. $P = p/l_1$ denotes the distance from the common focus to the terminal face of the fourth lens; $D = d/l_1$ is the distance between the two central lenses. The focal length is given by $F = f/l_1$.

It is also possible to obtain a similar solution with a symmetrical quadruplet. It possesses the desired properties only for a single pair of conjugate points on the axis Oz, which severely limits its usefulness.

10.5 A CHAIN OF N IDENTICAL CROSSED QUADRUPOLE LENSES

10.5.1 Introduction

If we want to channel a beam of high energy particles over a long distance, we can employ a series of N quadrupole lenses, provided a lateral stability condition which is identical to the condition which is imposed on the optical system of an ordinary periscope is fulfilled. The transverse dimensions of the beam must always remain smaller than the apertures of the

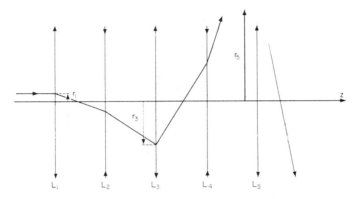

FIG. 112. A poorly calculated system of N lenses alternately excited. The pencil is lost after the fifth lens.

lenses, and hence smaller than the radius of the bore, a. The same problem crops up in all periodic optical systems, whether they be formed from N identical convergent lenses or, as in this case, from $\frac{1}{2}N$ convergent lenses with $\frac{1}{2}N$ divergent lenses interleaved between them. We shall only consider the case in which the lenses are all identical (both in length and excitation). A thorough examination of more general cases shows that, to within the degree of complication considered, the conclusions are unaltered. Figure 112 shows the evolution of a beam which is incident parallel to the axis in a poorly chosen system; the particles are lost at the walls after the fifth lens.

To obtain the condition for lateral stability, we shall argue first of all in terms of an optical system consisting of thick convergent lenses, and use the formalism of elementary geometrical optics; afterwards, we shall introduce the matrix formalism which proves to be particularly fruitful. The thick lens systems which we have in mind may be either ordinary convergent lenses or quadrupole doublets. In the latter case, the positions of the principal planes and the foci, measured relative to the beginning of the channel, will be different in the two planes XOz and YOz; if we suppose the succession to be infinitely long, however, we can consider that the basic unit is identical in the two planes.

10.5.2 The Behaviour of a Ray through a Characteristic Section of the Chain

The regularly repeated cell of the pattern will consist of a thick convergent system which is defined by its object and image principal planes, P_0 and P_i, and by its object and image foci, F_0 and F_i; the constants of the system will thus be

$$\overrightarrow{P_0 F_0} = f_0, \qquad \overrightarrow{P_i F_i} = f_i, \qquad \overrightarrow{P_0 P_i} = \alpha.$$

We have $f_0 = -f_i$, and for convergent lenses, $f_i > 0$; we write $f = f_i = |f_0|$. The position of one cell or "element" relative to its successor will be fixed by the distance $a + b$ between the focus $F_{i,m}$ and the focus $F_{0,m+1}$, so that the length of the element (the geometrical period) is (see Fig. 113):

$$L = a + b + 2f + d.$$

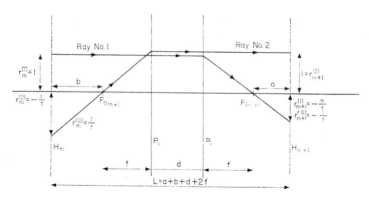

Fig. 113. The paths of two particular rays in an elementary cell of the m-th rank (the cell is equivalent to a thick lens with the same foci and principal planes).

Each element is bounded by two planes H_m and H_{m+1} which are at a distance a from F_i and b from F_0 respectively.

$$\overrightarrow{H_m F_{0,m}} = b, \qquad \overrightarrow{F_{i,m+1} H_{m+1}} = a.$$

We suppose that the distance of the ray from the axis, r_m, and its slope r'_m at the entry to the m-th element are known; the corresponding quantities at the exit, in the plane H_{m+1}, will be linearly related to the values at the entry:

$$r_{m+1} = \alpha r_m + \beta r'_m,$$
$$r'_{m+1} = \gamma r_m + \delta r'_m.$$

It is easy to relate α, β, γ and δ to the known optical quantities f, d, a and b by applying these equations to the two rays usually employed in optical constructions.

The first is parallel to the axis at H_m, and a distance $r_m = 1$ from it; it leaves the element with a slope $r_{m+1} = -\dfrac{1}{f}$, passing through the image focus $F_{i,m}$. The second ray is symmetrical to the first with respect to the symmetry plane of the thick lens—it passes through $F_{0,m}$ and has a slope $r'_m = +\dfrac{1}{f}$ at the entry; it leaves parallel to the axis and a distance $r_{m+1} = 1$ from it (Fig. 112).

We have therefore

$$\text{(i)} \quad r'_m = 0, \qquad r_m = 1, \qquad r'_{m+1} = \frac{1}{f}, \qquad r_{m+1} = -\frac{a}{f},$$

$$\text{(ii)} \quad r'_m = \frac{1}{f}, \qquad r_m = \frac{b}{f}, \qquad r'_{m+1} = 0, \qquad r_{m+1} = 1,$$

so that

$$\alpha = -\frac{a}{f}, \qquad \beta = \frac{f^2 - ab}{f}, \qquad \gamma = -\frac{1}{f}, \qquad \delta = -\frac{b}{f}.$$

The transfer relations thus become:

$$r_{m+1} = -\frac{a}{f} r_m + \frac{f^2 - ab}{f} r'_m,$$

$$r'_{m+1} = -\frac{1}{f} r_m - \frac{b}{f} r'_m.$$

We eliminate the slopes r', and obtain the following recurrence relation:

$$r_{m+2} - \frac{a+b}{f} r_{m+1} + r_m = 0,$$

the general solution of which is a linear combination, with constant coefficients, of the two fundamental solutions $e^{m\omega}$ and $e^{-m\omega}$:

$$r_m = A e^{m\omega} + B e^{-m\omega}.$$

The constants A and B are determined by the initial conditions r_0 and r'_0 at the beginning of the chain of lenses. From the recurrence relation we can calculate ω; by substituting for r_m, and dividing by e^m, we obtain

$$e^{2\omega} - \frac{a+b}{f} e^{\omega} + 1 = 0,$$

or

$$\frac{e^{\omega} + e^{-\omega}}{2} = \frac{a+b}{2f},$$

the solution of which is

$$\cosh \omega = \frac{a + b}{2f} \quad \text{if} \quad \frac{a + b}{2f} > 1,$$

or

$$\cos \mu = \frac{a + b}{2f} \quad \text{if} \quad \frac{a + b}{2f} \leq 1 \quad \text{with} \quad \omega = j\mu.$$

The outermost trajectory of the beam defined by r_m will be stable only when ω is imaginary, so that

$$r_m = A \cos m\mu + B \sin m\mu.$$

r_m is then always bounded, irrespective of the values of A and B. Otherwise, when ω takes real values, r_m increases with m without limit for every set of initial conditions for which $A \neq 0$.

We therefore obtain a simple relation between the value of the focal length and the distance between the two foci F_i and F_0:

$$a + b < 2f.$$

We have thus expressed the condition in terms of the basic lens parameters.

10.5.3 The Matrix Representation

The form of the stability condition obtained above is highly suggestive, but nevertheless, it is preferable to obtain a more abstract but still more general expression by employing the matrix representation. The method which we shall use allows us to find the stability condition in cases where the individual units of the structure consist of several different elements— convergent or divergent lenses, regions of field-free space, etc. For a motif or "cell" which consists of three separate parts, for example, we have

$$\| M \| = \| T_3 \| \ \| T_2 \| \ \| T_1 \|,$$

so that

$$\begin{pmatrix} r_{m+1} \\ r'_{m+1} \end{pmatrix} = \begin{pmatrix} \alpha & \beta \\ \gamma & \delta \end{pmatrix} \begin{pmatrix} r_m \\ r'_m \end{pmatrix},$$

where

$$\| M \| = \begin{pmatrix} \alpha & \beta \\ \gamma & \delta \end{pmatrix}.$$

The expressions given above for α, β, γ and δ show that

$$|M| = \alpha \delta - \beta \gamma = 1.$$

(This is a perfectly general property of all the elementary matrices $\| T_n \|$ which we use.)

The stability condition $a + b < 2f$ here takes the form

$$\alpha + \delta < 2,$$

or

$$\text{Trace} \ \| M \| < 2.$$

(The trace of a matrix is the sum of the elements on the principal diagonal.)

10.5.4 The Case of N Quadrupole Lenses

The stability condition takes a particularly simple form when the lenses (represented by their equivalent rectangular model) are in contact. (If the distance d between the lenses is not zero, the expression which is obtained is more complicated, but is still obtained by writing Trace $\|M\| < 2$.)

The motif of length $2L$ is now composed of a convergent lens and a divergent lens each of length L. The corresponding elementary matrices are then respectively:

$$\|T_C\| = \begin{pmatrix} \cos \beta L & \frac{1}{\beta} \sin \beta L \\ -\beta \sin \beta L & \cos \beta L \end{pmatrix},$$

$$\|T_D\| = \begin{pmatrix} \cosh \beta L & \frac{1}{\beta} \sinh \beta L \\ \beta \sinh \beta L & \cosh \beta L \end{pmatrix}.$$

The transfer matrix for a region of field-free space D is, we recall,

$$\|T_G\| = \begin{pmatrix} 1 & D \\ 0 & 1 \end{pmatrix}.$$

Writing $\beta L = x$, the matrix $\|M\|$ is therefore given by

$$\|M\| = \begin{pmatrix} \cosh x & \frac{1}{\beta} \sinh x \\ \beta \sinh x & \cosh x \end{pmatrix} \begin{pmatrix} \cos x & \frac{1}{\beta} \sin x \\ -\beta \sin x & \cos x \end{pmatrix},$$

or

$$\|M\| = \begin{pmatrix} \cosh x \cos x - \sinh x \sin x & \frac{1}{\beta} (\cosh x \sin x + \sinh x \cos x) \\ \beta (\sinh x \cos x - \sin x \cosh x) & \sinh x \sin x + \cosh x \cos x \end{pmatrix}.$$

The periscopic stability condition is thus

$$2 \cosh x \cos x \leq 2,$$

or

$$\cosh x \leq \sec x,$$

which gives

$$0 < \beta L \leq 1 \cdot 873.$$

When $\beta L > 1 \cdot 873$, the system is over-focused and the beam is rapidly lost at the walls.

Within the channel, the trajectory will display a certain periodicity, of length Λ, which is equal to twice the distance between the two successive points at which $r = 0$; the shortest value of Λ will be obtained when $\beta L = 1 \cdot 873$, when it is equal to twice the period, $p = 2L$, of the structure itself.

10.6 APPLICATIONS OF QUADRUPOLE LENSES

We have already mentioned two common uses of these lenses, namely, to focus into an image plane the particles which are emitted at an object plane, which is usually the exit diaphragm of an accelerator or the plane of a target which is being bombarded by some primary radiation, and to channel a beam over a long distance, the only requirement being that the trajectories shall be stable; here it is not a faithful image that interests us.

The same channelling principle has been applied to the circular accelerators which have recently been constructed; the principle was described in 1952 by Courant, Livingstone and Snyder for strong-focusing synchrotons. Each elementary magnet of the synchroton plays two roles; the trajectory has to be curved in such a way that the mean path is circular, and the particles have to be focused in the horizontal and the vertical planes. These two operations are combined by employing hyperbolic pole-pieces, between which the radial gradient is constant.

Another application of a similar kind, though in theory more simple, was first suggested by Blewett (1952) and later studied by Bernard (1953a). Here, small quadrupole lenses are inserted into the tubes which separate the accelerating gaps in a linear ion accelerator in order to cancel the defocusing effect of the high frequency accelerating field. We can briefly analyse the stability conditions of the whole system by supposing that each accelerating gap behaves like a divergent thin lens, the focal length of which is known as soon as the conditions which produce acceleration have been fixed (the field, the phase angle, the value of the high frequency). If we assume that the velocity of the particles varies very little as they pass through a gap, and if we place one lens in each tube, we obtain a certain periodicity p, which may be produced, for example, by:

—the second half of a convergent quadrupole lens Q_1 in the tube preceding the first gap;

—a divergent lens L_1 in the first gap;

—a divergent quadrupole lens Q_2 which is identical to Q_1 in the tube following the first gap;

—a lens L_2 identical to L_1;

—and, finally, the first half of a convergent quadrupole lens Q_3 identical to Q_1 (see Fig. 114a).

The stability condition provides the value of the minimum gradient which is necessary if the accelerator is to function well (see, for example, the study by Teng, 1954). The lower the energy of the particles, the greater must the convergence of the lenses be; we cannot even consider using electrostatic lenses, as the potentials involved would be too high. In a magnetic version which has recently been constructed in Geneva at CERN in a linear proton accelerator, gradients of the order of 10 kilogauss cm^{-1} are obtained at the point of injection (with an initial energy of 500 keV) in pulsed operation; when the energy lies between 10 and 50 MeV, the gradient which is necessary varies between about 2000 and 500 G cm^{-1}, and the quadrupole lenses can be supplied steadily as the heat which is dissipated in the windings is now sufficiently low.

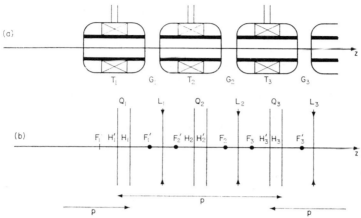

Fig. 114a. Focusing in a linear ion accelerator. (a) Tubes in which the motion of the particles is left undisturbed, save by a single quadrupole lens in each tube, alternately convergent and divergent. (b) The optical elements of the period; L_1 and L_2 are fictitious divergent lenses which lie in the gaps.

In conclusion, we should point out that quadrupole lenses can equally well be used to focus molecules and atoms which possess a magnetic moment proportional to the applied field, as Vauthier discovered (1949). The focusing force is then proportional to the distance from the axis in every radial direction, and the lens behaves like an axially symmetrical lens (Keller, 1957; Bennewitz and Paul, 1954; and Bennewitz, Paul and Schlier, 1955), and finally, several authors have shown that electric quadrupole lenses supplied with a high frequency potential provide a mass filter with an extremely good resolution (Paul and Raether, 1955; and Paul, Reinhard and von Zahn, 1958).

10.7 THE ABERRATIONS OF QUADRUPOLE LENSES

10.7.1 Aperture Aberrations

The vital aberration here is the aperture aberration. In the Gaussian image plane, the aberration figure depends upon the symmetry of the system; in the case of stigmatic operation (Burfoot, 1954), it may be (Fig. 114b):

FIG. 114b. The basic aberration figures for systems with four symmetry planes. (a) Circle. (b) Star. (c) Rosette. (d) An example of the previous figures superimposed.

(a) a circle, as in axially symmetrical lenses;
(b) a star with four points;
(c) a rosette;
(d) or, more commonly, a composite figure formed from a superposition of the three elementary aberrations. Three coefficients are necessary to describe the aberration (whereas one alone, C_s, is sufficient when the lenses are axially symmetric).

When the system is pseudo-stigmatic, we again obtain these same shapes, but with different magnifications in the X and Y directions (Septier, 1960). Finally, when the system is astigmatic, the effect of the aberrations on the focal lines has been studied in detail both for magnetic and electrostatic lenses (Septier, 1958; Septier and van Acker, 1960. See Fig. 6b).

Four coefficients are required to describe the transverse aberration in the Gaussian image plane. We write

$$\triangle x = C_x x_a^3 + C_{xy} x_a y_a^2,$$
$$\triangle y = C_y y_a^3 + C_{yx} y_a x_a^2. \tag{10.A}$$

(For a stigmatic or pseudo-stigmatic system, we should find that $C_{xy} = C_{yx}$.) Here, x_a and y_a denote the distance from the axis in the plane $z = z_a$ of the entrance pupil and $x(z)$, $y(z)$ are trajectories travelling in the two orthogonal planes of symmetry.

These trajectories $x(z)$ and $y(z)$ can be expressed in terms of two particular trajectories $g(z)$ and $h(z)$, defined by

$$g_x(z_0) = g_y(z_0) = h_x(z_a) = h_y(z_a) = 1$$

and

$$g_x(z_a) = g_y(z_a) = h_x(z_0) = h_y(z_0) = 0.$$

We find (Fig. 114c)

$$\begin{cases} x(z) = x_0 g_x(z) + x_a h_x(z) \\ y(z) = y_0 g_y(z) + y_a h_y(z), \end{cases}$$

where $z = z_0$ denotes the object plane and $z = z_a$, the aperture plane. At $z = z_i$, we have

$$x'(z_i) = x_a h'_x(z_i) = \alpha,$$
$$y'(z_i) = y_a h'_y(z_i) = \beta,$$

if we consider a particle from an axial point source S. We can also express $\triangle x$ and $\triangle y$ in terms of α and β, which are easier to measure experimentally:

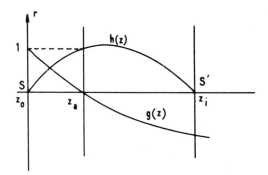

FIG. 114c. Definition of the particular trajectories, $g(z)$ and $h(z)$.

$$\triangle x = C'_x \alpha^3 + C'_{xy} \alpha \beta^2, \qquad \triangle y = C'_y \beta^3 + C'_{yx} \beta \alpha^2.$$

If S'_x and S'_y coincide, $C'_{xy} = C'_{yx}$ even if the magnifications M_x and M_y are different.

As in the case of rotationally symmetric systems, the aberrations may be calculated by either of two methods: the trajectory method or the characteristic function method, developed and applied by Hawkes (1963, 1965) in the general case of systems with two symmetry planes (rectilinear orthogonal systems).

Hawkes' theory also allows us to calculate all the aberration coefficients and we do indeed recover, as a special case, the formulae listed in Chapter 7 for rotationally symmetric systems.

10.7.2 The Trajectory Method

Into the equations of motion we now introduce firstly the transverse components of the velocity and secondly the higher order terms in the field expansion in the fringe field zone (Septier, 1961a). In the plane that is initially convergent, we obtain the following equations:

Magnetic case

$$X'' + \beta^2 kX = -\beta^2 \left\{ kX \frac{3X'^2 + Y'^2}{2} - kX'Y'Y - k'XYY' \right.$$
$$\left. - \frac{k''}{12} X(X^2 + 3Y^2) \right\} = -\beta^2 P(z).$$

Electrostatic case   (10.B)

$$x'' + \beta^2 kx = -\beta^2 \left\{ kx(x'^2 + y'^2) + \beta^2 k^2 x(x^2 - y^2) \right.$$
$$\left. - \frac{k'x'}{2}(x^2 - y^2) - \frac{k''x^3}{6} \right\} = -\beta^2 Q(z).$$

In the initially divergent plane, we obtain identical equations if the following exchanges are performed:

$$\beta^2 \to -\beta^2 \quad X \rightleftharpoons Y \quad X' \rightleftharpoons Y'$$
$$x \rightleftharpoons y \quad x' \rightleftharpoons y'$$

For a given system, the function $\beta^2 k(z)$ is known everywhere on the axis Oz, and can be represented by simple analytic functions, usually bell-shaped fields (Bernard and Hue, 1956; Grivet and Septier, 1960; Septier, 1961a). With the aid of an electronic computer, we can then either (i) integrate the so-called "third order equations" and calculate the final discrepancy between their solutions and those of the "first order equations" (obtained by setting the right hand sides equal to zero) or alternatively (ii) we can calculate the differences, $\varepsilon(z)$ and $\eta(z)$, between the Gaussian and "third order" trajectories throughout the whole system. For this, we replace such quantities as X, Y, X' and Y' by the first order solutions, obtained in a preliminary calculation. $P(z)$ and $Q(z)$ are then known functions.

If the rectangular model ($K = $ const.), which yields very simple first order solutions, is used, the aberrations cannot at first sight be calculated by this method, for the terms in $k'(z)$ and $k''(z)$ have to be neglected in the equations. Efforts have been made, nevertheless, to take them into account (Bernard and Hue, 1956). It seems, however, that by carrying out partial integrations on the general equations, expressions for the aperture aberration coefficients not containing k' and k'' can be obtained (Dymnikov et al., 1965c); these can then be integrated directly using the simple solutions afforded by the rectangular model.

We can also calculate the aberration coefficients of each lens and hence the total aberration coefficients for a system consisting of a number of lenses (Deltrap, 1964a, 1964b).

Using this trajectory method, Septier (1961a), Strashkevich (1964) and Dymnikov et al. (1964a) have succeeded in determining the aperture aberrations of electrostatic and magnetic single lenses and doublets. More recently, Meads (1966) has written a programme for the IBM 7094 computer with which the aberration coefficients can be calculated and an optimum system designed (by varying the geometrical and electrical parameters). Dhuicq and Septier (1966b) have calculated the aperture aberration coefficients of a quadruplet equivalent to a round lens (the characteristics of which have been given in § 10.4.4): for the magnetic case,

$$\frac{C_x}{f} = 39, \quad \frac{C_{xy}}{f} = \frac{C_{yx}}{f} \simeq 72, \quad \frac{C_y}{f} \simeq 1.4 \times 10^5$$

and for the electrostatic case

$$\frac{C_x}{f} = 12, \quad \frac{C_{xy}}{f} = \frac{C_{yx}}{f} \simeq 550, \quad \frac{C_y}{f} \simeq 0.5 \times 10^5$$

for

$$f = 0.8 \text{ mm.}$$

They have shown that it is possible to cancel these aberrations completely by employing three octopole lenses, judiciously disposed within the system.

Deltrap (1964a, 1964b) has measured the aberration coefficients of a system consisting of four mixed magnetic lenses (each a combined quadrupole and octopole) together with a round lens and has shown that it is possible to reduce the aberration coefficients considerably. Septier and Dhuicq (1964) have performed similar experiments using a symmetric triplet of unit magnification. These two latter systems are stigmatic only for a privileged pair of conjugate points.

10.7.3 The Characteristic Function Method (Hawkes, 1963, 1965a, 1965b)

Setting out from the expression for Fermat's principle, as applied in electron optics,

$$\delta \int n \, ds = 0,$$

in which n denotes the refractive index

$$n = \Phi^{*1/2} - \sqrt{\frac{e}{2m_0}} A \cdot s,$$

Hawkes transforms the variational equation into

$$\delta \int m \, dz = 0,$$

in which

$$m = \{ \Phi^{*1/2} (1 + x'^2 + y'^2)^{1/2} - \eta (A_x x' + A_y y' + A_z) \}, \qquad (10.\text{C})$$

where

$$\eta = \sqrt{\frac{e}{2m_0}} .$$

The function $\Phi(x, y, z)$ and the components of the vector $A(x, y, z)$ can be expressed as power series in x and y, while the variable z figures only in the coefficients. On substituting these series into m and collecting up, we obtain

$$m = m^{(0)} + m^{(2)} + m^{(4)} + \cdots + m^{(2n)} + \cdots, \qquad (10.\text{D})$$

in which $m^{(2n)}$ contains all the terms of degree $2n$.

If S denotes the optical path-length between two points in the planes z_1 and z_2, and S^* an infinitesimally close optical path, we have

$$S = \int_{z_1}^{z_2} m \, dz,$$

$$S^* = S + \delta S.$$

If the two rays begin and end at the same points, we have

$$\delta S = 0,$$

which implies

$$\frac{dp}{dz} = \frac{\partial m}{\partial x}, \quad \frac{dq}{dz} = \frac{\partial m}{\partial y},$$

where we have written $p = \partial m / \partial x'$, $q = \partial m / \partial y'$.

The function $V_{1,2} = \int_{z_1}^{z_2} m \, dz$ is Hamilton's point characteristic function, and we have

$$\left.\begin{aligned}
\delta V_{1,2} &= p_2 \delta x_2 + q_2 \delta y_2 - p_1 \delta x_1 - q_1 \delta x_1 \\
p_1 &= -\frac{\partial V_{1,2}}{\partial x_1} \quad q_1 = -\frac{\partial V_{1,2}}{\partial y_1}, \\
p_2 &= \frac{\partial V_{1,2}}{\partial x_2} \quad q_2 = \frac{\partial V_{1,2}}{\partial y_2}.
\end{aligned}\right\} \quad (10.E)$$

If m is replaced by a perturbed function, m^*,

$$m^* = m + m^{(p)},$$

we obtain (Sturrock, 1951)

$$V_{1,2}^{(p)} = \int_{z_1}^{z_2} m^{(p)} \, dz$$

and if in addition we impose the conditions

$$x_1^{(p)} = y_1^{(p)} = x_2^{(p)} = y_2^{(p)} = 0,$$

we can deduce the perturbations of p and q ($p^{(p)}$ and $q^{(p)}$) from formulae resembling (10.E) except that the perturbation $V_{1,2}^{(p)}$ replaces $V_{1,2}$.

The Gaussian (first order) properties of a system will be described by the term

$$m = m^{(2)},$$

the third order geometrical aberrations by

$$m^{(p)} = m^{(4)},$$

and the chromatic aberration by

$$m^{(p)} = \frac{\partial m^{(2)}}{\partial \Phi}.$$

We obtain $V_{1,2}^{(p)}$ by integrating $m^{(p)}$ with respect to z; to be able to perform this integration, we replace the quantities x, y, x' and y' by their unperturbed first order values, obtained from $m = m^{(2)}$. We can then calculate the quantities $p_c^{(p)}$, $q_c^{(p)}$, $x_c^{(p)}$ and $y_c^{(p)}$ in an arbitrary current plane of abscissa $z = z_c$ in terms of $V_{oc}^{(p)}$ and $V_{ac}^{(p)}$ (these two quantities correspond to integrations from the object point z_0 to z_c and from the aperture z_a to z_c, respectively).

In a system of quadrupoles possessing two planes in which the trajectories remain plane, we have

$$m^{(2)} = \tfrac{1}{2}C(x'^2 + y'^2) + \tfrac{1}{2}(\bar{C}_x x^2 + \bar{C}_y y^2)$$
$$p = Cx', \quad q = Cy' \tag{10.F}$$

(C, $\bar{C}_x$ and $\bar{C}_y$ are functions of z.) The equations of motion are now

$$\frac{d}{dz}(Cx') = \bar{C}_x x, \quad \frac{d}{dz}(Cy') = \bar{C}_y y, \tag{10.G}$$

and have general solutions

$$x(z) = x_0 g_x(z) + x_a h_x(z),$$
$$y(z) = y_0 g_y(z) + y_a h_y(z). \tag{10.H}$$

It is easy to show from equations (10.G) that the quantities

$$k_x = C(g_x h_x' - g_x' h_x)$$
$$k_y = C(g_y h_y' - g_y' h_y)$$

are constants. We then find

$$k_x x_c^{(p)} = h_{xc} \frac{\partial V_{ac}^{(p)}}{\partial x_0} - g_{xc} \frac{\partial V_{oc}^{(p)}}{\partial x_a},$$

$$k_x p_c^{(p)} = C\left(h_{xc}' \frac{\partial V_{ac}^{(p)}}{\partial x_0} - g_{xc}' \frac{\partial V_{oc}^{(p)}}{\partial x_a} \right), \tag{10.I}$$

with two analogous expressions for $y_c^{(p)}$ and $q_c^{(p)}$.

Setting out from equations (10.H) and (10.I), and taking the symmetry into account, we can express x_c and y_c as functions of x_0, y_0, x_a and y_a, and the corresponding slopes x_0', y_0', x_a' and y_a'; this enables us to calculate the cross-section of the beam at every plane $z = z_c$ (and in particular, at the Gaussian image plane) and to ascertain the various aberration coefficients.

In order to calculate the aperture aberration coefficients, Hawkes obtains the expressions for $m^{(2)}$ and $m^{(4)}$, using the expansions of $\Phi(x, y, z)$

(scalar electrostatic potential) and of A_x, A_y and A_z given by Glaser (1956); we recall that the expression for the vector potential A chosen by Glaser differs from the (more common) one that has been given in Chapter 3.

In an orthogonal system with a straight axis, consisting of round electrostatic lenses together with (electrostatic or magnetic) quadrupoles and octopoles, oriented in such a way that there are two privileged planes zOx and zOy in which the trajectories remain plane, round magnetic lenses may be included; the function $B(z)$ must, however, not overlap any of the other functions of the type $\Phi(z)$. In a magnetic lens, the equations for x and y are independent only if we use a rotating coordinate system, the rate of rotation of which is given (see Chapter 5) by

$$\frac{d\theta}{dz} = \frac{\eta}{2} \frac{B(z)}{\{\Phi^*(z)\}^{1/2}};$$

the trajectories are then "uncoupled" over two curved orthogonal surfaces defined by $\theta(z)$ and $\theta(z) + \pi/2$. If the round magnetic lens is placed between two series of systems with plane orthogonal surfaces, the planes *behind* the magnetic lens will have to be twisted through an angle

$$\theta_{1,2} = \int_{z_1}^{z_2} \frac{d\theta}{dz} dz$$

with respect to the planes *in front of* the magnetic lens. This situation never arises in practice, for $\theta_{1,2}$ would vary with the excitation of the round lens. A system consisting of a round lens followed by quadrupoles and octopoles has, on the other hand, been widely used with a view to correcting the spherical aberration of a round electron microscope objective (see §7.7).

Hawkes' major contribution in this domain is to have put the expressions for the aperture aberration coefficients into forms that can be used directly, in several special cases:

(i) systems of magnetic and electrostatic quadrupoles and octopoles (without round lenses);

(ii) systems containing only one type of quadrupole and octopole lens (magnetic or electrostatic);

(iii) systems with rotational symmetry.

Full expressions for all these coefficients are to be found in Hawkes (1965a, b). We give as an example only those relating to a purely electrostatic stigmatic system (a quadruplet, for example), using the rectangular model approximation ($k(z) = 1$ inside the lenses). Each lens, labelled i, will be represented by a function $D(z)$:

$$D(z) = D_i = \text{const}.$$

(or with the notation of §10.2, we should have $D_i = 4\Phi_i/a^2$) and to obtain C_x, C_y and C_{xy} in the Gaussian image plane, we must calculate g_x, g_y, h_x and h_y; in the Gaussian image plane, $z = z_G$, we have $g_x = g_y = g$ and $h_x = h_y = 0$. If z_{ia} and z_{ib} denote the abscissae of the entry and exit planes of the i-th lens, we have

$$kC_x = \left(\frac{g}{12\Phi_0^{3/2}}\right)\sum_i \int_{z_{ia}}^{z_{ib}} D_i^2 h_x^4 \, dz - \left(\frac{g}{24\Phi_0^{1/2}}\right)\sum_i [D_i h_x^3 h_x']_{z_{ia}}^{z_{ib}},$$

$$kC_y = \left(\frac{g}{12\Phi_0^{3/2}}\right)\sum_i \int_{z_{ia}}^{z_{ib}} D_i^2 h_y^4 \, dz + \left(\frac{g}{24\Phi_0^{1/2}}\right)\sum_i [D_i h_y^3 h_y']_{z_{ia}}^{z_{ib}},$$

$$kC_{yx} = kC_{xy} = -\left(\frac{g}{8\Phi_0^{3/2}}\right)\sum_i \int_{z_{ia}}^{z_{ib}} \frac{D_i^2}{h_x^2 h_y^2} \, dz + \left(\frac{g}{32\Phi_0^{1/2}}\right)\sum_i 3D_i[\{h_x^2(h_y^2)' - (h_x^2)'h_y^2\}_{z_{ia}}$$

$$- \{h_x^2(h_y^2)' - (h_x^2)'h_y^2\}_{z_{ib}}].$$

We recall that in the Gaussian image plane, z_G, $k_x = k_y = k$:

$$k = \sqrt{\Phi_0}(gh' - g'h) = gh' \quad \text{at } z = z_G \text{ (no distortion)}.$$

In conclusion, it is possible to show that the aperture aberration coefficients of a system of quadrupoles and octopoles, or of a round lens followed by such a system, can be cancelled by a judicious choice of the electrical and geometrical parameters.

10.7.4 Chromatic Aberration

The achromatic system. By superimposing an electrostatic potential $\Phi(x, y, z)$ and a magnetic potential $\Phi_m(x, y, z)$ within a single lens, in which the electrodes are turned through 45° with respect to the magnetic poles (to make the system orthogonal), we can design an achromatic mixed lens, as first Kel'man and Yavor (1961, 1963) and later Septier (1963) have shown; the converging power suffers, however. This interesting property has been demonstrated experimentally (Kel'man et al., 1963; Yavor et al., 1964). The presence of the electrostatic lens means that such systems can only be used for relatively low energies.

We reconsider the equations of motion in a mixed quadrupole system (in the non-relativistic case), using the rectangular approximation:

$$x'' + \left(\frac{K_E}{2\Phi_0} + \frac{eK_M}{mv}\right)x = 0.$$

The electric field gradient is denoted by K_E and the magnetic field gradient by K_M; the equivalent lengths of the magnetic and electrostatic contributions are supposed identical. We may rewrite this

$$x'' + \left(\frac{eK_E}{mv^2} + \frac{eK_M}{mv}\right)x = 0$$

or

$$x'' + A(v)x = 0,$$

so that the system will be achromatic if $\partial A(v)/\partial v = 0$, which implies

$$2\frac{eK_E}{mv^2} + \frac{eK_M}{mv} = 0.$$

Since $\beta_E^2 = eK_E/mv^2$ and $\beta_M^2 = eK_M/mv$, the condition to be fulfilled is

$$\beta_E^2 = -\tfrac{1}{2}\beta_M^2$$

and the equations of motion will then be of the form

$$x'' + (\tfrac{1}{2}\beta_M^2)x = 0,$$
$$y'' - (\tfrac{1}{2}\beta_M^2)y = 0.$$

The convergence is thus half as strong as that of the magnetic lens acting alone.

Using the method set out in the preceding section, Hawkes (1965c) has found it possible to calculate the chromatic aberration of rectilinear orthogonal systems (and in particular, of these mixed lenses) in a completely general way, together with the aperture aberration of achromatic lenses. In addition, Septier (1963) and later Dymnikov et al. (1964b, 1965b) have shown that the aperture aberration remains of the same order of magnitude as that of electrostatic or magnetic lenses.

PRISM OPTICS

THE foregoing chapters have been devoted to the study of optical systems with straight axes that have the same effect on charged particles as glass lenses have on photons. We have, moreover, considered only beams that consist of a single type of particle (electrons or identical ions) and are monochromatic (or quasi-monochromatic). Any small differences between the energies or velocities of the various particles are embraced within the "chromatic aberration" of the lenses.

In particle optics, there is a whole range of situations in which we have to analyse the different components of a heterogeneous beam. We then use deflector systems with dispersive properties and by analogy with glass optics, we call such systems "prisms". We shall find that these prisms also possess valuable focusing properties; in fact they form another class of thick optical systems, for which asymptotic cardinal elements can be defined. The straight optic axis of the electron lens is here replaced by a curved axis coinciding with a particular "mean trajectory" (or "equilibrium trajectory") which lies in a plane of symmetry or antisymmetry of the system.

Prisms consist of a pair of electrodes (in the electrostatic case) or of poles (in the magnetic case), which create an electric field E or a magnetic field B directed in such a way that the force exerted on a particle is always normal to the direction in which the particle is travelling, at least in the privileged ("median") plane containing the curved axis.

In its simplest form, a prism consists of a limited region of space containing a homogeneous electric or magnetic field: a plane condenser or a pair of parallel poles very close together. We should use such a system to obtain fairly weak angular deflexions (the beam of electrons in an oscilloscope is deflected in this way for example, see Chapter 12). In order to obtain larger deflexions and a high dispersion, it is better to use long systems with a large radius of curvature in which—for obvious reasons of constructional simplicity and ease of calculation—the mean trajectory (or curved axis) is a segment of a circle centred at O and of radius R.

Electrostatic prisms will thus consist of portions of cylindrical, spherical (or even toroidal) condensers and magnetic prisms of circular sector magnets. The angle at the centre of the prism, ϕ, is then equal to the deflexion imposed on the mean trajectory of the beam in the median plane.

This median plane, which contains the curved optic axis, is a plane of symmetry in the electrostatic prism and of antisymmetry in the magnetic prism; the scalar potentials in the two cases therefore have different expressions. We shall study the two types of prisms in turn, first examining in detail the first order optical properties (Gaussian optics) and then briefly considering the aberrations.

10A.1 MAGNETIC PRISMS

10A.1.1 Potential and Field

We consider a magnet consisting of two identical poles in the form of circular sectors of angle ϕ, centre C and mean radius r_0, terminated by the plane faces AB and $A'B'$ (Fig. 115a(a)). The gap has a plane of mechanical symmetry (Fig. 115a(b)). The magnetic field is constant along an arbitrary circle (C) of radius r lying in this symmetry plane (the "median" plane)—on (C_0) we have $B = B_0$—and we assume to begin with that $B = 0$ outside the end faces and that $B = $ constant along (C) for $0 \leqslant \theta \leqslant \phi$

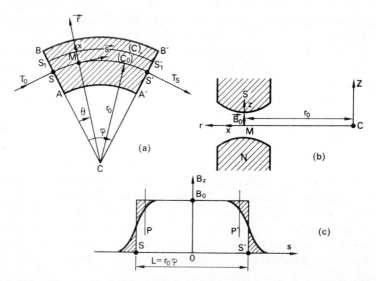

FIG. 115a. Diagram of a magnetic prism. (a) The mean trajectory (C_0) in the median plane. (b) Section by a vertical plane. (c) Rectangular model: real pole faces at P and P', magnetic virtual faces at S and S' (equal areas).

(Fig. 115a(c)). This is the rectangular model approximation that has already been used for quadrupole lenses. For reasons of symmetry, the field B is parallel to the axis CZ in the median plane. In reality, the fringing fields at the edges extend beyond the poles, and the rectangular model obtained by setting the areas under the curves equal:

$$B_0 L = r_0 \int_{-\infty}^{\infty} B_z(\theta) \, d\theta,$$

is slightly longer than the circular arc (C_0):

$$L > r_0 \phi_{\text{mech}},$$

where ϕ_{mech} denotes the angle of the real sectors. We must therefore replace the real sector by a magnetic sector of angle $\phi = L/r_0 > \phi_{\text{mech}}$; the difference $\triangle\phi = \phi - \phi_{\text{mech}}$ depends upon the width of the gap, D, and on the geometry of the poles and coils, but it is always true that

$$0 \cdot 5D < r_0 \triangle\phi < 0 \cdot 7D.$$

We shall assume henceforward that the faces $AB, A'B'$ are the hypothetical magnetic faces corresponding to the angle $\phi = L/r_0$. It is easy to determine L from measured values of the magnetic field.

Instead of the standard cylindrical coordinates r, θ, Z, with axis CZ, it is more convenient to use the reduced dimensionless coordinates x, z, θ, defined by

$$x = \frac{r - r_0}{r_0}, \quad z = \frac{Z}{r_0}, \quad \theta = \frac{s}{r_0}. \tag{10A.1}$$

(C_0) then plays the part of a curved axis, which extends outside the sector as two straight lines T_0 and T_S (see Fig. 115a(a)); ds is an element of length along this axis, (C_0).

In the gap, the scalar magnetic potential Φ_m corresponding to the induction B satisfies Laplace's equation,

$$\nabla^2 \Phi_m = 0. \tag{10A.2}$$

Taking the rotational symmetry into account $(\partial/\partial\theta = 0)$ and introducing the new variables x and z, equation (10A.2) becomes

$$\frac{\partial^2 \Phi_m}{\partial x^2} + \frac{1}{1+x} \frac{\partial^2 \Phi_m}{\partial x} + \frac{\partial^2 \Phi_m}{\partial z^2} = 0. \tag{10A.3}$$

In the symmetry plane of the system, which we shall call the "horizontal plane" H, Φ_m is a function of x only. We can therefore write it in the form

$$\Phi(0, x) = \Phi_0(x) = \Phi_0(1 - nx + bx^2 - cx^3 + dx^4 - \cdots). \tag{10A.4}$$

Outside the plane H, Φ_m can be expanded as a polynomial containing only odd powers of z:

$$\frac{\Phi_m}{\Phi_0} = -\sum_{n=0}^{\infty} A_{2n+1}(x)z^{2n+1}. \tag{10A.5}$$

Substituting equation (10A.5) into (10A.3), we obtain a recurrence relation between the terms $A_{2n+1}(x)$ and their x-derivatives, from which we can calculate the first few terms of (10A.5). We obtain, for example,

$$A_3(x) = -\tfrac{1}{6}\{(2b-n)+(n+2b-6c)x \\ -(n+2b+3c-12d)x^2+\cdots\}. \tag{10A.6}$$

The first term, $A_1(x)$, corresponds to $n=0$ and is given by equation (10A.4). We finally obtain, up to terms of fourth order in x and z,

$$\frac{\Phi_m}{\Phi_0} = (-1+nx-bx^2+cx^3)z-\{(n-2b)-(n+2b-6c)x\}\frac{z^3}{6}+\cdots. \tag{10A.7}$$

Writing

$$B_z = -\frac{\partial \Phi_m}{\partial Z} = -\frac{1}{r_0}\frac{\partial \Phi_m}{\partial z},$$

$$B_x = -\frac{\partial \Phi_m}{\partial r} = -\frac{1}{r_0}\frac{\partial \Phi_m}{\partial x}, \tag{10A.8}$$

$$B_0 = \Phi_0/r_0,$$

we arrive at the expressions for the two components of the field B:

$$\frac{B_z}{B_0} = (1-nx+bx^2-cx^3)+\tfrac{1}{2}z^2\{(n-2b)-x(n+2b-6c)\}+\cdots,$$

$$\frac{B_x}{B_0} = (-n+2bx-3cx^2)z-(n+2b-6c)\frac{z^3}{6}+\cdots \tag{10A.9}$$

Only the first terms in these expansions are retained when we study the "first order" optical properties, so that

$$B_z = (1-nx)B_0 = B_0\left(1-n\frac{r-r_0}{r_0}\right),$$

$$B_x = -B_0nz = -B_0n\frac{Z}{r_0}. \tag{10A.10}$$

The coefficient n is called the "index" of the field; it is proportional to the transverse gradient of B_z in the vicinity of r_0:

$$n = -\frac{r_0}{B_0}\frac{\partial B_z}{\partial r} = -\frac{1}{B_0}\frac{\partial B_z}{\partial x}. \tag{10A.11}$$

When $n=0$, we have $B_z = B_0$ and $B_x = 0$ so that the magnetic field is homogeneous throughout the sector.

10A.1.2 First Order Trajectory Equations (Gaussian Optics)

We first observe that one of the zero order trajectories is the circle (C_0); this curve is the path of a particle of charge e and momentum $p_0 = mv_0 = \sqrt{2em_0V_0^*}$ (with $V_0^* = V_0(1 + eV_0/2m_0c^2)$: the relativistic potential) chosen such that

$$B_0r_0 = p_0/e, \qquad (10A.12)$$

injected along T_0, normal to the entry face AB at S. For non-relativistic particles (electrons at a few keV or ions up to a few hundred keV), we have $V_0^* \simeq V_0$, and so

$$B_0r_0 = \sqrt{\frac{2m_0}{e}}\, V_0. \qquad (10A.13)$$

Particles of momentum $p_1 \neq p_0$ can then also follow circular paths (C) of radius $r_1 = r_0(1 + x_1)$, satisfying the equation

$$B(x_1)r_1 = p_1/e.$$

We now study the trajectories of particles *of momentum* p_0, injected near to the curved axis (C_0), $x, z \ll 1$, with small slopes ($x'^2, y'^2 \ll 1$). We know that in a static magnetic field, the initial energy of the particle does not change, since the magnetic force

$$f_m = ev \times B \qquad (10A.14)$$

is always normal to the direction of the trajectory. We thus deduce that $m = \text{const.}$ and $v = \text{const.} = v_0$. Furthermore, since v_z and v_x are both much smaller than v_θ, we may write

$$v_\theta = v_0\left\{1 - \left(\frac{v_z}{v_0}\right)^2 - \left(\frac{v_x}{v_0}\right)^2\right\}^{-1/2} \simeq v_0,$$

and so

$$v_\theta = r\frac{d\theta}{dt} \simeq r_0\left(\frac{d\theta}{dt}\right)_0 = v_0$$

or

$$\frac{d\theta}{dt} = \frac{1}{1+x}\left(\frac{d\theta}{dt}\right)_0 = \frac{\omega}{1+x}. \qquad (10A.15)$$

ω is known as the "cyclotron frequency" of the particle in the field B_0. From equation (10A.12), we have

$$\omega = eB_0/m. \qquad (10A.16)$$

In the particular coordinate system that we have selected (Fig. 115a), the set of vectors (r, Z, s) are mutually orthogonal. The equations of motion for r and Z then take the following form, when we project onto the corresponding axes:

$$\frac{d^2r}{dt^2} - r\left(\frac{d\theta}{dt}\right)^2 = -\frac{e}{m} rB_z \frac{d\theta}{dt},$$

$$\frac{d^2Z}{dt^2} = \frac{e}{m} rB_r \frac{d\theta}{dt}. \tag{10A.17}$$

Replacing B_z and B_r by the expressions given in equation (10A.10), we obtain

$$\frac{d^2x}{dt^2} + \omega^2(1-n)x = 0,$$

$$\frac{d^2z}{dt^2} + \omega^2 nz = 0. \tag{10A.18}$$

We can eliminate time with the aid of the formula $dt = d\theta/\omega$, and we find

$$\frac{d^2x}{d\theta^2} + (1-n)x = 0, \tag{10A.19}$$

$$\frac{d^2z}{d\theta^2} + nz = 0. \tag{10A.20}$$

For $0 < n < 1$, the solutions of equations (10A.19) and (10A.20) can be expressed in terms of circular functions: the prism is convergent in both the x- and z-directions (that is, in both the horizontal plane H in which the deflexion occurs and in the plane that is customarily called the "vertical plane", V, containing the axes Oz and Os). If $n = 0$ (homogeneous field), there is no focusing in V, but there is again a focusing effect in H. When $n < 0$, the plane V will be a divergent plane and H will be a convergent plane. We have already encountered such a situation in the chapter on strong-focusing lenses. Finally, we observe that the two equations (10A.19) and (10A.20) are identical when $n = \frac{1}{2}$, so that the convergence is the same in H and V.

10A.1.3 Optical Properties of the Prism with Entry and Exit Faces Normal to the Mean Trajectory (C_0)

10A.1.3.1   The general case

The solutions of equations (10A.19) and (10A.20) are of the form

$$x = A_H \cos(\sqrt{1-n}\,\theta) + B_H \sin(\sqrt{1-n}\,\theta),$$

$$z = A_V \cos(\sqrt{n}\,\theta) + B_V \sin(\sqrt{n}\,\theta).$$

The constants A and B can easily be expressed as a function of the initial conditions at the entry face, and we obtain the coordinates (of position and slope) of the trajectory in the plane of the exit face, after traversing an angular path of length $\theta = \phi$, in the form

$$x_S = x_0 \cos(\sqrt{1-n}\,\phi) + \frac{x_0'}{\sqrt{1-n}} \sin(\sqrt{1-n}\,\phi),$$
$$\tag{10A.21}$$
$$x_S' = -x_0 \sqrt{1-n} \sin(\sqrt{1-n}\,\phi) + x_0' \cos(\sqrt{1-n}\,\phi),$$

$$z_S = z_0 (\cos \sqrt{n}\,\phi) + \frac{z_0'}{\sqrt{n}} \sin(\sqrt{n}\,\phi),$$
$$\tag{10A.22}$$
$$z_S' = -z_0 \sqrt{n} \sin(\sqrt{n}\,\phi) + z_0' \cos(\sqrt{n}\,\phi),$$

where $x' = dx/d\theta = dr/ds$ and $z' = dz/d\theta = dZ/ds$. From equations (10A.21) and (10A.22), we can define the transfer matrices of the prism, $\|H\|$ and $\|V\|$, in the planes H and V respectively. We write

$$\begin{pmatrix} x_S \\ x_S' \end{pmatrix} = \|H\| \begin{pmatrix} x_0 \\ x_0' \end{pmatrix},$$

with

$$\|H\| = \begin{pmatrix} \cos(\sqrt{1-n}\,\phi) & \dfrac{1}{\sqrt{1-n}} \sin(\sqrt{1-n}\,\phi) \\ -\sqrt{1-n} \sin(\sqrt{1-n}\,\phi) & \cos(\sqrt{1-n}\,\phi) \end{pmatrix}$$
$$\tag{10A.23}$$
$$= \begin{pmatrix} H_{11} & H_{12} \\ H_{21} & H_{22} \end{pmatrix}$$

and

$$\begin{pmatrix} z_S \\ z_S' \end{pmatrix} = \|V\| \begin{pmatrix} z_0 \\ z_0' \end{pmatrix}$$

with

$$\|V\| = \begin{pmatrix} \cos(\sqrt{n}\,\phi) & \dfrac{1}{\sqrt{n}} \sin(\sqrt{n}\,\phi) \\ -\sqrt{n} \sin(\sqrt{n}\phi) & \cos(\sqrt{n}\,\phi) \end{pmatrix}$$
$$= \begin{pmatrix} V_{11} & V_{12} \\ V_{21} & V_{22} \end{pmatrix}.$$
$$\tag{10A.24}$$

We recall that all distances in real space are measured in units of r_0. From $\|H\|$ and $\|V\|$, we can determine the focal lengths, f_H and f_V and the positions of the foci and principal planes (Fig. 115b). The cardinal elements are symmetrical with respect to the plane $\theta_1 = \phi/2$.

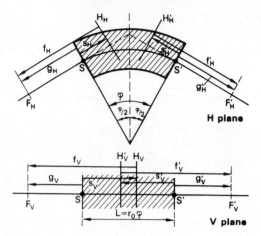

FIG. 115b. Positions of the foci and principal planes (with respect to the magnetic faces).

Returning to the unscaled variables, we can easily show that

$$f'_H = -f_H = \frac{r_0}{\sqrt{1-n}\,\sin(\sqrt{1-n}\,\phi)},$$

$$g'_H = -g_H = \frac{r_0}{\sqrt{1-n}\,\tan(\sqrt{1-n}\,\phi)}, \qquad (10A.25)$$

$$s'_H = -s_H = \frac{r_0}{\sqrt{1-n}}\tan(\tfrac{1}{2}\sqrt{1-n}\,\phi),$$

$$f'_V = -f_V = \frac{r_0}{\sqrt{n}\,\sin(\sqrt{n}\,\phi)},$$

$$g'_V = -g_V = \frac{r_0}{\sqrt{n}\,\tan(\sqrt{n}\,\phi)}, \qquad (10A.26)$$

$$s'_V = -s_V = \frac{r_0}{\sqrt{n}}\tan(\tfrac{1}{2}\sqrt{n}\,\phi).$$

10A.1.3.2 First special case: $n = 0$

The matrices $\|H\|$ and $\|V\|$ now take the simple form

$$\|H\| = \begin{pmatrix} \cos\phi & \sin\phi \\ -\sin\phi & \cos\phi \end{pmatrix}, \quad \|V\| = \begin{pmatrix} 1 & \phi \\ 0 & 1 \end{pmatrix}. \qquad (10A.27)$$

In the plane V, the prism is equivalent to a drift space of length $L = r_0\phi$, and has no effect on the trajectories. In the plane H, it behaves like a thick lens, convergent if $\phi < \pi$ and divergent if $\pi < \phi < 2\pi$. We then have

$$f'_H = -f_H = r_0 \operatorname{cosec} \phi,$$

$$g'_H = -g_H = r_0 \cot \phi, \qquad (10A.28)$$

$$s'_H = -s_H = r_0 \tan(\phi/2).$$

The diagram of Fig. 115c(a) shows that the principal planes lie at P, the point at which T_0 and T_S intersect; we can therefore replace the prism by a thin lens of focal length of $f = r_0 \operatorname{cosec} \phi$ at P and two drift spaces $SP = PS' = r_0 \tan(\phi/2)$. The length of this optical system is greater than that of the real trajectory ($L = r\phi$), however, and it is thus better to consider the thick lens of length L with crossed principal planes (Fig. 115c(b)). For $\phi = 90°$, the foci lie at S and S' and we have $f'_H = r_0$.

THE CORRESPONDENCE BETWEEN OBJECT AND IMAGE. We consider an object P, a distance p from S, and its image Q, distant $S'Q = q$ from S'. In reduced coordinates, p/r_0 and q/r_0, the complete transfer matrix between P and Q becomes

$$\|T\| = \begin{pmatrix} 1 & q/R \\ 0 & 1 \end{pmatrix} \begin{pmatrix} H_{11} & H_{12} \\ H_{21} & H_{22} \end{pmatrix} \begin{pmatrix} 1 & p/R \\ 1 & 0 \end{pmatrix} = \begin{pmatrix} T_{11} & T_{12} \\ T_{21} & T_{22} \end{pmatrix}. \quad (10A.29)$$

For a ray leaving P ($x_P = 0$) at slope $x'_p \neq 0$, we see that at Q,

$$x_Q = x'_P T_{12} = 0,$$

so that

$$T_{12} = 0. \qquad (10A.30a)$$

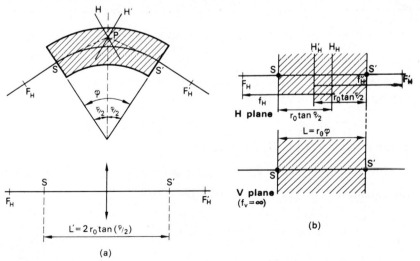

FIG. 115c. The special case of the homogeneous field prism (faces normal to (C_0)). (a) Thin lens representation. (b) Representation in terms of thick centred systems.

Equation (10A.30a) is the condition that P and Q be conjugate points. $T_{12} = 0$ implies that

$$\tan \phi = -\frac{(p+q)/r_0}{1 - pq/r_0^2}.$$ (10A.30b)

From Fig. 115d, we see that since $CS = CS' = r_0$, we have

$$\tan \phi = -\frac{\tan \phi_P + \tan \phi_Q}{1 - \tan \phi_P \tan \phi_Q}$$

$$= -\tan (\phi_P + \phi_Q),$$

and hence

$$\phi + \phi_P + \phi_Q = \pi.$$ (10A.31)

Thus the points P, C and Q are collinear (Barber's rule). Once P has been selected, and knowing T_0 and T_S, we have only to produce PC to obtain Q. *Stigmatic operation* is obtained when

$$p = q = r_0 \cot (\phi/2).$$

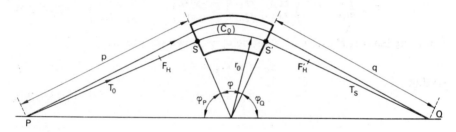

FIG. 115d. Correspondence between object and image in a homogeneous field prism (faces normal to (C_0)).

LINEAR MAGNIFICATION. Consider now a trajectory for which $x_p \neq 0$. Knowing that $T_{12} = 0$, we have

$$x_Q = T_{11} x_P.$$

The linear magnification, G_H, is given by

$$G_H = T_{11} = \cos \phi - q/R \sin \phi.$$ (10A.32a)

Using equation (10A.31), we can easily show that

$$G_H = -\frac{\cos \phi_P}{\cos \phi_Q}$$

$$= -\frac{CQ}{CP}.$$ (10A.32b)

If $p = q$, we have $G_H = -1$.

10A.1.3.3 Second special case: $n = \frac{1}{2}$

For this value of n, equations (10A.19) and (10A.20) are identical and the prism is *equivalent to a stigmatic, centred, rotationally symmetric system* (apart from the deflexion through an angle ϕ), for particles of the same velocity v. We now have

$$f' = -f = r_0\,\sqrt{2}\,\operatorname{cosec}(\phi/\sqrt{2}),$$

$$g' = -g = r_0\,\sqrt{2}\,\cot(\phi/\sqrt{2}). \tag{10A.33}$$

If $\phi = \pi/\sqrt{2} \simeq 127°$, the foci are situated at S and S', and $f' = r_0\sqrt{2}$. Symmetrical operation ($p = q$) is obtained for $|p-g| = |f|$, or

$$p = q = r_0\,\sqrt{2}\,\cot\left(\frac{1}{2}\frac{\phi}{\sqrt{2}}\right). \tag{10A.34}$$

We then have $G_H = G_V = -1$.

In the general case ($p \neq q$), the relation between conjugates takes the form

$$\tan(\phi/\sqrt{2}) = -\frac{(p+q)/r_0\sqrt{2}}{1 - pq/2r_0^2}$$

or

$$\frac{q}{r_0} = \frac{(p/r_0)\,\sqrt{2} + 2\tan(\phi/\sqrt{2})}{(p/r_0)\tan(\phi/\sqrt{2}) - \sqrt{2}}. \tag{10A.35}$$

Barber's rule still applies, provided we consider a fictitious system of angle $\phi' = \phi/\sqrt{2}$, of the same radius r_0, and divide p and q by $\sqrt{2}$. The corresponding points P' and Q' will be collinear with C.

For a given angle ϕ and the same radius of curvature r_0, a prism of index $n = \frac{1}{2}$ is *less convergent* in the H plane than the corresponding prism with homogeneous field, but has the advantage of convergence in the V plane.

10A.1.4 Method of Obtaining a Field of Index n

If the material of which the poles are composed has a very high permeability ($\mu \to \infty$), the surfaces of the poles are equipotentials. If each of the poles is excited by a coil of N turns carrying a current I, the magnetic potential will be given by $\pm\Phi_1 = \pm\mu_0 NI$. When the width of the gap is much less than r_0, we can use equation (10A.9) and write down the equation in x and z that gives the profile of the pole held at potential Φ_1:

$$-\frac{\Phi_1}{\Phi_{0m}} = (1 - nx + bx^2 - cx^3)z + \{(n-2b) - (n+2b-6c)x\}\frac{z^3}{3}.$$

At $x = 0$, the reduced width of the gap is $2h/r_0$. To first order, therefore,

$$-\Phi_1 = \frac{h}{r_0} \Phi_0 = B_0 h = \mu_0 NI \qquad (10A.36)$$

and

$$z = \frac{h}{r_0} (1 - nx)^{-1}. \qquad (10A.37)$$

In a first approximation, the field with index n will therefore be obtained with poles of hyperbolic cross-section, defined by equation (10A.37). In order to simplify their mechanical construction, the hyperbola is usually replaced by its tangent at the point $(x = 0, z = h)$; the pole pieces are then portions of cones of revolution, with meridian

$$z = \frac{h}{r_0} (1 + nx). \qquad (10A.38)$$

When $n = \frac{1}{2}$,

$$z_{(1/2)} = \frac{h}{r_0} (1 + x/2) \qquad (10A.39)$$

and the apex of the cone, C', is at $x = -2$, that is at a point distant $2r_0$ from (C_0) (Fig. 115e). The slope of the tangent is then

$$\tan \alpha = h/2r_0. \qquad (10A.40)$$

The index $n = \frac{1}{2}$ is only obtained in a narrow zone on either side of (C_0) and the prism must only be used with beams of small diameter.

In the special case for which $n = 2b$ (for example, $n = \frac{1}{2}$ and $b = \frac{1}{4}$), a rigorous solution of Laplace's equation, $\nabla^2 \Phi_m = 0$, can be found (Bosi, 1965, 1967):

$$-\frac{\Phi_m}{\Phi_0} = z\{1 - n \log (1 + x)\} . \qquad (10A.41)$$

The equation of the pole is then given by

$$z = -\frac{\Phi_1/\Phi_0}{1 - n \log (1 + x)} = -\frac{h}{1 - n \log (1 + x)}. \qquad (10A.42)$$

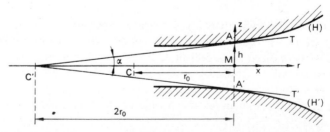

FIG. 115e. The pole cross-section that gives $n = \frac{1}{2}$ (H, H': hyperbolae; T, T': tangents at A, A').

This expression coincides with equation (10A.37) in first order, and at $x = 0$ equation (10A.36) is satisfied exactly.

10A.1.5 Properties of the Prism with Inclined Entry and Exit Faces

So far, we have assumed the mean trajectory $T_0 T_S$ to be normal to the entry and exit faces of the prism; the same is true to a first approximation of the other trajectories, in Gaussian optics. Let us now suppose that the entry and exit faces are turned (about S and S') through angles α and β respectively, with respect to their initial orientation. *By convention*, the angle of rotation is taken to be positive when the normal to the tilted face is outside the mean trajectory $T_0(C_0)T_S$ (Fig. 115f). For simplicity, we only consider here the case of the prism with homogeneous magnetic field ($n = 0$), and it is, in fact, extremely rare for prisms to be constructed with tilted faces and $n \neq 0$. To a first approximation, however, the results are true for any value of n.

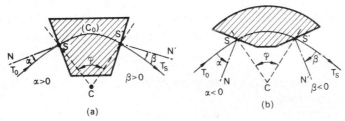

FIG. 115f. Tilted faces: definition of the signs of α and β.

10A.1.5.1 Trajectories in the median plane

Trajectories that arrive at the inclined entry face at a distance $r = r_0(1+x)$ with $x > 0$ ("external" trajectories) will experience the action of B_z over a distance L' which is less than $L = r_0\phi$ (when $\alpha > 0$); they will therefore be deflected through an angle $\phi_{ext} < \phi$. For "internal" trajectories ($x < 0$), on the other hand, we have $L' > L$ and hence $\phi_{int} > \phi$. The rotation of the faces will thus modify the convergence in the H plane. In order to calculate the effect of a rotation of the entry face through an angle α, we can reason as follows (Enge, 1967): when the face AB is rotated to $A_1 B_1$, a small prism SAA_1 of angle α (Fig. 115g) is added to the original prism (P), and in it the field B_z is in the same sense as in (P); simultaneously, a small prism SSB_1, also of angle α, is removed from (P). The new prism with tilted face is thus equivalent to the prism (P) provided that we *add* to the latter a double prism of angle α consisting of two symmetrical parts, SAA_1 and SBB_1. In SAA_1, the field B_z has the same ampli-

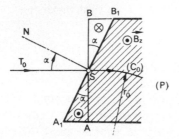

FIG. 115g. Passage from the prism (P)
with normal entry face (AB) to the prism
with inclined face (A_1B_1).

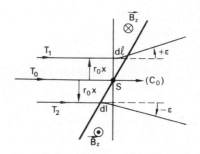

FIG. 115h. Deflexions of the incident
rays in the H plane.

tude and sense as the main field in (P), but in SBB_1, the field has the same
amplitude as that in (P) but is in the opposite direction.

If x is always much smaller than unity, the paths of the beams in the
small prisms will be short, so that to a first approximation the double
prism will be equivalent to a thin divergent lens, which causes only a
refraction (that is, a change of slope) ϵ, without affecting the position x.
A trajectory T_1, parallel to T_0 (Fig. 115h), travels a distance

$$dl = r_0 x \tan \alpha \qquad (10\text{A}.43)$$

in the prism. The radius of curvature of the trajectory is equal to r_0, since
$|B| = B_0$, and hence

$$r_0 \epsilon = r_0 x \tan \alpha,$$

so that

$$\epsilon = x \tan \alpha.$$

The trajectory T_2 experiences an equal and opposite refraction, $-\epsilon$. The
double prism is thus equivalent to a thin divergent lens of focal length

$$f_{EH} = -\frac{r_0 x}{\epsilon} = -r_0 \cot \alpha. \qquad (10\text{A}.44)$$

If $\alpha < 0$, the lens will be convergent.

Similarly the exit double prism will be equivalent to a lens of focal length

$$f_{SH} = -r_0 \cot \beta \quad (\beta > 0 \text{ or} < 0). \qquad (10\text{A}.45)$$

The transfer matrix of a thin lens of convergence $c = 1/f$ is of the form

$$\|T\| = \begin{pmatrix} 1 & 0 \\ -c & 1 \end{pmatrix}.$$

The total transfer matrix of the prism (P') with inclined end-faces is therefore given in the H plane by the product of three matrices:

$$\|H\| = \begin{pmatrix} 1 & 0 \\ \tan\beta & 1 \end{pmatrix} \begin{pmatrix} \cos\phi & \sin\phi \\ -\sin\phi & \cos\phi \end{pmatrix} \begin{pmatrix} 1 & 0 \\ \tan\alpha & 1 \end{pmatrix},$$

which yields

$$\|H\| = \begin{pmatrix} \dfrac{\cos(\phi-\alpha)}{\cos\alpha} & \sin\phi \\ -\dfrac{\sin\{\phi-(\alpha+\beta)\}}{\cos\alpha\cos\beta} & \dfrac{\cos(\phi-\beta)}{\cos\beta} \end{pmatrix}. \qquad (10A.46)$$

In the symmetrical situation for which $\alpha = \beta$, $\|H\|$ simplifies to

$$\|H\| = \begin{pmatrix} \dfrac{\cos(\phi-\alpha)}{\cos\alpha} & \sin\phi \\ -\dfrac{\sin(\phi-2\alpha)}{\cos^2\alpha} & \dfrac{\cos(\phi-\alpha)}{\cos\alpha} \end{pmatrix}. \qquad (10A.47)$$

When $\phi = \alpha + \beta$, the entry and exit planes are parallel; the prism (P') then behaves as a "parallel plate" and in the H plane, its convergence is zero, since $H_{21} = 0$. If $\alpha + \beta < \phi$, the prism is convergent in the H plane, and we have

$$f'_H = -f_H = \frac{r_0 \cos\alpha \cos\beta}{\sin\{\phi-(\alpha+\beta)\}},$$

$$g'_H = r_0 \frac{\cos\beta \cos(\phi-\alpha)}{\sin\{\phi-(\alpha+\beta)\}}, \qquad (10A.48)$$

$$g_H = -r_0 \frac{\cos\alpha \cos(\phi-\beta)}{\sin\{\phi-(\alpha+\beta)\}}.$$

$g_H = -g'_H$ when α and β are equal.

10A.1.5.2 Trajectories in the vertical plane

In the prism (P), where $\alpha = \beta = 0$, there are zones in the vicinity of the faces in which the field falls rapidly from $B = B_0$ to $B = 0$; the field lines become curved outside the prism (Fig. 115i). A component $B_\theta \neq 0$ appears, which has virtually no effect on the particles since B_θ is parallel to the velocity up to terms in second order (it must be taken into account in calculating the aberrations). Because of the symmetry, we still have $B_r = 0$, but if the faces are tilted, the field lines remain plane and normal to the surfaces of the poles; we thus find that B_r no longer vanishes. We thus write (in unscaled coordinates Z and s)

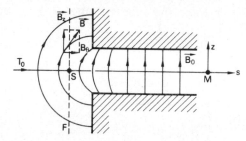

FIG. 115i. The appearance of the lines of force of B in the fringing field zone near the entry face. F: magnetic virtual face.

$$\frac{d^2Z}{dt^2} = \frac{e}{m} vB_r \quad \text{with} \quad v = ds/dt.$$

$$\frac{d^2Z}{ds^2} = -\frac{B_r}{r_0 B_0}. \tag{10A.49}$$

We do not know the expression for B_r, but we do know that

$$\text{curl } B = 0,$$

so that

$$\frac{\partial B_r}{\partial Z} = \frac{\partial B_z}{\partial r}. \tag{10A.50}$$

Integrating over the fringing field region, we obtain

$$\frac{\partial}{\partial Z}\int B_r \, ds = \frac{\partial}{\partial r}\int B_z \, ds.$$

In the (r, s) plane, we have characterized the action of this fringing field by that of the equivalent double prism containing the homogeneous field B_0 and at the entry, we have

$$\int B_z \, ds = B_0 r_0 x \tan \alpha \tag{10A.51}$$

according to equation (10A.43). Since $\partial r = r_0 \partial x$ and $\partial Z = r_0 \partial z$, we obtain

$$\frac{\partial}{\partial z}\int B_r \, ds = B_0 \tan \alpha$$

or

$$\int B_r \, ds = B_0 z \tan \alpha. \tag{10A.52}$$

In the vertical plane, an incident ray is thus refracted by an amount

$$\frac{dZ}{ds} = -\frac{1}{r_0 B_0}\int B_r \, ds = -\frac{Z}{r_0} \tan \alpha. \tag{10A.53}$$

The double prism is therefore equivalent to a thin lens in the V plane, of focal length

$$f_{EV} = -\frac{Z}{dZ/ds} = r_0 \cot \alpha = -f_{EH}. \tag{10A.54}$$

Similarly, at the exit plane we have

$$f_{SV} = r_0 \cot \beta = -f_{SH}. \tag{10A.55}$$

The relations (10A.44, 10A.45) and (10A.54, 10A.55) remain valid to a first approximation when $n \neq 0$.

A tilted face thus has the same effect as a thin strong-focusing lens: divergent in the H plane and convergent in the V plane (or conversely, depending on the sign of α or β). This brings out the reason for this modification to the end faces when $n = 0$: the prism with homogeneous field, which is much the easiest to construct mechanically, becomes convergent in the V plane and remains convergent in the H plane. These features were foreseen by Cotte as early as 1938. The transfer matrix $\|V\|$ therefore becomes

$$\|V\| = \begin{pmatrix} 1 & 0 \\ -\tan \beta & 1 \end{pmatrix} \begin{pmatrix} 1 & \phi \\ 0 & 1 \end{pmatrix} \begin{pmatrix} 1 & 0 \\ -\tan \alpha & 1 \end{pmatrix}$$

$$\tag{10A.56}$$

$$= \begin{pmatrix} 1 - \phi \tan \alpha & \phi \\ \phi \tan \alpha \tan \beta - (\tan \alpha + \tan \beta) & 1 - \phi \tan \beta \end{pmatrix}.$$

We therefore have

$$f_V' = -f_V = \frac{r_0}{\tan \alpha + \tan \beta - \phi \tan \alpha \tan \beta},$$

$$g_V' = \frac{r_0(1 - \phi \tan \alpha)}{\tan \alpha + \tan \beta - \phi \tan \alpha \tan \beta}, \tag{10A.57}$$

$$g_V = -\frac{r_0(1 - \phi \tan \beta)}{\tan \alpha + \tan \beta - \phi \tan \alpha \tan \beta}.$$

10A.1.5.3 Stigmatic operation

With a homogeneous field prism with tilted faces, we can achieve stigmatic operation but, unlike the case $n = \frac{1}{2}$, the system is stigmatic for only one particular pair of conjugate planes, since the cardinal elements (f and g) are different in the planes H and V.

In the case of a symmetric system ($\alpha = \beta$, $p = q$), we can easily calculate the values of $\tan \alpha$ and p/r_0 for which this type of operation obtains.

We establish the overall transfer matrices $\|T_H\|$ and $\|T_V\|$ (see equation 10A.29) and then write down the conjugacy relation for each:

$$T_{H12} = 0, \tag{10A.58}$$

$$T_{V12} = 0. \tag{10A.59}$$

From equation (10A.59), we derive

$$\tan \alpha = r_0/p. \tag{10A.60}$$

Substituting (10A.60) into (10A.58), we finally obtain

$$p/r_0 = 2 \cot (\phi/2),$$

$$\tan \alpha = \tfrac{1}{2} \tan (\phi/2). \tag{10A.61}$$

The magnification G is -1 in both planes.

EXAMPLE: *The prism with deflexion* $\pi/2$. The matrices $\|H\|$ and $\|V\|$ take the form

$$\|H\| = \begin{pmatrix} \tan \alpha & 1 \\ -(1 - \tan^2 \alpha) & \tan \alpha \end{pmatrix},$$

$$\|V\| = \begin{pmatrix} 1 - \dfrac{\pi}{2} \tan \alpha & \dfrac{\pi}{2} \\ -\tan \alpha \left(2 - \dfrac{\pi}{2} \tan \alpha\right) & 1 - \dfrac{\pi}{2} \tan \alpha \end{pmatrix}, \tag{10A.62}$$

and equation (10A.61) gives

$$\tan \alpha = \tfrac{1}{2} \quad (\text{or} \quad \alpha \simeq 26°34'), \tag{10A.63}$$

$$p = 2r_0.$$

A device incorporating a stigmatic prism of this type ($\phi = \pi/2$) has been successfully used by Castaing and Slodzian (1962a, b) to transport the image in ion microscopy and to act as a mass selector simultaneously.

10A.1.6 The Effect of the Fringing Fields

10A.1.6.1 In the H plane

In a real magnet, the trajectories do not enter the region containing the field B_z abruptly; a particle of momentum p_0 is progressively deflected, and the radius of curvature of the trajectory decreases from $r = \infty$ to $r = r_0$ (if $B = B_0$). The trajectory that follows the circle (C_0) inside the prism does not coincide with T_0 and T_S outside, but corresponds instead to the trajectories T'_0 and T'_S (Fig. 115j). A more detailed analysis shows

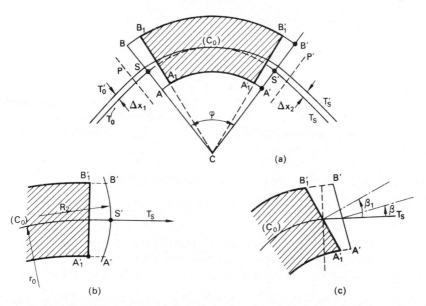

FIG. 115j. The effect of the fringing field in the H plane. (a) Shift of the trajectories. A_1B_1 and $A_1'B_1'$ are the mechanical faces and $AB,A'B'$ are the magnetic faces that form the edges of the rectangular model; in reality, the fringing fields extend to PP'. (b) Curvature of the magnetic face. (c): misorientation of the magnetic virtual face when the mechanical face is tilted, $\beta < \beta_1$.

that T_0' and T_S' are parallel to T_0 and T_S respectively, but are slightly shifted towards the centre of curvature C by an amount $\triangle x_1$. The effect can be corrected by slightly altering the relative positions of the prism and the ideal path selected initially. With the new curved axis $T_0'(C_0)T_S'$, the formulae derived above are directly applicable between the virtual magnetic faces AB and $A'B'$. Furthermore, the virtual magnetic faces obtained from measurements of the magnetic field distribution by the method suggested in §10A.1.1 are slightly curved, since the dimensions of the poles in the radial direction are finite, and are convex towards the outside of the prism (Fig. 115k). This curvature introduces aberrations.

10A.1.6.2 In the vertical plane

When the entry and exit faces $(AB, A'B')$, which are assumed to be of infinite extent in the radial direction, are inclined $(\alpha, \beta \neq 0)$, it can be shown by calculating the trajectories rigorously (Enge, 1964) that the convergence of the prism in the V plane is slightly weaker than that predicted by the formulae given above. This error can be compensated by increasing the angles α, β of the mechanical faces to values α', β' slightly larger than those given by the theory. When the width of the poles is small, on the other hand (Fig. 115k), the rotation of the magnetic face is less than that of the mechanical face, and this effect becomes greater as r_0 is made smaller. The angle. β_1, of the mechanical faces must therefore be adjusted in such a way that the desired rotation β of the magnetic faces is achieved. Accurate magnetic measurements are necessary when designing a prism with tilted faces. Alternatively, the magnet may be equipped with rotatable mechanical faces; this is particularly convenient when rigorously stigmatic operation is required.

10A.1.7 The Dispersive Properties of a Magnetic Prism

10A.1.7.1 Trajectories and transfer matrices for particles of momentum $p_0 + \triangle p$

Hitherto, we have been considering particles of charge e, all having the same momentum $p_0 = mv_0 = eB_0r_0$. We now consider a particle incident along the trajectory T_0, entering the prism at S, but having momentum

$$p = p_0 + \triangle p = p_0(1 + \triangle p/p_0) = p_0(1 + \delta),$$

where $\delta \ll 1$. In a prism of arbitrary index n, the circular equilibrium trajectory corresponding to this momentum will be a circle (C) of radius $r = r_0 + dr$, such that

$$(r_0 + dr)B_z = p_0(1 + \delta)/e \qquad (10A.64)$$

and

$$B_0r_0(1 + x)(1 - nx) = p_0(1 + \delta)/e,$$

$$x = \frac{\delta}{1 - n}. \qquad (10A.65)$$

With respect to the *new curved axis* (C), defined by these conditions, the initial values where the particle enters the prism are as follows:

$$x_0 = -\delta/(1 - n), \quad x_0' = 0.$$

The transfer matrix $\|H\|$ of the prism is again given by equation (10A.23). The trajectory is therefore of the form

$$x_{(C)} = x_0 \cos(\sqrt{1 - n}\,\theta), \qquad (10A.66)$$
$$x_{(C)}' = -x_0\sqrt{1 - n} \sin(\sqrt{1 - n}\,\theta).$$

If we now measure the abscissa, x, with respect to the curved axis (C_0) corresponding to momentum p_0, we find at the exit of the prism

$$x_S = \{1 - \cos(\sqrt{1 - n}\,\phi)\}\frac{\delta}{1 - n},$$

$$x_S' = \frac{\delta}{\sqrt{1 - n}} \sin(\sqrt{1 - n}\,\phi). \qquad (10A.67)$$

We can thus extend the matrix $\|H\|$ to include the effect on x_S and x'_S of a variation $\triangle p/p_0 = \delta$; for $\alpha = \beta = 0$, we may write

$$\begin{pmatrix} x_S \\ x'_S \\ \delta \end{pmatrix} = \|H\| \begin{pmatrix} x_0 \\ x'_0 \\ \delta \end{pmatrix} = \begin{pmatrix} H_{11} & H_{12} & H_{13} \\ H_{21} & H_{22} & H_{23} \\ H_{31} & H_{32} & H_{33} \end{pmatrix} \begin{pmatrix} x_0 \\ x'_0 \\ \delta \end{pmatrix}, \tag{10A.68}$$

with

$$H_{11} = \cos\left(\sqrt{1-n}\,\phi\right), \quad H_{12} = \frac{1}{\sqrt{1-n}} \sin\left(\sqrt{1-n}\,\phi\right)$$

$$H_{13} = \frac{1}{1-n}\{1 - \cos\left(\sqrt{1-n}\,\phi\right)\}$$

$$H_{21} = -\sqrt{1-n} \sin\left(\sqrt{1-n}\,\phi\right), \quad H_{22} = \cos\left(\sqrt{1-n}\,\phi\right) \tag{10A.69}$$

$$H_{23} = \frac{1}{\sqrt{1-n}} \sin\left(\sqrt{1-n}\,\phi\right)$$

$$H_{31} = H_{32} = 0, \quad H_{33} = 1$$

When the entry and exit faces are tilted $(\alpha, \beta \neq 0)$, we find

$$H_{11} = \cos\left(\sqrt{1-n}\,\phi\right) + \frac{1}{\sqrt{1-n}} \tan\alpha \sin\left(\sqrt{1-n}\,\phi\right)$$

$$\begin{aligned} H_{21} = {} & \cos\left(\sqrt{1-n}\,\phi\right)(\tan\alpha + \tan\beta) \\ & - \sqrt{1-n} \sin\left(\sqrt{1-n}\,\phi\right)\left(1 - \frac{1}{1-n}\tan\alpha\tan\beta\right) \end{aligned} \tag{10A.70}$$

$$H_{22} = \cos\left(\sqrt{1-n}\,\phi\right) + \frac{1}{\sqrt{1-n}} \tan\beta \sin\left(\sqrt{1-n}\,\phi\right)$$

$$H_{23} = \frac{1}{\sqrt{1-n}} \sin\left(\sqrt{1-n}\,\phi\right) + \frac{1}{1-n}\{1 - \cos\left(\sqrt{1-n}\,\phi\right)\}.$$

$H_{12}, H_{13}, H_{31}, H_{32}$ and H_{33} are unchanged (see 10A.69).

When $n = 0$ (homogeneous field), $\|H\|$ simplifies:

$$\|H\| \atop (n=0) = \begin{pmatrix} \dfrac{\cos(\phi-\alpha)}{\cos\alpha} & \sin\phi & 1-\cos\phi \\[2mm] -\dfrac{\sin\{\phi-(\alpha+\beta)\}}{\cos\alpha\cos\beta} & \dfrac{\cos(\phi-\beta)}{\cos\beta} & \sin\phi + (1-\cos\phi)\tan\beta \\[2mm] 0 & 0 & 1 \end{pmatrix}.$$

$$\tag{10A.71}$$

The transfer matrix in the V plane is unaltered, so far as the first order properties are concerned at least.

10A.1.7.2 Dispersion

10A.1.7.2.1 Definitions. The dispersive power of a prism—its ability to separate (in the image plane) particles of slightly different momenta originating in the same object point on the axis ($x_0 = 0$) by an amount x_i (Fig. 115k)—is characterized by a dimensionless quantity $\mathscr{D}_p$:

$$\mathscr{D}_p = x_i \frac{p_0}{\Delta p} = \frac{x_i}{\delta}. \qquad (10\text{A}.72)$$

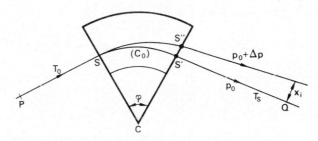

FIG. 115k. Dispersion of the prism.

Writing $\overline{SP} = p$ and $\overline{S'Q} = q$, the total transfer matrix between P and Q takes the form

$$\|T\| = \begin{pmatrix} 1 & q/r_0 & 0 \\ 0 & 1 & 0 \\ 0 & 0 & 1 \end{pmatrix} \|H\| \begin{pmatrix} 1 & p/r_0 & 0 \\ 0 & 1 & 0 \\ 0 & 0 & 1 \end{pmatrix}$$

$$= \begin{pmatrix} T_{11} & T_{12} & T_{13} \\ T_{21} & T_{22} & T_{23} \\ T_{31} & T_{32} & T_{33} \end{pmatrix}.$$

Since P and Q are conjugates,

$$T_{12} = 0,$$

and hence

$$\frac{p}{r_0} H_{11} + \frac{q}{r_0} H_{22} + \frac{pq}{r_0^2} H_{21} + H_{12} = 0, \qquad (10\text{A}.73)$$

where H_{ij} are the elements of $\|H\|$. In the image plane Q, we have

$$x_i = x_0 T_{11} + x_0' T_{12} + \delta T_{13},$$

which reduces to

$$x_i = \delta T_{13}.$$

We therefore find

$$\mathscr{D}_p = T_{13} = H_{13} + \frac{q}{r_0} H_{23}. \qquad (10\text{A}.74)$$

We recall that the linear magnification in the H plane is given by

$$G = T_{11} = H_{11} + \frac{q}{r_0} H_{21}. \tag{10A.75}$$

Examining the matrices $\|H\|$ given in equation (10A.69), corresponding to systems with no inclination of the entry and exit faces, we see that

$$H_{13} = \frac{1}{1-n}(1 - H_{11}), \, H_{23} = H_{12} = -\frac{1}{1-n} H_{21},$$

and hence

$$\mathscr{D}_p = H_{13} + \frac{q}{r_0} H_{23} = \frac{1}{1-n}\left\{1 - \left(H_{11} + \frac{q}{r_0} H_{21}\right)\right\}$$

or

$$\mathscr{D}_p = \frac{1}{1-n}(1 - G_H). \tag{10A.76}$$

This extremely useful relation was proved analytically by H. Bruck (1958).

10A.1.7.2.2 Special cases. We assume symmetric conditions ($p = q$, $G = -1$).

(i) Homogeneous field prism ($n = 0$). When the faces are normal to T_0 ($\alpha = \beta = 0$), we have $p = q = r_0 \cot(\phi/2)$ and we find that

$$\mathscr{D}_p = 2. \tag{10A.77}$$

When the faces are tilted and $\alpha = \beta$, $\tan\alpha = \frac{1}{2}\tan(\phi/2)$ and $p = q = 2\cot(\phi/2)$. We find

$$\mathscr{D}_p = 4. \tag{10A.78}$$

The dispersion is thus doubled by the use of inclined faces.

(ii) Prisms of non-zero index. For $\alpha = \beta = 0$, we obtain

$$\mathscr{D}_p = \frac{2}{1-n},$$

so that for $n = 1/2$,

$$\mathscr{D}_p = 4. \tag{10A.79}$$

In all these special cases, $\mathscr{D}_p$ is independent of ϕ.

10A.1.7.2.3 The non-dispersive system. When we wish simply to deflect a beam consisting of particles of different momenta and focus it bodily without dispersion, a prism (or a more complicated optical system) for which the term T_{13} of the total transfer matrix vanishes must be employed. With $n = 0$ and $\alpha = \beta = 0$, this condition implies $\phi = 2\pi$ so that there would be no overall deflection. With $\alpha \neq \beta$ ($n = 0$), however, non-dispersive prisms can be obtained. Figure 115 l, for example, shows

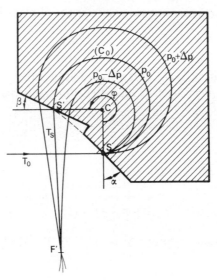

FIG. 115 l. Example of a single achromatic prism with triple focusing (in H, V and momentum).

a solution to this problem (net deflection of 90°) given by Enge (1967): a beam incident parallel to T_0 is focused in both the H and the V planes, and also in energy, at the image focus F'. The parameters of the system are then $\phi = 270°$, $\alpha = 45°$, $\beta = 32°40'$, $p = \infty$ and $q = 2\cdot74\,r_0$.

10A.1.7.2.4 Mass dispersion. When the incident beam consists of ions of different masses, all carrying the same charge e and accelerated through the same potential V_0, the different masses can be separated with the aid of a prism. A mass dispersion, $\mathscr{D}_m$, may be defined by the relation

$$\mathscr{D}_m = \frac{m_0}{\delta m}, \tag{10A.80}$$

where δm is the difference in mass between two ions of similar mass in the vicinity of the mass m_0. If $\delta m \ll m_0$, the relation $mv = p = \sqrt{2emV_0}$ implies

$$\frac{\delta p}{p_0} = \frac{1}{2}\frac{\delta m}{m_0},$$

and hence

$$\mathscr{D}_m = \tfrac{1}{2}\mathscr{D}_p. \tag{10A.81}$$

10A.1.7.2.5 Theoretical resolving powers R_p and R_m. In a prism free of aberrations, all particles of the same momentum p_0, from a slit source (object) of reduced width a_0/r_0 in the H plane and situated at P, will pass through the exit slit of reduced width a_i/r_0 situated at Q (the image of P);

we have

$$a_i = |G_H| a_0.$$

In a first approximation, the particles of momentum $p_i = p_0 + \triangle p$ will be focused over an image of the same width a_i but situated at a distance $x_i = \mathscr{D}_p(\triangle p/p_0)$ from the axis (Fig. 115m).

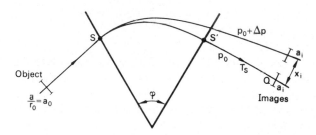

FIG. 115m. Images when the particles have different momenta.

If we wish to record particles of momentum p_0 and those of momentum $p_0 + \triangle p_i$ simultaneously on a double collector, we can establish the minimum value of $\triangle p$, $\triangle p_{min}$, such that no particle of momentum $p < p_0 - \triangle p_{min}$ or $p > p_0 + \triangle p_{min}$ passes through the image slit of width a_i lying on the axis. We then have $a_i = x_i$ and

$$\mathscr{D}_p \frac{\triangle p_{min}}{p_0} = |G_H| a_0.$$

The theoretical resolving power, in the first order approximation, is then defined by

$$R_p = \frac{p_0}{\triangle p_{min}} = \frac{\mathscr{D}_p}{|G_H| a_0}. \qquad (10A.82)$$

If, on the other hand, we have a single collector behind a slit of reduced width $s_0 = s/r_0$ placed on the axis (and this case occurs more frequently), we may pass particles of different momenta across this slit (by varying B_0); particles of momentum p_0, focused into an image of width a_i, cannot all be completely separated from particles of momentum $p_0 + \triangle p$ unless (Fig. 115m)

$$x_i = a_i + s_o$$

or

$$\mathscr{D}_p \frac{\triangle p_i}{p_0} = |G_H| a_0 + s_0.$$

Hence,

$$R_p = \frac{\mathscr{D}_p}{|G_H| a_0 + s_0} \qquad (10A.83)$$

since the electric signal recorded by the collector C is trapezoidal and its width is $a_i + s_0$.

When a source of given width a is used, it will be advantageous to increase $\mathcal{D}_p$, the ratio $\mathcal{D}_p/|G_H|$ and the radius r_0 and to reduce s, since in real space we have

$$R_p = \frac{r_0 \mathcal{D}_p}{|G_H| a + s} \qquad (10A.84)$$

(where $s = r_0 s_0$ and $a = r_0 a_0$).

Formulae (10A.83) and (10A.84) are used to calculate the theoretical resolving power in first order (which is valid, as we shall see, when the angular aperture of the beam is very small: less than 10^{-2}, say).

EXAMPLES. For a homogeneous field prism, in the symmetrical case ($|G| = 1$), we find:

$$\alpha = \beta = 0 \quad R_p = \frac{2r_0}{a+s},$$

$$\alpha = \beta \neq 0 \quad R_p = \frac{4r_0}{a+s}.$$

For $n = \frac{1}{2}$, $\alpha = \beta = 0$, in the symmetrical case:

$$R_p = \frac{4r_0}{a+s}.$$

When the prism is used for mass separation, we can define a corresponding resolving power:

$$R_m = \frac{m_0}{\Delta m}$$

and, from equation (10A.81), we have

$$R_m = \tfrac{1}{2} R_p. \qquad (10A.85)$$

In reality, the width of the image in the H plane is not equal to $a_i = G_H a_0$, for the aberrations of the prism must be taken into account, especially the second order aberrations, which are the most important.

10A.1.8 Aberrations of Magnetic Prisms

We have obtained the first order optical properties with the aid of a rectangular field model (taking edge effects into account when necessary). We have, however, neglected

   (i) terms of higher order in the expansions of B_z and B_x;
  (ii) transverse velocities v_z and v_x;

(iii) the effect of the component B_θ that is present in the stray fields and which affects v_z and v_x;

(iv) the curvature of the virtual magnetic faces.

The dominant aberration terms that appear when we take all these factors into account are of second order, and they have been extensively studied both for the homogeneous field prism ($n = 0$) with normal and tilted faces and for the case where the index $n \neq 0$. Just as for lenses, the most important aberration, the defect that limits the resolving power of the prism, is the aperture aberration, or more exactly the aperture aberration in the H plane. Unlike round lenses, however, for which the (third order) spherical aberration cannot be removed, the second order aperture aberration of magnetic prisms can be eliminated in a relatively simple way. Some attention has also been devoted to the aberrations of higher order.

The aperture aberration in the H plane can be calculated easily in the simple case of the 180° prism (Fig. 115n). The trajectory (C_0), centred on C and of radius r_0, arrives at A' (the Gaussian image). The trajectories (C_1) and (C_2) corresponding to particles of the same momentum p_0 but initially inclined to (C_0) at angles $\pm\gamma_0$ intersect CA at A'', between C and A'. We have

$$AA'' = 2r_0 \cos \gamma_0 \simeq 2r_0(1 - \gamma_0^2/2).$$

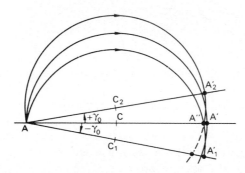

FIG. 115n. Second order aberration in a homogeneous field prism with $\phi = 180°$.

In this special case, therefore, the aberration $A'A''$ is given by

$$A'A'' = r_0\gamma_0^2. \tag{10A.86}$$

The reader will find a detailed study of the second order aberrations in the articles by Hintenberger (1948, 1949, 1951), Kerwin (1949), Ikegami (1958), Tasman and Boerboom (1959), Wachsmuth, Tasman and Boerboom (1959), Tasman, Boerboom and Wachsmuth (1959), Bretscher (1960), Brown, Belbeoch and Bounin (1964) and Enge (1967). Here, we confine

ourselves to the principal results, usually obtained by applying a perturbation method to the trajectories; we have already used this technique
in the sections on round lens aberrations.

We first consider an ideal prism of angle ϕ, limited at S and S' by two
plane faces F and F' (the virtual magnetic faces), in which the mean
trajectory is (C_0). A second trajectory falls on the entry face F with initial
conditions x_1, x_1', z_1, z_1', δ. The corresponding quantities in the plane F' of
the exit face can be obtained to a first order approximation by means of
the transfer matrices $\|H\|$ and $\|V\|$. In reality, however, the trajectory
intersects F' at a point $x_2 + \triangle x_2$, $z_2 + \triangle z_2$ with slope $x_2' + \triangle x_2'$, $z_2' + \triangle z_2'$ (Fig.
115o). The differences $\triangle x_2 \ldots \triangle z_2'$ may be expanded, and if we consider
only terms of second order, we obtain

$$\triangle x_2 = a_1 x_1^2 + a_2 x_1 x_1' + a_3 x_1'^2 + a_4 z_1^2$$
$$+ a_5 z_1 z_1' + a_6 z_1'^2 + a_7 x_1 \delta + a_8 x_1' \delta + a_9 \delta^2,$$
$$\triangle z_2 = c_1 x_1 z_1 + c_2 x_1' z_1 + c_3 x_1 z_1' + c_4 z_1'^2$$
$$+ c_5 z_1 \delta + c_6 z_1' \delta. \tag{10A.87}$$

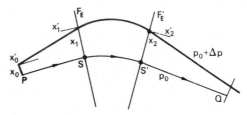

FIG. 115o. Definition of the notation: ideal prism limited at F_E and F_E'.

The expansions for $\triangle x_2'$ and $\triangle z_2'$ are similar to those for $\triangle x_2$ and $\triangle z_2$ except
that the coefficients $a_1 \ldots a_9$ and $c_1 \ldots c_6$ are replaced by $b_1 \ldots b_9$ and
$d_1 \ldots d_6$ respectively. A consequence of the symmetry about the median
plane is that there are only thirty second order aberration coefficients; the
terms independent of δ are the geometrical aberrations (20 coefficients).

For a flat monochromatic beam ($z_1 = z_1' = \delta = 0$) emerging from the point
P on T_0, the aperture aberration term is given by $\triangle x_2 = a_3 x_1'^2$ in the exit
plane of the prism.

All the coefficients a_k, b_k, c_k and d_k can be calculated by integrating
the equations of motion, in which second order terms arising from B_z,
B_x and the velocities v_x and v_z have been retained.

As the second stage, we calculate the trajectories in the entry and
exit zones more accurately. For this, we assume that the magnetic entry
and exit faces passing through S and S' (F_E and F_E') are curved (with
radii of curvature R_1 and R_2) and tilted at angles α and β (these are the

angles at S and S'). The real ray (Fig. 115p(a)), which impinges on the plane of the flat face F at x_F with slope x'_F crosses F_E at x_E (with slope $x'_E = x'_F$); it then follows a curved path (C). If the prism were in fact bounded by the face F, the trajectory (C) would correspond to a particle emerging from P' and impinging on F at x_i with slope x'_1. The parameters x_1 and x'_1 are then calculated as a function of x_E, x'_E, α and R_1.

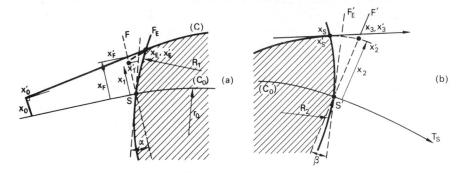

FIG. 115p. Fringing field zone, with tilted curved faces; (a) entry; (b) exit.

When the particle leaves the magnet (Fig. 115p(b)), the trajectory intersects the real face F'_E at x_S (slope x'_S) and then continues in a straight line to F', which it crosses at x_3 with slope x'_3 (Fig. 115p(b)). If the prism ended at F', the curve (C) would cut this face at x_2 (with slope x'_2). It is possible to calculate x_S and x'_S as a function of x_2 and x'_2.

The first calculation, on the prism with normal plane faces FF', allows us to relate the entry parameters, x_1 and x'_1, to the exit parameters, x_2 and x'_2, however; it is therefore possible to express x_S and x'_S as a function of x_E and x'_E. The same procedure can be followed in the vertical plane for z_S and z'_S. In first order, we recover the transfer matrices $\|H\|$ and $\|V\|$ that we have met above (these will contain supplementary terms involving the curvatures $1/R_1$ and $1/R_2$ of the entry and exit faces).

The values in the Gaussian image plane, situated at a distance q from S, are then calculated from x_S, x'_S, z_S and z'_S. We know that

$$x_E = x_0 + \frac{p}{r_0} x'_0, \quad x'_E = x'_0,$$

$$z_E = z_0 + \frac{p}{r_0} z'_0, \quad z'_E = z'_0$$

$$(10A.88)$$

(the suffix 0 corresponds to the object at P, distance p from the magnet) and at Q we have

$$x_i = x_S + \frac{q}{r_0} x'_S.$$

$$(10A.89)$$

Equation (10A.89) may be put in the form

$$x_i = G_H x_0 + M_1 x_0' + M_2 \delta + \triangle x_i,$$

$$(10A.90)$$

$$z_1 = G_V z_0 + N_1 z_0' + \triangle z_i,$$

where $\triangle x_i$ and $\triangle z_i$ are second order terms.

In the Gaussian image plane, the relation expressing the fact that P and Q are conjugates in the H plane implies that $M_1 = 0$. When there is double focusing in first order, we have $M_1 = 0$ and $N_1 = 0$ simultaneously. The dispersion is given by M_2. The term $\triangle x_i$ is of the form

$$\triangle x_i = M_{11} x_0^2 + M_{12} x_0 x_0' + M_{13} x_0'^2 + M_{14} z_0^2 + M_{15} z_0 z_0'$$

$$(10A.91)$$

$$+ M_{16} z_0'^2 + N_{11} x_0 \delta + N_{12} x_0' \delta + N_{13} \delta^2.$$

The M_{1k} correspond to geometrical aberrations and the N_{1j} to chromatic aberrations.

In devices (such as mass or velocity spectrographs) in which a high resolving power is required in the H plane, sources (objects) that are very narrow in the x-direction are employed; the terms in x_0^2 and $x_0 x_0'$ can then be neglected. Furthermore, the optics can be designed so that z_0 is also small, and that the incident beam is parallel to T_0 in the V plane ($z_0' = 0$); the principal aberration term is then given by

$$\triangle x_i \simeq M_{13} x_0'^2. \qquad (10A.92)$$

The coefficient M_{13} depends on the radii of curvature, R_1 and R_2, of the faces of the prism; these curvatures can be selected in such a way that the coefficient vanishes (Hintenberger, 1949). In the case of the prism with homogeneous field, we arrive at an expression of the form

$$M_{13} = r_0 \left(\frac{\gamma_1}{R_1} + \frac{\gamma_2}{R_2} \right) - (C_1 + C_2)$$

with

$$\gamma_1 = \frac{1}{\cos^3 \alpha \, \{1 + (r_0/p + \tan \alpha)^2\}^{3/2}},$$

$$\gamma_2 = \frac{1}{\cos^3 \beta \, \{1 + (r_0/q + \tan \beta)^2\}^{3/2}},$$

$$(10A.93)$$

$$C_1 = \frac{r_0^2}{p^2} \frac{r_0/p + 3 \tan \alpha}{\{1 + (r_0/p + \tan \alpha)^2\}^{3/2}},$$

$$C_2 = \frac{r_0^2}{q^2} \frac{r_0/q + 3 \tan \beta}{\{1 + (r_0/q + \tan \beta)^2\}^{3/2}};$$

p and q are related by the first order conjugacy condition. If the entry and exit faces are plane $(R_1 = R_2 = \infty)$, we cannot achieve $M_{13} = 0$ except for particular values of p and q:

$$\frac{r_0}{p} = -3 \tan \alpha \quad \text{and} \quad \frac{r_0}{q} = -3 \tan \beta. \tag{10A.94}$$

In the symmetric case $(p = q, \alpha = \beta)$, we should then have

$$\tan \alpha = -\tfrac{1}{2} \tan (\phi/2),$$

$$\frac{p}{r_0} = \tfrac{2}{3} \cot (\phi/2). \tag{10A.95}$$

The faces must thus be tilted *towards the interior* of the prism, which increases the horizontal convergence in H but renders the V plane divergent. (We recall that a doubly convergent system is obtained for $\tan \alpha = \tfrac{1}{2} \tan (\phi/2), p/r_0 = 2 \cot (\phi/2).$)

More generally, with $\alpha, \beta \neq 0$ but again with $p = q$, we obtain

$$r_0 \left(\frac{1}{R_1} + \frac{1}{R_2}\right) = 2 \frac{r_0^2}{p^2} \left(\frac{r_0}{p} + 3 \tan \alpha\right) \cos^3 \alpha. \tag{10A.96}$$

The radii R_1 and R_2 and the tilt α can then be chosen arbitrarily, subject to this condition.

For prisms in which $n \neq 0$, relations of the form (10A.93) or (10A.96) are again obtained, and it is possible to remove M_{13} as before.

If we now consider rays from a point object P lying in the median plane $(x_0 = z_0 = 0)$, we have

$$\triangle x_i = M_{13} x_0'^2 + M_{16} z_0'^2 \tag{10A.97}$$

(neglecting the chromatic terms). The first term represents the aperture aberration (broadening of the image), which is independent of the height z_i of the point at which the trajectory falls on the Gaussian image plane. The second term, on the other hand, vanishes when $z_i = 0$, and is only present outside the median plane H; it corresponds to curvature of the image (which in first order would be a segment of an infinitesimally thin straight line). In first order, we have

$$z_i \simeq K z_0',$$

so that

$$\triangle x_i \propto z_i^2.$$

The image becomes a parabolic segment, symmetrical about the median plane, and $M_{13} x_0'^2$ wide.

RESOLVING POWER. If we take only the principal (aperture) aberration term into account, the resolving power will be given by

$$R_p = \frac{\mathcal{D}_p}{|G|a_0 + s_0 + |M_{13}|x_0'^2} \qquad (10A.98)$$

and

$$R_m = \tfrac{1}{2} R_p.$$

EXAMPLE: *The prism with deflection* $\pi/2$. The matrices $\|H\|$ and $\|V\|$ take the form

$$R_p = \frac{2r_0}{a + s + r_0 x_0'^2}.$$

Unsymmetrical conditions $(G_H \neq -1)$. It can be shown (Bruck, 1958) that

$$|M_{13}| = \tfrac{1}{2}(|G_H| + 1/G_H^2), \qquad (10A.99)$$

and hence

$$R_p = \frac{r_0(1 + |G_H|)}{a + s + \tfrac{1}{2} r_0(|G_H| + 1/G_H^2)x_0'^2}. \qquad (10A.100)$$

10A.2 ELECTROSTATIC PRISMS

10A.2.1 Description

An electrostatic prism (Herzog, 1934) consists of a portion of a condenser with two metal electrodes, having an axis of revolution Z and a plane of mechanical *and electrical* symmetry. This symmetry plane will be called the horizontal or H plane, as before. The electrodes coincide with part of the surface of a torus of revolution with axis $\overrightarrow{CZ}$ and they therefore have a double curvature: in the H plane and in a vertical plane $(\mathbf{r}, \mathbf{Z})$. The angle of the prism (the angle of deflexion) is ϕ and the mean radius of curvature in the H plane is r_0.

Figure 115q shows a section through the prism by the H plane, and by the radial plane $\theta = 0$. The radii of curvature of the electrodes in the H plane are denoted by r_i and r_e and those in the radial plane by R_i and R_e. We restrict the discussion to the simple case in which the signs of R_i and R_e are the same and the corresponding centres of curvature are coincident at C_1. We also assume that the separation of the electrodes, $\triangle r = R_e - R_i = r_e - r_i$, is small in comparison with r_i and R_i.

The electrodes are held at potentials $\pm V_1$, symmetric with respect to earth $(V = 0)$; the mean equipotential surface, which intersects H along a circle (C_0) of radius r_0, corresponds to $V = 0$. If we take the origin of potential at the source which is emitting accelerated particles at voltage Φ_0, the potential on (C_0) will be equal to Φ_0 (and the electrode voltages

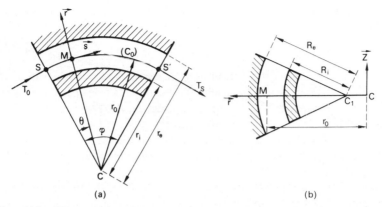

(a)   (b)

FIG. 115q. Electrostatic prism. (a) Section by a horizontal plane H. (b) Section
by a radial vertical plane.

will be $\Phi_0 \pm V_1$). With our assumption that $(r_e - r_i)/r_0 \ll 1$, we have $r_0 \simeq$
$(r_e + r_i)/2$ and $R_0 \simeq (R_e + R_i)/2$.

As a first approximation, we assume that the electric field vanishes outside the prism and is constant within it, for $0 < \theta < \phi$, when we follow one
of the family of circles (C) of radius r centred on $\overrightarrow{CZ}$. In the H plane, this
field is wholly radial. We shall thus be using a rectangular model, whereby
the prism is bounded not by the edges of the real electrodes but by virtual
electric faces situated outside the prism. In order to obtain the distance l
between the edges of the electrodes and this virtual face (Fig. 115r), the
real distribution of E_r in the H plane must be established, either by calculation or by means of an electrolytic tank. When (earthed) metal slits or
diaphragms are placed at the entry and exit of the prism, the stray fields
will be altered; the distribution $E_r(\theta)$ must be determined for the complete

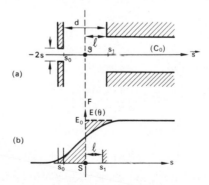

FIG. 115r. Entry zone of an electrostatic prism with a diaphragm. (a) Position of
the electric face Γ with respect to the electrodes and the diaphragm (separation l).
(b) Determination of the limit of the rectangular model (equal areas).

system of electrodes and diaphragms, therefore. The distance l satisfies the empirical expression

$$l \simeq 0 \cdot 4b + 0 \cdot 2d + 0 \cdot 5s \qquad (10A.101)$$

(Wollnik and Ewald, 1965), in which $2b$ represents the separation of the electrodes, d the distance between the electrodes and the diaphragm and $2s$ the width of the opening in the diaphragm in the H plane. This formula is valid for $d \leqslant 2b$ and for plane diaphragms.

If the curvatures r_0 and R_0 are equal, the condenser is called a spherical condenser; if $R_0 = R_i = R_e = \infty$, the condenser is cylindrical, with $\overrightarrow{CZ}$ as axis.

We shall first consider only particles travelling in the vicinity of (C_0) with small slopes dr/ds or dZ/ds (Gaussian optics), as we did in the case of magnetic prisms; we also limit the discussion to prisms whose entry and exit faces lie in planes intersecting along $\overrightarrow{CZ}$.

10A.2.2 Potential and Field

In the ideal prism (rectangular model), these quantities are independent of θ in the range $0 < \theta < \phi$.

We adopt as before the system of reduced coordinates

$$x = \frac{r - r_0}{r_0}, \quad z = \frac{Z}{r_0}, \quad \theta = \frac{s}{r_0},$$

where s is the arc-length along the curved axis (C_0). In the neighbourhood of (C_0) in the H plane, the radial field E_r may be written in the form

$$E_r = E_0(1 - nx + bx^2 - \cdots), \qquad (10A.102)$$

where

$$n = -\frac{r_0}{E_0}\left(\frac{\partial E_r}{\partial r}\right)_{z=0}$$

is the index of the field and

$$b = \frac{1}{2}\frac{r_0^2}{E_0}\left(\frac{\partial^2 E_r}{\partial r^2}\right)_{z=0}.$$

Taking $\Phi = \Phi_0$ on (C_0), the potential is given by

$$\Phi(x, 0) - \Phi_0 = -E_0 r_0 (x - \tfrac{1}{2} nx^2 + \tfrac{1}{3} bx^3 - \cdots). \qquad (10A.103)$$

The coefficients n and b depend on the radii of curvature, r_0 and R_0, of the equipotential surface $\Phi(x, z) = \Phi_0$. Since we know that $\Phi(x, z)$ must satisfy Laplace's equation and that, because of the symmetry about the plane $z = 0$, it can be expanded in the form

$$\frac{\Phi(x,z) - \Phi_0}{E_0 r_0} = \sum_n A_{2n}(x) z^{2n}, \tag{10A.104}$$

we can show (Ewald and Liebl, 1955; Albrecht, 1956) that

$$\frac{\Phi(x,z) - \Phi_0}{E_0 r_0} = -\{x - \tfrac{1}{2} n x^2 + \tfrac{1}{3} b x^3$$
$$-\tfrac{1}{2}(1-n)z^2 + \tfrac{1}{2}(1+n-2b)xz^2 + \cdots\}. \tag{10A.105}$$

For the field components, we obtain

$$E_r(x,z) = E_0 \{1 - nx + bx^2 + \tfrac{1}{2}(1+n-2b)z^2 + \cdots\},$$
$$E_z(x,z) = E_0 \{-(1-n)z + (1+n-2b)xz + \cdots\}, \tag{10A.106}$$

where

$$n = 1 + r_0/R_0,$$

$$b = 1 + \frac{r_0}{R_0} + \frac{r_0^2}{2R_0^2}(1 + R_0'), \tag{10A.107}$$

$$R_0' = \left(\frac{dR}{dr}\right)_{r=r_0, z=0}.$$

Halting the field expansions at second order terms, we find

$$\Phi(x,z) = \Phi_0 - E_0 r_0 \left[x - \frac{1}{2}\left(1 + \frac{r_0}{R_0}\right)x^2 \right.$$
$$+ \frac{1}{3}\left\{1 + \frac{r_0}{R_0} + \frac{r_0^2}{2R_0^2}(1 + R_0')\right\}x^3 + \frac{1}{2}\frac{r_0}{R_0}z^2$$
$$\left. - \frac{1}{2}\left\{\frac{r_0}{R_0} + \frac{r_0^2}{R_0^2}(1 + R_0')\right\}xz^2 \right],$$

$$E_r(x,z) = E_0 \left[1 - \left(1 + \frac{r_0}{R_0}\right)x + \left\{1 + \frac{r_0}{R_0} + \frac{r_0^2}{2R_0^2}(1 + R_0')\right\}x^2 \right.$$
$$\left. - \left\{\frac{r_0}{R_0} + \frac{r_0^2}{R_0^2}(1 + R_0')\right\}\frac{z^2}{2} \right], \tag{10A.108}$$

$$E_z(x,z) = E_0 \left[\frac{r_0}{R_0}z - \left\{\frac{r_0}{R_0} + \frac{r_0^2}{R_0^2}(1 + R_0')\right\}xz\right].$$

For the study of the first order properties, we simply write

$$E_r(x,z) = E_0(1 - nx) = E_0\left\{1 - \left(1 + \frac{r_0}{R_0}\right)x\right\},$$

$$E_z(x,z) = E_0(n-1)z = E_0\frac{r_0}{R_0}z. \tag{10A.109}$$

A positive field E_0 is created by applying a potential difference $2V_1$ across the electrodes in such a way that the outer electrode is negative with respect to Φ_0. In two special cases (the spherical condenser and the cylindrical condenser), $\Phi(r, z)$ is given for every point in space by simple analytical expressions. For the *spherical prism* ($n = 2$):

(a) $$V_1 = \Phi_0 r_0\left(\frac{1}{r_i} - \frac{1}{r_e}\right) = \Phi_0 r_0 \frac{r_e - r_i}{r_e r_i},$$

(b) $$r_0 = \frac{2 r_i r_e}{r_i + r_e}, \qquad (10A.110)$$

(c) $$\Phi(0, r) = \Phi_0\left(\frac{2r_0}{r} - 1\right),$$

and for the *cylindrical prism* ($n = 1$):

(a) $$V_1 = \Phi_0 \log (r_e/r_i),$$

(b) $$r_0 = r_i r_e, \qquad (10A.111)$$

(c) $$\Phi(0, r) = \Phi_0\{1 - 2 \log (r/r_0)\}.$$

If $r_e - r_i \ll r_0$, we can regard r_0 as approximately equal to $(r_i + r_e)/2$ in both cases.

10A.2.3 Trajectories

We consider only the non-relativistic situation; this implies
$$m = m_0, \quad dm/dt = 0, \quad v_0 = \sqrt{2e\Phi_0/m}$$

(where m_0 is the rest mass of the particle). When a particle of mass m, charge e and energy $W_0 = e\Phi_0 = \frac{1}{2}mv_0^2$, travelling in the H plane, enters the prism at S, it will follow the circle (C_0) if the value of E_0 is such that

$$\frac{mv_0^2}{r_0} = -e\bar{E}_0, \qquad (10A.112)$$

and hence

$$E_0 r_0 = -2\Phi_0. \qquad (10A.113)$$

The velocity of the particle will not remain equal to v_0 except on (C_0), since $\Phi(r, z) \neq \Phi_0$. More generally, a particle incident with velocity

$$v = v_0(1 + \beta), \quad \beta \ll 1 \qquad (10A.114)$$

will have a velocity

$$v = v_0\left\{1 + \beta - x + \frac{1}{2}\frac{r_0}{R_0}(x^2 - z^2) + x\beta\right\} \qquad (10A.115)$$

at a point (x, z); in a first order approximation, this reduces to

$$v = v_0(1 + \beta - x). \tag{10A.116}$$

In cylindrical coordinates, the equations of motion take the form

(a) $$m\frac{d^2r}{dt^2} = mr\left(\frac{d\theta}{dt}\right)^2 + eE_r,$$

(b) $$m\frac{d^2z}{dt^2} = eE_z, \tag{10A.117}$$

(c) $$m\frac{d}{dt}\left(r^2\frac{d\theta}{dt}\right) = 0.$$

Equation (10A.117c) may be integrated immediately, and gives an expression for $d\theta/dt$

$$\frac{d}{dt}\left\{r_0^2(1+x)^2\frac{d\theta}{dt}\right\} = \frac{d}{dt}\{r_0(1+x)v_\theta\} = 0$$

(v_θ denotes the azimuthal velocity). Hence

$$r_0^2(1+x)^2\frac{d\theta}{dt} = r_0(1+x)v_\theta = A = \text{const.} \tag{10A.118}$$

The constant A is easily obtained from the initial conditions (x_1, x_1', z_1, z_1') at $\theta = 0$:

$$A = r_0(1+x_1)(v_\theta)_1, \tag{10A.119}$$

and

$$(v_\theta)_1 = v_1\{1 - (x_1'^2 + z_1'^2)\}^{1/2}$$
$$\simeq v_1\{1 - \tfrac{1}{2}(x_1'^2 + z_1'^2)\}. \tag{10A.120}$$

In first order

$$(v_\theta)_1 = v_1 = v_0(1 + \beta - x_1)$$

so that

$$A = r_0v_0(1+x_1)(1+\beta-x_1)$$
$$\simeq r_0v_0(1+\beta). \tag{10A.121}$$

A is independent of x_1 and hence, on combining equations (10A.118) and (10A.121), we obtain

$$\frac{d\theta}{dt} = \frac{v_0}{r_0}(1 + \beta - 2x). \tag{10A.122}$$

To the second order approximation, we should find (Ewald and Liebl, 1965)

$$\frac{d\theta}{dt} = \frac{v_0}{r_0}\left\{1 + \beta - 2x - 2x\beta + 3x^2 - x_1^2 + 2\beta x_1\right.$$

$$\left. - \frac{1}{2}(x_1'^2 + z_1'^2) + \frac{1}{2}\frac{r_0}{R_0}(x_1^2 - z_1^2)\right\}. \tag{10A.123}$$

Substituting for E_r and for $d\theta/dt$ from equation (10A.122) into equation (10A.117a) gives

$$\frac{d^2r}{dt^2} = \frac{v_0^2}{r_0}(1 + \beta - 2x)^2(1 + x) + \frac{e}{m}E_0(1 - nx),$$

so that with the aid of equation (10A.112), we find

$$\frac{d^2x}{dt^2} = \frac{v_0^2}{r_0^2}\left\{(1 + x)(1 + \beta - 2x)^2 - (1 - nx)\right\}$$

or

$$\frac{d^2x}{dt^2} = \frac{v_0^2}{r_0^2}\left\{(n - 3)x + 2\beta\right\}. \tag{10A.124}$$

Finally, from equation (10A.117b) we arrive at

$$\frac{d^2z}{dt^2} = \frac{e}{m}E_0(n - 1)z = -\frac{v_0^2}{r_0}(n - 1)z,$$

and hence

$$\frac{d^2z}{dt^2} = \frac{v_0^2}{r_0^2}(1 - n)z. \tag{10A.125}$$

On eliminating the time from equations (10A.123) and (10A.124), we obtain the equations of the paraxial trajectories in reduced coordinates:

(a)
$$\frac{d^2x}{d\theta^2} + (3 - n)x = 2\beta,$$

$$\tag{10A.126a}$$

(b)
$$\frac{d^2z}{d\theta^2} + (n - 1)z = 0.$$

For a monochromatic beam of particles, these equations simplify still further to

(a) $x'' + (3 - n)x = 0,$

(b) $z'' + (n - 1)z = 0.$ (10A.126b)

REMARK. It can be shown (Bruck, 1958) that if the particle velocities are relativistic, the trajectory equations become

$$\frac{d^2x}{d\theta^2} + \left(3 - n - \frac{v^2}{c^2}\right)x = \frac{d^2x}{d\theta^2} + \left(2 - \frac{r_0}{R_0} - \frac{v^2}{c^2}\right)x = 0,$$

$$\tag{10A.127}$$

$$\frac{d^2z}{d\theta^2} + (n - 1)z = \frac{d^2z}{d\theta^2} + \frac{r_0}{R_0}z = 0.$$

We note that $2\beta = 2dv/v_0 = d\Phi/\Phi_0$ is the relative *energy* variation in the beam.

Equations (10A.126) show that the particles oscillate about the curved axis (C_0), provided that $3 - n > 0$ and $n - 1 > 0$. The motion in the H plane and that in the vertical or V plane $(\mathbf{Z}, \mathbf{s})$ are uncoupled, exactly as in the case of the magnetic prism.

We set

$$\omega_H^2 = 3 - n,$$
$$\omega_V^2 = n - 1 \qquad\qquad (10A.128)$$

$(\omega_H$ and ω_V may be imaginary); the solutions of equations (10A.126) then take the following form, expressed in terms of the initial conditions, at $\theta = 0$:

$$x = x_1 \cos(\omega_H \theta) + \frac{x_1'}{\omega_H} \sin(\omega_H \theta),$$

$$\qquad\qquad (10A.129)$$

$$z = z_1 \cos(\omega_V \theta) + \frac{z_1'}{\omega_V} \sin(\omega_V \theta).$$

The values of n, ω_H and ω_V for the three types of prism commonly used are tabulated below, for non-relativistic particles; the relativistic values are related to these by

$$(\omega_H^2)_{\text{rel}} = \omega_H^2 - v^2/c^2,$$

$$\qquad\qquad (10A.130)$$

$$(\omega_V^2)_{\text{rel}} = \omega_V^2.$$

Condenser	Toroidal	Spherical	Cylindrical
n	$1 + r_0/R_0$	2	1
ω_H	$(2 - r_0/R_0)^{1/2}$	1	$\sqrt{2}$
ω_V	$(r_0/R_0)^{1/2}$	1	0

10A.2.4 First Order Optical Properties; Transfer Matrices

When the incident beam is not monoenergetic, the equation of motion in the H plane can be obtained by integrating the first of equations (10A.126a); the solution when the right hand side is zero (10A.126b) is already known, and hence the full solution is

$$x = A \cos(\omega_H \theta) + B \sin(\omega_H \theta) + \frac{1}{3 - n}\frac{\Delta\Phi}{\Phi_0}. \qquad (10A.131)$$

Introducing the initial conditions at $\theta = 0$, we find

$$x = x_1 \cos (\omega_H\theta) + \frac{x_1'}{\omega_H} \sin (\omega_H\theta) + \frac{1}{\omega_H^2}\{1 - \cos (\omega_H\theta)\} \frac{\triangle\Phi}{\Phi_0},$$

$$x' = -x_1\omega_H \sin (\omega_H\theta) + x_1' \cos (\omega_H\theta) + \frac{1}{\omega_H} \sin (\omega_H\theta) \tag{10A.132a}$$

and

$$z = z_1 \cos (\omega_V\theta) + \frac{z_1'}{\omega_V} \sin (\omega_V\theta),$$

$$z' = -z_1\omega_V \sin (\omega_V\theta) + z_1' \cos (\omega_V\theta). \tag{10A.132b}$$

These equations enable us to write down the transfer matrices $\|H\|$ and $\|V\|$ and hence to express the trajectory parameters at $\theta = \phi$ in terms of those at $\theta = 0$:

$$\begin{pmatrix} x_2 \\ x_2' \\ \dfrac{\triangle\Phi}{\Phi_0} \end{pmatrix} = \|H\| \begin{pmatrix} x_1 \\ x_1' \\ \dfrac{\triangle\Phi}{\Phi_0} \end{pmatrix},$$

$$\begin{pmatrix} z_2 \\ z_2' \end{pmatrix} = \|V\| \begin{pmatrix} z_1 \\ z_1' \end{pmatrix},$$

with

$$\|H\| = \begin{pmatrix} \cos (\omega_H\phi) & \dfrac{1}{\omega_H} \sin (\omega_H\phi) & \dfrac{1 - \cos (\omega_H\phi)}{\omega_H^2} \\ -\omega_H \sin (\omega_H\phi) & \cos (\omega_H\phi) & \dfrac{\sin (\omega_H\phi)}{\omega_H} \\ 0 & 0 & 1 \end{pmatrix} \tag{10A.133}$$

and

$$\|V\| = \begin{pmatrix} \cos (\omega_V\phi) & \dfrac{\sin (\omega_V\phi)}{\omega_V} \\ -\omega_V \sin (\omega_V\phi) & \cos (\omega_V\phi) \end{pmatrix}. \tag{10A.134}$$

Unlike the case of magnetic prisms, the *mass* of the particles does not occur in the foregoing equations; we shall therefore define the dispersive properties of the prism in terms of *energy* and not velocity. Particles of the same charge but with different masses that have been accelerated through the same potential Φ will follow the same path through the prism: electrostatic prisms cannot be used to separate particles of different masses emitted by an ion source. They can, however, be used as *velocity analysers* when all the particles are of the same type. In this case, we have

$$\frac{\triangle v}{v_0} = \frac{1}{2} \frac{\triangle\Phi}{\Phi_0}. \tag{10A.135}$$

The optical elements and the energy dispersion, $\mathscr{D}_e$, can be calculated from equations (10A.133) and (10A.134). Returning to real coordinates (see Fig. 115s), we obtain:

in the H plane

$$f'_H = -f_H = \frac{r_0}{\sin(\omega_H \phi)},$$

$$\omega_H = \sqrt{3-n}; \qquad (10A.136)$$

$$g'_H = -g_H = \frac{r_0}{\tan(\omega_H \phi)},$$

in the V plane

$$f'_V = -f_V = \frac{r_0}{\sin(\omega_V \phi)},$$

$$\omega_V = \sqrt{n-1}. \qquad (10A.137)$$

$$g'_V = -g_V = \frac{r_0}{\tan(\omega_V \phi)},$$

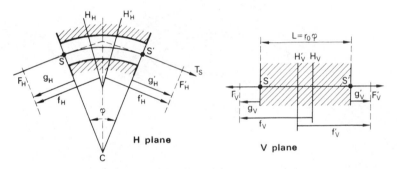

FIG. 115s. Positions of the foci and principal planes.

The electrostatic prism, like the magnetic prism, is equivalent to two centred systems of length $L = r_0\phi$, the cardinal elements of which are given by equations (10A.136) and (10A.137). If ω_H and ω_V are real, the two systems are convergent. If, on the other hand, ω_H (or ω_V) is imaginary, the circular functions must be replaced by hyperbolic functions, and the corresponding system is divergent. We recall that the term r_0/R_0 that appears in n can be negative, if R_0 is negative (the electrodes would then be convex in the $(\mathbf{Z}, \mathbf{r})$ plane towards $\overrightarrow{CZ}$; in Fig. 115q, $R_0 > 0$). Summarizing the various possibilities, we have:

$n = 1 + r_0/R_0 < 1 \quad (R_0 < 0) \qquad \begin{cases} H \text{ plane convergent} \\ V \text{ plane divergent,} \end{cases}$

$n = 1 \text{ (cylindrical condenser)} \qquad \begin{cases} H \text{ plane convergent} \\ V \text{ plane equivalent to a drift space,} \end{cases}$

$1 < n < 3$	both planes convergent,
$n = 3$ $(r_0/R_0 = 2)$	$\begin{cases} H \text{ plane equivalent to a drift space} \\ V \text{ plane convergent,} \end{cases}$
$n > 3$	$\begin{cases} H \text{ plane divergent} \\ V \text{ plane convergent.} \end{cases}$

The properties of the *cylindrical prism* are thus similar to those of the magnetic prism with homogeneous field, since $f_V = g_V = \infty$. We have $g'_H = g_H = 0$, however, so that the foci are at S and S', for tan $(\sqrt{2}\ \phi) = 0$ or $\phi_1 = \pi/2\sqrt{2}$. By joining two prisms of angle ϕ_1, therefore, we can focus particles from a point source at S at the point S'. The image of S will be a segment of a straight line (in first order) since there is no convergence in the V plane, and the total deflection will then be (Fig. 115t)

$$\phi = 2\phi_1 = \pi/\sqrt{2} = 127°2'. \tag{10A.138}$$

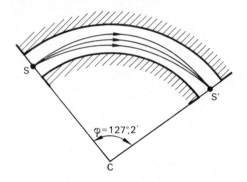

FIG. 115t. Cylindrical 127° prism; focusing at S and S'.

This is true when the source at S, in the electric face of the prism, is in fact the crossover of a beam formed by an optical system placed in front of the prism. If, on the other hand, we attempt to define an "object" by means of a metal diaphragm placed at S, the electric face will move closer to the electrodes and we shall no longer have $\phi = 127°2'$; the exact position of the image must then be calculated with the aid of equation (10A.136).

For the cylindrical prism, equations (10A.136) become

$$f'_H = r_0/\sin(\sqrt{2}\ \phi),$$
$$g'_H = r_0/\tan(\sqrt{2}\ \phi). \tag{10A.139}$$

THE SPHERICAL PRISM is equivalent to a rotationally symmetric optical system, since $\omega_H = \omega_V = 1$ (like the magnetic prism with index $n = \frac{1}{2}$); the system is stigmatic for an infinite number of pairs of conjugate planes.

We have

$$f'_H = r_0/\sin\phi,$$
$$g'_H = r_0/\tan\phi. \tag{10A.140}$$

Particles from S can be focused at S', if $\phi = 180°$, but we now have a point image (if aberrations are neglected).

TOROIDAL PRISM. For $1 < n < 3$, symmetric stigmatic operation can again be achieved in certain cases, but for only one pair of conjugate planes.

10A.2.5 Dispersion; Theoretical Resolving Power (First Order Approximation)

The dispersion, $\mathcal{D}_e$, is defined, in the Gaussian image plane, which lies at a reduced distance q/r_0 from S, by

$$\mathcal{D}_e = \frac{\triangle x_i}{\triangle\Phi/\Phi_0} \tag{10A.141}$$

or

$$\mathcal{D}_e = \triangle x_2 + \frac{q}{r_0}\triangle x'_2,$$

so that

$$\mathcal{D}_e = \left\{1 - \cos(\omega_H\phi) + \frac{q}{r_0}\omega_H\sin(\omega_H\phi)\right\}\Big/\omega_H^2. \tag{10A.142}$$

It can be shown (Bruck, 1958) that we again obtain the simple relation

$$\mathcal{D}_e = \frac{1 - G_H}{\omega_H^2}. \tag{10A.143}$$

From the relation between conjugates in the H plane, we can calculate q/r_0 when the object distance, p/r_0, is known; we have merely to write down the total transfer matrix $\|T\|$ between the object P and its image Q and then set

$$T_{12} = 0, \tag{10A.144}$$

which gives

$$\tan(\omega_H\phi) = -\frac{\{(p+q)/r_0\}\omega_H}{1 - (pq/r_0^2)\,\omega_H^2}. \tag{10A.145}$$

CYLINDRICAL PRISM. With symmetric operation, we have

$$\frac{p}{r_0} = \frac{q}{r_0} = \frac{1}{\omega_H}\cot(\omega_H\phi/2) = \frac{1}{\sqrt{2}}\cot(\sqrt{2}\phi/2), \tag{10A.146}$$
$$G_H = -1$$

and

$$\mathscr{D}_e = 1. \tag{10A.147}$$

If $p \neq q$, $G_H \neq -1$ and we find

$$\mathscr{D}_e = \tfrac{1}{2}(1 - G_H). \tag{10A.148}$$

$\mathscr{D}_e$ is thus independent of ϕ and of r_0. The *real separation*, d_c, between particles of energies Φ_0 and $\Phi_0(1 + \triangle\Phi/\Phi_0)$ in the Gaussian image plane does depend on r_0 and is given by

$$d_c = r_0 \triangle x_i = r_0 \mathscr{D}_e \frac{\triangle\Phi}{\Phi_0} = \tfrac{1}{2} r_0 (1 - G_H) \frac{\triangle\Phi}{\Phi_0}. \tag{10A.149}$$

SPHERICAL PRISM. From equation (10A.143) we obtain

$$\tan \phi = \frac{(p+q)/r_0}{1 - pq/r_0^2}. \tag{10A.150}$$

The points P and Q (and G_H which is equal to G_V) can be determined graphically by Barber's rule (see the section on the magnetic prism with homogeneous field). In the symmetric case,

$$p/r_0 = q/r_0 = \cot (\phi/2),$$

$$G_H = G_V, \tag{10A.151}$$

$$\mathscr{D}_e = 2. \tag{10A.152}$$

The dispersion is therefore twice as large as that of the cylindrical prism. In the general case ($p \neq q$), we have

$$\mathscr{D}_e = (1 - G_H) \tag{10A.153}$$

and

$$d_s = r_0(1 - G_H) \frac{\triangle\Phi}{\Phi_0}. \tag{10A.154}$$

If we do not take the aberrations into account, we can define a theoretical first order resolving power, $R_e = \Phi_0/\triangle\Phi$; the width of the image of an object a_0 is given by $a_i = |G|a_0$. If a slit of width s_0 is placed on the axis at Q, we have

$$R_e = \frac{r_0 \mathscr{D}_e}{|G|a_0 + s_0}. \tag{10A.155}$$

In reality, the practical resolving power will be poorer than the value predicted by equation (10A.155), since allowance must be made for the aberrations, especially the second order aperture aberration.

10A.2.6 Focusing in the Vertical Plane with a Cylindrical Prism

The cylindrical prism is the most common in practice, since it is easy to construct. The convergence is zero in the vertical plane, however, as we have already found, and this often limits the "clarity" of the system. It is in fact very easy to transform a cylindrical prism into a prism with non-zero convergence in the V plane: we have only to arrange that the equipotential surface $\Phi(x, z) = \Phi_0$ and the neighbouring surfaces have a positive vertical curvature, $1/R$. This is achieved by placing two supplementary circular plane electrodes, E_1 and E_2, above and below the electrodes of the condenser and holding them at the same potential, which is not equal to Φ_0. If the field E_0 is positive, the outer electrode is held at $\Phi_0 - V_1$ and the inner electrode at $\Phi_0 + V_1$. In order to obtain $R_0 > 0$, we apply to E_1 and E_2 voltages that are slightly negative with respect to Φ_0. The electrolytic tank yields R_0 as a function of the applied voltage for a given electrode geometry. In Fig. 115u(a), the form of the equipotential surfaces in a cylindrical condenser is shown. If we require $R_0 = \infty$, we must have $b \ll c$ (where $2b$ is the distance between the electrodes and c is their height). Figure 115u(b) shows the equipotentials when a potential $-V_2$ is applied to the supplementary electrodes. (In this figure, the potentials are measured with respect to earth—as they are in practice when we apply a voltage to a prism—(C_0) corresponds to $\Phi = 0$.)

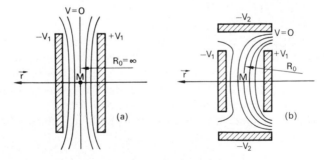

Fig. 115u. The conversion of a cylindrical prism into a toroidal prism (a) Equipotential surfaces before modification. (b) Form of the equipotentials after application of a negative potential to the supplementary electrodes.

10A.2.7 Aberrations of Electrostatic Prisms

If we retain second and third order terms in the expansion for $\Phi(x, z)$ and second order terms in v, we can first of all calculate the second order aberrations in an ideal prism, bounded by plane virtual electric faces, using the rectangular model approximation (Ewald and Liebl, 1957;

Liebl and Ewald, 1959). The equations giving the second order approximation to the variations of x, x', z and z' through the stray fields are then derived, after which the aberration coefficients of the real prism can be calculated (Wollnik and Ewald, 1965; Wollnik, 1965). Finally, the aberrations can be calculated in the Gaussian image plane (Wollnik, 1967).

10A.2.7.1 Aberrations of the ideal prism

The second order trajectory equations, in the form

$$x'' + \omega_H^2 x - 2\delta = f(x, x', z, z', \delta) \qquad (10A.156)$$

are integrated; the right-hand side contains only terms of second order, and into it are substituted the (first order) solutions for x, x', z and z', obtained by integrating the corresponding homogeneous equation, for $0 < \theta < \phi$. Thus in the exit face of the prism (suffix 2), for example, equation (10A.156) gives

$$x_2 = (x_2)_1 + \triangle x_2,$$
$$\qquad (10A.157)$$
$$x_2' = (x_2')_1 + \triangle x_2',$$

with

$$(x_2)_1 = H_{11}x_1 + H_{12}x_1' + H_{13}\frac{\triangle\Phi}{\Phi_0},$$
$$\qquad (10A.158)$$
$$(x_2')_1 = H_{21}x_1 + H_{22}x_1' + H_{23}\frac{\triangle\Phi}{\Phi_0}.$$

(Suffix 1 characterizes quantities measured in the entry face.) Thirty coefficients are again necessary to describe the departures $\triangle x_2$, $\triangle x_2'$, $\triangle z_2$ and $\triangle z_2'$ completely (including the chromatic aberrations arising from an energy spread $\triangle\Phi/\Phi_0 = \delta$). These aberrations are given by expressions similar to those for magnetic prisms (see equation 10A.87). For the aperture aberration coefficient a_3, for example (Wollnik, 1965), which is defined by

$$(\triangle x_2)_{ap} = a_3 x_1'^2,$$

we find

$$a_3 = \omega_H^{-4}\left[\left\{4\left(\frac{r_0}{R_0} - 1\right) - 2b/3 + \omega_H^2\right\}\{1 - \cos(\omega_H\phi)\}\right.$$
$$\left. - \left\{2\left(\frac{r_0}{R_0} - 1\right) - b/3\right\}\sin^2(\omega_H\phi)\right], \qquad (10A.159)$$

where

$$b = -\frac{1}{2E_0 r_0}\left(\frac{\partial^2\Phi(x, z)}{\partial x^2}\right)_{x=z=0}.$$

We observe that $a_3 = 0$ for $\omega_H \phi = 2\pi$ (and the coefficients of the terms in x_1^2 and $x_1 x_1'$ vanish at the same time), but this property cannot be exploited for in these conditions the dispersion $\mathscr{D}_e$ vanishes.

SPECIAL CASES: *Spherical prism* ($r_0 = R_0, \omega_H^2 = 1, b = 3$):

$$(a_3)_{\text{sph}} = \sin^2 \phi - (1 - \cos \phi). \tag{10A.160}$$

If we wish to focus S at S' ($\phi = 180°$), we have

$$(a_3)_{\text{sph},180°} = -2.$$

Cylindrical prism ($R_0 = \infty, \omega_H^2 = 2, b = 1$):

$$(a_3)_{\text{cyl}} = \tfrac{1}{12}\{7 \sin^2 (\sqrt{2} \phi) - 8(1 - \cos (\sqrt{2} \phi))\} \tag{10A.161}$$

For a prism for which $\phi = 127°2'$, we have

$$(a_3)_{\text{cyl},127°} = -4/3.$$

10A.2.7.2 Passage through the fringing fields

If we consider the passage through the real fringing fields, we see that, in the H plane, the trajectory (C_0) corresponds to an incident trajectory T_0', normal to the virtual entry face that we defined earlier and parallel to T_0 but shifted from it a distance $\triangle x_E$ (Fig. 115v). Outside the ideal prism,

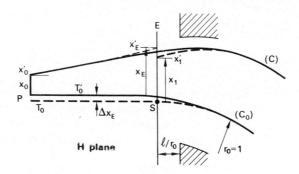

FIG. 115v. Influence of the fringing field at the entry. E: Electric face. (The H plane is represented.)

we use T_0' as curved axis and inside the prism, the circle (C_0). A trajectory from a point object ($x_0, x_0', z_0 = z_0' = 0$) will strike the entry face E at x_E with slope x_E' and will then follow the trajectory (C) inside the prism. (C) appears to come from a point with coordinate x_1 (with respect to (C_0)) and slope x_1', lying in E.

In order to obtain an actual deflection ϕ, the object must therefore be shifted a distance $\triangle x_E$ (as in the case of magnetic prisms, see §10A.1.8).

A second order calculation shows (Wollnik and Ewald, 1965) that

$$x_1 = x_E + \tfrac{1}{2} x_E^2 + \cdots,$$

(10A.162)

$$x_1' \simeq x_E' - \frac{1}{2R_1}(x_E^2 - z_E^2) - x_E x_E' + x_E l/3 r_0.$$

(l denotes the gap between the electrodes and the virtual electric face.) We here assume that the electric face E has a radius of curvature R_1 (Fig. 115w) in a vertical plane (Z, s).

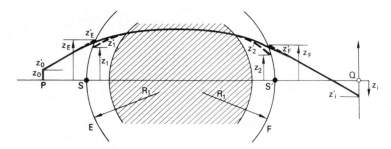

FIG. 115w. Notation. (V plane.)

In the vertical plane, we have

$$z_1 = z_E + \cdots,$$

(10A.163)

$$z_1' = z_E' + \frac{1}{R_1} x_E z_E.$$

At the exit face F, we can likewise calculate the corrected values x_F, x_F', z_F and z_F' as a function of x_2, x_2', z_2 and z_2', the values corresponding to an ideal prism of angle ϕ. (For simplicity, we assume that the exit face has the same radius of curvature, R_1, as the entry face.) We have

$$x_F = x_2 - \tfrac{1}{2} x_2^2,$$

$$x_F' = x_2' - \frac{1}{2R_1}(x_2^2 - z_2^2),$$

(10A.164)

$$z_F = z_2 + \cdots,$$

$$z_F' = z_2' + \frac{1}{R_1} x_2 z_2.$$

From the values at F, we can calculate the corresponding values (x_i, x_i', z_i, z_i') in the Gaussian image plane in the H plane, situated a reduced distance q/r_0 from S'.

The departures $\triangle x_i$, $\triangle x'_i$, $\triangle z_i$ and $\triangle z'_i$ can therefore be calculated as a function of x_0, x'_0, z_0, z'_0 and δ, and the corresponding aberration coefficients can be extracted. Of particular interest is the departure $\triangle x_i$ in the H plane from which we can calculate the energy resolution. According to equations (10A.162) and (10A.164), $\triangle x_i$ will depend upon R_1, so that a partial correction at least of the spherical aberration will be possible. Up to the present time, however, no expression for the aberrations in the Gaussian image plane has been published; less work has been devoted to electro-static prisms than to magnetic prisms.

10A.2.7.3 Resolving power

We can only give an approximate value for the resolution R_e in a prism in which the image of S is formed at S' (neglecting all the other aberration terms in comparison with the aperture aberration term $a_3 x_1'^2$), since to a first approximation $\triangle x_2$ is given by equation (10A.159). We add a term $r_0 \triangle x_2$ to the denominator of equation (10A.155).

SPHERICAL CONDENSER ($\phi = 180°$, $|G_H| = 1$). In real coordinates,

$$R_e = \frac{r_0 \mathcal{D}_e}{|G|a_0 + s_0 + |a_3|x_1'^2 r_0} \tag{10A.165}$$

or

$$R_e = \frac{2r_0}{a_0 + s_0 + 2r_0 x_1'^2} \tag{10A.166}$$

$$= \frac{2}{(a_0 + s_0)/r_0 + 2x_1'^2}.$$

If, for example, we have

$$r_0 = 5 \text{ cm,}$$

$$a_0 = 0 \cdot 25 \text{ mm,}$$

$$s_0 = 0 \cdot 25 \text{ mm (circular holes as object and image),}$$

then

$$R_e = \frac{2}{10^{-2} + 2x_1'^2} \simeq 200\,(1 - 200\,x_1'^2).$$

The aberration has virtually no effect so long as $x'_1 < 10^{-2}$ rad.

CYLINDRICAL CONDENSER ($\phi = 127°2'$, $|G_H| = 1$):

$$R_e = \frac{r_0}{a_0 + s_0 + \frac{4}{3}r_0 x_1'^2} = \frac{1}{(a_0 + s_0)/r_0 + \frac{4}{3}x_1'^2}. \tag{10A.167}$$

With the same values of r_0, a_0 and s_0 (slits as object and image), we find
$R_e \simeq 100\,(1 - 133\,x_1'^2)$.

10A.2.7.4 Direct calculation of the aberrations in the image plane

With the advent of fast computers, it has become possible to calculate the aberrations and the resolution directly, by integrating the general equations of motion from the object plane to the image plane; the values of the potential employed are either measured or calculated by solving Laplace's equation with the computer. The trajectories have recently been calculated in this way for beams of slow electrons, taking the space charge into account (Ballu, 1968; Barat, 1968). These calculations show that for beams of angular aperture $\alpha > 10^{-2}$ in a cylindrical 127° prism, it is better to use entry and exit diaphragms in the form of inclined "horns" (Fig. 115x(a)) instead of plane diaphragms (Fig. 115x(b)). The potential distribution is disturbed less in the stray field zone with inclined planes.

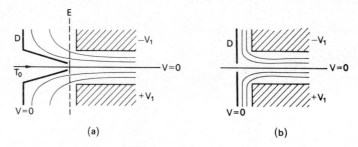

(a)   (b)

FIG. 115x. Examples of diaphragms (screens) and the appearance of the equipotentials in the H plane. (a) Plane diaphragm. (b) Inclined planes (thin walls).

TABLE 1

SOME USEFUL CONSTANTS

(in R.M.K.S. units)

Dielectric constant of free space:
$$\epsilon_0 = \frac{10^{-9}}{36\pi} 8\cdot855 \times 10^{-12} \text{ F m}^{-1}$$

Permeability of free space:
$$\mu_0 = \frac{4\pi}{10^7} = 1\cdot257 \times 10^{-6} \text{ henry m}^{-1}$$

Velocity of light in free space:
$$c = 2\cdot998 \times 10^8 \simeq 3 \times 10^8 \text{ m sec}^{-1}$$

Charge of the electron or proton:
$$e = 1\cdot602 \times 10^{-19} \text{ coulomb (C)}$$

Rest mass of the electron:
$$m_0 = 9\cdot107 \times 10^{-31} \text{ kg}$$

Rest mass of the proton:
$$M_0 = 1\cdot672 \times 10^{-27} \text{ kg}$$

Ratio e/m_0 (electron):
$$\frac{e}{m_0} = 1\cdot759 \times 10^{11} \text{ C kg}^{-1}$$

Ratio e/M_0 (proton):
$$\frac{e}{M_0} = 0\cdot958 \times 10^8 \text{ C kg}^{-1}$$

Ratio $\dfrac{m_0}{M_0}$:
$$\frac{m_0}{M_0} = 1836$$

Relativistic correction factors;
electron:
$$\frac{e}{m_0 c^2} = 1\cdot957 \times 10^{-6} \text{ V}^{-1}$$

proton:
$$\frac{e}{M_0 c^2} = 1\cdot066 \times 10^{-9} \text{ V}^{-1}$$

Boltzmann's constant:
$$k = 1\cdot380 \times 10^{-23} \text{ J deg.}^{-1}$$

Reciprocal $1/k$:
$$\frac{1}{k} = 11{,}600 \text{ deg. eV}^{-1}$$

Planck's constant:
$$h = 6\cdot624 \times 10^{-34} \text{ J sec}^{-1}$$

Electron volt:
$$1 \text{ eV} = 1\cdot602 \times 10^{-19} \text{ J}$$

Velocity of a 1 eV electron:
$$v_e = \sqrt{\frac{2e}{m_0}} = 5\cdot932 \times 10^5 \text{ m sec}^{-1}$$

Velocity of a 1 eV proton:
$$v_H = 1\cdot384 \times 10^4 \text{ m sec}^{-1}$$

TABLE 2

SOME USEFUL FORMULAE

(The numerical values are given in R.M.K.S. units for electrons)

---

(1) *Low velocities* ($v \ll c$)

$$v = \sqrt{\frac{2e}{m}} \cdot \sqrt{V} = 0.5932 \times 10^6 \sqrt{V}$$

$$p = m v \qquad\qquad = 5.425 \times 10^{-25} \sqrt{V}$$

$$B \varrho = \frac{m v}{e} \qquad\quad = 5.685 \times 10^{-12} v$$

$$B \varrho = \sqrt{\frac{2 m V}{e}} \quad = 3.372 \times 10^{-6} \sqrt{V}$$

$$v = \frac{e}{m} B \varrho \qquad = 1.759 \times 10^{11} (B \varrho)$$

$$\lambda = h/p \qquad\quad = \frac{12.26}{\sqrt{V}} \times 10^{-10}$$

or

$$\lambda = \frac{12.26}{\sqrt{V}} \text{Å}$$

v in m sec^{-1}; B in teslas; ρ in metres; V in volts; m in kg; e in coulombs.

(2) *Relativistic velocities*

$$\varepsilon = \frac{e}{2 m_0 c^2} \qquad = 0.978 \times 10^{-6}$$

$$V^* = V(1 + \varepsilon V)$$

$$\frac{m}{m_0} = \frac{1}{\sqrt{1 - \frac{v^2}{c^2}}} = 1 + 2 e V^{\cdot}$$

$$p = m v \qquad\qquad = 2 m_0 c \sqrt{e V(1 + \varepsilon V)}$$

$$\qquad\qquad\qquad\quad = \sqrt{2 m_0 e V(1 + \varepsilon V)}$$

$$\beta = v/c \qquad\qquad = 2 \frac{\sqrt{\varepsilon V(1 + \varepsilon V)}}{1 + 2 \varepsilon V}$$

$$B \varrho = \frac{p}{e} \qquad\qquad = 2 \frac{m_0 c}{e} \sqrt{\varepsilon(1 + \varepsilon V)}$$

$$\lambda = \frac{h}{p} \qquad\qquad = \frac{h}{2 m_0 c} \frac{1}{\sqrt{\varepsilon(1 + \varepsilon V)}}$$

TABLE 3

FUNDAMENTAL CHARACTERISTICS OF FREE ELECTRONS IN MOTION

In the following table, the following quantities are tabulated to four significant figures:

V – The potential necessary to raise the kinetic energy of an electron initially at rest to a value E. V is expressed in volts.

V^* – The potential, relativistically corrected in such a way that the relativistic value of the momentum is obtained by simply replacing V by V^* in the non-relativistic expression. V and V^* are related by the equation:

$$V^* = V + \frac{e}{2 m_0 c^2} V^2,$$

and the relativistic momentum p has the form:

$$p = m_0 c \sqrt{\frac{2 E}{m_0 c^2} + \left(\frac{E}{m_0 c^2} \right)^2} = (2 e m_0 V^*)^{1/2}.$$

$B\rho$ – The product of the magnetic induction B, measured in teslas (1 tesla = 10^4 gauss) and the radius of curvature of the electron trajectory ρ, measured in meters, in a plane perpendicular to the field (1 tesla $\cdot$ m = 10^6 gauss $\cdot$ cm).

λ – The de Broglie wavelength of the electron in the absence of a magnetic field; λ is expressed in metres (1 m = 10^{10} Å).

(a) Thermal and photoelectric velocities
(0·1 to 10 eV)

V volts	V^* volts	$B\rho$ tesla · m	λ m
0·1	0·1000		
0·2	0·2000		
0·3	0·3000	$1\cdot846 \times 10^{-6}$	$2\cdot240 \times 10^{-9}$
0·4	0·4000	2·132	1·940
0·5	0·5000	2·385	1·734
0·6	0·6000	2·616	1·583
0·7	0·7000	2·821	1·469
0·8	0·8000	3·024	1·371
0·9	0·9000	3·184	1·293
1	1·0000	3·372	1·226
2	2·0000	4·769	$8\cdot671 \times 10^{-10}$
3	3·0000	5·840	7·080
4	4·0000	6·744	6·132
5	5·0000	7·540	5·484
6	6·0000	8·260	5·006
7	7·0000	8·922	4·635
8	8·0000	9·538	4·336
9	9·0000	10·12	4·088
10	10·00	10·66	3·878

(b) Vacuum tube region (10 to 10^4 eV)

V volts	V^* volts	$B\rho$ tesla · m	λ m
10	10·00	10·66	$3\cdot878 \times 10^{-10}$
20	20·00	15·08	2·742
30	30·00	18·47	2·239
40	40·00	21·32	1·939
50	50·00	23·84	1·734
100	100·0	33·72	1·226
150	150·0	41·30	1·001
200	200·0	47·69	$8\cdot670 \times 10^{-11}$
250	250·0	53·32	7·755
300	300·0	58·41	7·079
500	500·2	75·42	5·483
1000	1001	106·7	3·876
2500	2506	168·8	2·449
5000	5024	239·0	1·730
7500	7555	293·1	1·410
10000	10097	338·8	1·220

(c) X-ray tubes, microscopes, diffraction cameras
(10^4 to $2 \cdot 10^5$ eV)

V volts	V^* volts	$B\rho$ tesla · m	λ m
10000	10097	338·8	$1·220 \times 10^{-11}$
20000	20391	481·5	$8·588 \times 10^{-12}$
30000	30880	592·5	6·979
40000	41560	687·5	6·015
50000	52440	772·2	5·355
60000	63520	849·9	4·866
70000	74790	922·2	4·484
80000	86260	990·4	4·175
90000	97920	1055	3·918
100×10^3	$109·8 \times 10^3$	1117	3·701
120	134·0	1235	3·349
140	159·1	1345	3·073
160	185·0	1450	2·850
180	211·7	1551	2·665
200	239·1	1649	2·507

(d) β-ray devices and accelerators
(2×10^5 to 20×10^6 eV)

V volts	V^* volts	$B\rho$ tesla · m	λ m
2×10^5	$2·391 \times 10^5$	1649	$2·507 \times 10^{-12}$
3	3·880	2100	1·968
4	5·565	2515	1·644
5	7·446	2909	1·421
6	9·522	3290	1·257
8×10^5	$1·426 \times 10^6$	4027	1·026
1×10^6	1·978	4743	$8·718 \times 10^{-13}$
1·5	3·701	6487	6·374
2	5·914	8200	5·042
10	107·8	35020	1·180
20	411·4	68390	$6·046 \times 10^{-14}$

TABLE 4

THE LEGENDRE FUNCTIONS OF FRACTIONAL ORDER

$$P_n(\cos\theta) \leqq 0.1 \leqq n \leqq 2.0$$

θ	n = 0.1	0.2	0.3	0.4	0.5	0.6	0.7	0.8	0.9	1.0
5°	0.365201	−0.254581	−0.813813	−1.272544	−1.599553	−1.774742	−1.791031	−1.652930	−1.378654	−0.996444
10°	0.501717	0.005894	−0.453932	−0.847492	−1.150000	−1.343918	−1.420160	−1.378607	−1.227945	−0.984808
20°	0.638358	0.268268	−0.088558	−0.411669	−0.683193	−0.888957	−1.019355	−1.069887	−1.041354	−0.939693
30°	0.718190	0.423320	0.130467	−0.145688	−0.391682	−0.596024	−0.749809	−0.847187	−0.885632	−0.866025
40°	0.774511	0.534092	0.289416	0.051166	−0.170483	−0.366364	−0.528746	−0.651698	−0.734616	−0.766044
50°	0.817704	0.620144	0.414869	0.209624	0.012012	−0.170817	−0.332468	−0.467557	−0.574395	−0.642788
60°	0.852374	0.690081	0.518406	0.342882	0.169084	0.002434	−0.151995	−0.289652	−0.406673	−0.500000
70°	0.880955	0.748422	0.606031	0.456324	0.307261	0.157754	0.016269	−0.116849	−0.237227	−0.342020
80°	0.904886	0.797813	0.681210	0.557670	0.430035	0.301038	0.173516	0.050203	−0.066312	−0.113648
90°	0.925086	0.839927	0.746089	0.645288	0.539353	0.430189	0.319752	0.209982	0.102787	0.000000
100°	0.942171	0.875872	0.802069	0.721889	0.636309	0.546730	0.454374	0.360536	0.266528	0.173648
110°	0.956571	0.906416	0.850092	0.788227	0.721505	0.650659	0.576469	0.499745	0.421314	0.342020
120°	0.968597	0.932102	0.890814	0.845072	0.795249	0.741748	0.685007	0.625480	0.563646	0.500000
130°	0.978471	0.953322	0.924694	0.892752	0.857676	0.819665	0.778935	0.735715	0.692387	0.642788
140°	0.986362	0.970362	0.952059	0.931521	0.908821	0.884042	0.857275	0.828616	0.798169	0.766044
150°	0.992387	0.983428	0.973140	0.961544	0.948665	0.934528	0.919163	0.902601	0.884877	0.866025
160°	0.996635	0.992665	0.988095	0.982927	0.977168	0.970822	0.963895	0.956393	0.948322	0.939693
170°	0.999161	0.998171	0.997028	0.995735	0.994289	0.992699	0.990947	0.989050	0.987003	0.984808
180°	1.000000	1.000000	1.000000	1.000000	1.000000	1.000000	1.000000	1.000000	1.000000	1.000000

Table 4 (continued)

θ	n = 1·1	1·2	1·3	1·4	1·5	1·6	1·7	1·8	1·9	2·0
5°	−0·542651	−0·058480	+0·413527	+0·833237	1·165932	1·385435	1·476543	1·433916	1·265476	0·989351
10°	−0·672324	−0·318188	+0·047601	+0·395130	0·696918	0·930007	1·077683	1·136627	1·087486	0·954769
20°	−0·775450	−0·562976	−0·319350	−0·063159	0·186880	0·412909	0·599730	0·735204	0·811308	0·824533
30°	−0·792525	−0·672251	−0·514788	−0·331564	−0·135126	+0·061648	+0·246321	0·407651	0·536224	0·625000
40°	−0·756377	−0·705113	−0·617004	−0·498494	−0·357352	−0·202235	−0·042220	+0·113682	+0·260689	0·380236
50°	−0·679240	−0·680298	−0·648801	−0·587376	−0·500235	−0·392577	−0·270327	−0·139860	−0·007413	+0·119764
60°	−0·567487	−0·607957	−0·621232	−0·608117	−0·570350	−0·510520	−0·431952	−0·338567	−0·234713	−0·125000
70°	−0·428902	−0·496128	−0·542575	−0·567190	−0·571540	−0·555254	−0·520788	−0·468913	−0·402560	−0·324533
80°	−0·269688	−0·352628	−0·421019	−0·473792	−0·510309	−0·530315	−0·533992	−0·521924	−0·495078	−0·454769
90°	−0·096662	−0·185629	−0·265506	−0·335101	−0·393447	−0·439820	−0·473745	−0·495011	−0·503662	−0·500000
100°	+0·083160	−0·003726	−0·085870	−0·162209	−0·231828	−0·293895	−0·347694	−0·392684	−0·428461	−0·454769
110°	0·262707	+0·184212	+0·107352	+0·032911	−0·038356	−0·105747	−0·168608	−0·226363	−0·278485	−0·324533
120°	0·435048	0·369302	0·303277	0·237487	+0·172439	+0·108625	+0·046528	−0·013395	−0·070707	−0·125000
130°	0·593260	0·542960	0·491146	0·438457	0·385144	0·331534	0·277906	+0·224546	+0·171992	+0·119764
140°	0·732358	0·697230	0·660787	0·623160	0·584483	0·544892	0·504529	0·463536	0·422057	0·380236
150°	0·846085	0·825099	0·803106	0·780153	0·756288	0·731556	0·706009	0·679698	0·652678	0·625000
160°	0·930510	0·920782	0·910521	0·899731	0·888427	0·876616	0·864311	0·851520	0·838256	0·824533
170°	0·982463	0·979971	0·977330	0·974544	0·971609	0·968530	0·965306	0·961937	0·958424	0·954769
180°	1·000000	1·000000	1·000000	1·000000	1·000000	1·000000	1·000000	1·000000	1·000000	1·000000

This table has been kindly provided by M. Schelkunoff.

BIBLIOGRAPHY

Books of a more general nature are to be found listed in the Bibliography for Part 1. More specialized works, devoted to a single electron optical instrument, are mentioned in the appropriate section of Part 2.

BIBLIOGRAPHY FOR PART 1†

I. BASIC WORKS

(A) On Electron Optics

ARDENNE, M. VON (1940) *Elektronen Übermikroskopie*, Springer, Berlin.
ARDENNE, M. VON (1956, 1962–3) *Tabellen für Elektronenphysik und Übermikroskopie* (2 vols.) V.E.B., Berlin.
BROGLIE, L. DE (1950) *Optique électronique et corpusculaire*, Hermann, Paris.
BROGLIE, L. DE, C. MAGNAN et al. (1946) *L'Optique électronique*, Revue d'Optique, Paris.
BUSCH, H. and E. BRÜCHE (1937) *Beiträge zur Elektronenoptik*, Barth, Leipzig.
COSSLETT, V. E. (1946; 2nd ed., 1950) *Introduction to Electron Optics*, Oxford University Press.
DUPOUY, G. (1952; 2nd ed., 1961) *Eléments d'optique électronique*, Armand Colin, Paris.
FLÜGGE, S. (1956) ed. *Encyclopædia of Physics*, **33**, *Optics of corpuscles*, Springer, Berlin.
GABOR, D. (1945) *The Electron Microscope*, Hutton, London.
GLASER, W. (1952) *Grundlagen der Elektronenoptik*, Springer, Vienna.
HAINE, M. (1961) *The Electron Microscope—the present state of the art*, Spon, London.
JACOB, L. (1950) *An Introduction to Electron Optics*, Methuen, London.
KLEMPERER, O. (1953) *Electron Optics*, Cambridge University Press.
KLEMPERER, O. and M. E. BARNETT, (1970) *Electron Optics*, Cambridge University Press.
MALOFF, I. G., and D. W. EPSTEIN (1938) *Electron Optics in Television*. McGraw-Hill, New York.
OLLENDORFF, F. (1955) *Elektronik des Einzelelektrons*, Springer, Berlin.
PICHT, J. (1939, 3rd ed. 1963) *Einführung in die Theorie der Elektronenoptik*, Barth Leipzig. (First edition distributed by Edwards, Ann Arbor.)
PICHT, J. and J. HEYDENREICH (1966) *Einführung in die Elektronenmikroskopie*, Verlag Technik, Berlin.
PIERCE, J. R. (1949, 2nd ed. 1954) *Theory and Design of Electron Beams*, Van Nostrand, New York.
SEPTIER, A. (Ed.) (1967) *Focusing of Charged Particles*, Academic Press, New York.
STURROCK, P. A. (1955) *Static and Dynamic Electron Optics*, Cambridge University Press.
TSUKKERMAN, I. I. (1961) *Electron Optics in Television* (trans. Fenn) Pergamon.
ZWORYKIN, V. K., G. A. MORTON, E. G. RAMBERG, J. HILLIER and A. W. VANCE (1945) *Electron Optics and the Electron Microscope*, Wiley, New York.

(B) On Glass Optics

BORN, M. (1933) *Optik*, Springer, Berlin.
BORN, M., and E. WOLF (1959; 4th ed. 1970) *Principles of Optics*, Pergamon.

† A number of the references in the bibliography are to communications to scientific congresses. The text in each case is to be found in the Proceedings, which used always to be published with an appreciable delay. A list of these congresses, with their full titles, is to be found in Section IVB of Part 2. In the text, only the date and the place where the congress was held, and the number or page number of the communication are given.

CARATHÉODORY, C. (1937) *Geometrische Optik*, Springer, Berlin.
MARÉCHAL, A. (1952) *Imagerie géométrique—Aberrations*, Revue d'Optique, Paris.
SYNGE, J. L. (1936) *Geometrical Optics*, Cambridge University Press.

(C) On Field Calculations and Basic Techniques

DURAND, E. (1952) *Electrostatique et Magnétostatique*, Masson, Paris.
DURAND, E. (1964–1966) *Electrostatique* (3 vols.); (1968) *Magnetostatique*, Masson, Paris.
GRANIER, J. (1941) *Les champs physiques*, Dunod, Paris.
JEANS, J. (1948) *Electricity and Magnetism*, Cambridge University Press.
SCARBOROUGH, J. B. (1930) *Numerical Mathematical Analysis*, Johns Hopkins University Press, Baltimore.
SHAW, F. S. (1953) *Relaxation Methods*, Dover, New York.
SOUTHWELL, R. V. (1946) *Relaxation Methods*, Oxford University Press.
WEBER, E. (1951) *Electromagnetic Fields*, McGraw-Hill, New York.
WHITTAKER, E. T., and G. ROBINSON (1944) *The Calculus of Observations*, Blackie, London.

(D) On Luminescent Screens

GARLICK, G. F. J. (1949) *Luminescent Materials*, Oxford University Press.
KRÖGER, F. A. (1948) *Luminescence of Solids*, Elsevier, Amsterdam.
LEVERENZ, H. W. (1950) *Luminescence of Solids*, Wiley, New York.

II. ARTICLES

AHARANOV, Y. and D. BOHM (1959) *Phys. Rev.* **115**, 485.
ALBRECHT, R. (1956) *Z. Naturforsch.* **11a**, 156.
ARCHARD, G. D. (1954a) *Brit. J. Appl. Phys.* **5**, 294; (1954b) London, 97; (1955) *Proc. Phys. Soc. Lond.* B **68**, 156; (1958) B **72**, 135.
ARCHARD, G. D., T. MULVEY and D. P. R. PETRIE (1960) Delft, 51.
ARDENNE, M. VON (1938) *Z. Physik* **108**, 338; (1939) *Z. Physik* **112**, 744; (1941) *Z. Physik* **117**, 602.
ARNAL, R. (1953) *C. R. Acad. Sci. Paris* **237**, 308.
ARTCIMOVITCH, L. A. (1944) *Izvestia Akad. Nauk SSSR.* **8**, 313.
BALLU, Y. (1968) Thèse (Doct. Ing.), Orsay.
BARAT, M. and FRANÇOIS, R. (1968) *C. R. Acad. Sci. Paris.* **B266**, 1306.
BARBER, M. R., and K. F. SANDERS (1959) *J. Elect. Cont.* **7**, 465.
BARBIER (1953) *Ann. Radioél.* **8**, 111.
BARKER, A. N., H. O. W. RICHARDSON and N. FEATHER, (1950) *Research* **3**, 431.
BAS, E. (1954) *G.F.F. Mitteilungen* **10**, 17; (1954) London; No. 146; (1955) *Optik* **12**, 377.
BAS, E., and F. GAYDOU (1959) *Z. angew. Phys.* **11**, 370.
BAUER, H. D. (1965–6) *Optik* **23**, 596.
BECKER, H., A. WALLRAFF (1940) *Arch. Elektrotechn.* **34**, 115.
BEHNE, R. (1936) *Ann. der Phys.* **26**, 372.
BELL, J. S. (1953) *Nature Lond.* **171**, 167.
BENNEWITZ, H. G., and W. PAUL (1954) *Z. Physik* **139**, 489.
BENNEWITZ, H. G., W. PAUL and C. SCHLIER (1954) *Z. Physik* **139**, 489; (1955) **141**, 6.
BERNARD, M. Y. (1951a) *C.R.Acad. Sci. Paris* **233**, 289 and 1354; (1951b) *J. Phys. Radium* **12**, 761; (1951c) *C.R. Acad. Sci. Paris* **233**, 1438; (1952a), **234**, 606; (1952b), **235**, 1115; (1953a) *J. Phys. Radium* **14**, 451 and 531; (1953b) *C.R. Acad. Sci. Paris* **236**, 185 and 902; (1954) *Ann. Phys. Paris* **9**, 633.

BERNARD, M. Y., and P. EHINGER (1954) *Cahiers Phys.* **50**, 8.
BERNARD, M. Y., and P. GRIVET (1951) *C. R. Acad. Sci. Paris* **233**, 788; (1952a) *Ann. Radioél.* **6**, 3; (1952b) *J. Phys. Radium* **13**, 47.
BERNARD, M. Y., and J. HUE (1956) *C. R. Acad. Sci.* **243**, 1852.
BERTEIN, F. (1941) *C. R. Acad. Sci. Paris* **224**, 560; (1947a) **225**, 801; (1947b), **225**, 863; (1947c) *Ann. Radioél.* **2**, 379; (1948a) **3**, 49; (1948b) *J. Phys. Radium* **9**, 104; (1951) **12**, 595 and 25 A; (1952a) **13**, 41 A and 91 A; (1952b) *C. R. Acad. Sci. Paris* **234**, 417; (1953) *J. Phys. Radium* **14**, 235.
BERTEIN, F., H. BRUCK and P. GRIVET (1947) *Ann. Radioél.* **2**, 249.
BERTEIN, F., P. GRIVET and E. REGENSTREIF (1949) Delft, 86.
BERTEIN, F., and E. REGENSTREIF (1947) *C. R. Acad. Sci. Paris* **224**, 737; (1949) **228**, 1854.
BERTRAM, S. (1940) *Proc. Inst. Radio Engrs.* **28**, 418; (1942) *J. Appl. Phys.* **13**, 496.
BETH, R. A. (1965) Brookhaven National Lab. Report AADD-66; (1966) *ibid.* AADD-102.
BLEWETT, J. P. (1947) *J. Appl. Phys.* **18**, 968; (1952) *Phys. Rev.* **88**, 1197; (1959) *Brook-haven Natl. Lab.* JPB-13.
BLOKHINTSEV, D. I. (1964) *Quantum Mechanics*, Reidel, Dordrecht.
BLOOMER, R. N. (1957) *Brit. J. Appl. Phys.* **8**, 83.
BOERSCH, H. (1940) *Naturwiss.* **28**, 709; (1942a) *Z. Techn. Phys.* **23**, 129; (1942b) *Naturwiss.* **30**, 711; (1942c) **30**, 120; (1947) *Z. Naturforsch.* 2a, 615; (1948) *Experientia* **4**, 1; *Physik* (1951) *Z.* **130**, 517.
BOERSCH, H., and G. BORN (1958) Berlin, 35.
BORRIES, B. VON (1942a) *Z. Physik* **119**, 498; (1942b) *Z. angew. Photograph.* **4**, 42; (1944) *Z. Physik* **122**, 539 (1948); *Optik* **3**, 321 and 389.
BOSI, G. (1965) *Nucl. Instr. and Meth.* **33**, 68.
BOSI, G. (1967) *Nucl. Instr. and Meth.* **46**, 55.
BRACHER, A. (1950) Thesis, Birmingham.
BRAUCKS, F. (1958) *Optik* **15**, 242; (1959) *Optik* **16**, 304.
BREMMER, H. (1950) Paris, communication No. 5.
BRETSCHER, M. M. (1960) ORNL Report No. 2864.
BREWER, G. R. (1957) *J. Appl. Phys.* **26**, 7.
BRICKA, M., and H. BRUCK (1948) *Ann. Radioél.* **3**, 339.
BROADWAY, L. F., and A. F. PIERCE (1939) *Proc. Phys. Soc. Lond.* **51**, 335.
BROWN, K. L., R. BELBEOCH and P. BOUNIN (1964) *Rev. Sci. Instrum.* **35**, 481.
BRÜCHE, E. (1942) *Kolloid Z.* **100**, 192.
BRÜCHE, E., and H. JOHANNSON (1932) *Naturwiss.* **20**, 49 and 353.
BRÜCHE, E., and R. KNECHT (1934) *Z. Physik* **92**, 462.
BRUCK, H. (1947a) *C. R. Acad. Sci. Paris* **224**, 1628; (1947b) **224**, 1553; (1947c), **224**, 1818.
BRUCK, H. (1958) *Optique Corpusculaire*, C.D.U., Paris.
BRUCK, H., and P. GRIVET (1947) *C. R. Acad. Sci. Paris* **224**, 1768.
BRUCK, H., R. REMILLON and L. ROMANI (1948) *C. R. Acad. Sci. Paris* **226**, 650.
BRUCK, H., and L. ROMANI (1944) *Cahiers Phys.* **24**, 1.
BURFOOT, J. C. (1952) *Brit. J. Appl. Phys.* **3**, 22; (1953a) Thesis, Cambridge; (1953b) *Proc. Phys. Soc.* B **66**, 775; (1954) *Proc. Phys. Soc.* B **67**, 523.
BUSCH, H. (1927) *Arch. Elektrotechn.* **18**, 583.
CARLILE, R. N. (1957) *H. E. P. L.* 33.
CARTAN, L. (1937) *J. Phys. Radium* **8**, 111.
CASTAING, R. (1950) Paris, communication No. 21; (1951). Thèse, Paris; (1960) *Adv. in Electronics* **13**, 317.
CASTAING, R. and G. SLODZIAN (1962a) *J. Micr.* **1**, 395.
CASTAING, R. and G. SLODZIAN (1962b) *C. R. Acad. Sci. Paris* **255**, 1893.

CHANSON, P. (1947) *Ann. Phys. Paris* **2**, 333.

CHANSON, P., A. ERTAUD and C. MAGNAN (1945) *C. R. Acad. Sci. Paris* **220**, 770.

CHARLES, D. (1949) *Ann. Radioél.* **4**, 1, 33.

CHARLES, D., and A. SEPTIER (1948) *C. R. Acad. Sci. Paris* **226**, 2056.

CHOUMOFF, S., and J. LAPLUME (1957) *Rev. Tech. CFTH* **25**, 115.

CHRISTENSEN, V. (1959) *Techn. Mem.* No. 71, *Project Matterhorn*, Princeton.

CITRON, A., F. J. FARLEY, E. L. MICHAELIS and H. ØVERÅS (1959) Cern 59-8.

CLOGSTON, A. M., and H. HEFFNER (1954) *J. Appl. Phys.* **25** 436.

COTTE, M. (1938) *Ann. Phys. Paris* **10**, 333; (1939) *Rev. Gen. Electr.* **45**, 675; (1949) *C.R. Acad. Sci. Paris* **228**, 377.

COUCHET, G. (1951) *C.R. Acad. Sci. Paris* **233**, 1013; (1954) *Ann. Phys. Paris* **9**, 731.

COUCHET, G., M. GAUZIT and A. SEPTIER (1951) *C.R. Acad. Sci. Paris* **233**, 1087; (1952) *Bull. Micr. Appl.* **2**, 55.

COURANT, E. D., M. S. LIVINGSTON and H. S. SNYDER (1952) *Phys. Rev.* **88**, 1168; (1953) **91**, 202.

DAVISSON, C. J., and C. J. CALBICK (1931) *Phys. Rev.* **38**, 585; (1932) *Phys. Rev.* **42**, 580.

DELTRAP. J. H. M. (1964a) Thesis, Cambridge; (1964b) Prague, 45.

DELTRAP. J. H. M. and V. E. COSSLETT (1962) Philadelphia, KK-8.

DHUICQ, D. (1960) *C.R. Acad. Sci. Paris* **251**, 1989.

DHUICQ, D. and A. SEPTIER (1959) *C.R. Acad. Sci. Paris* **249**, 2001; (1966a) *C.R. Acad. Sci. Paris* **B263**, 280; (1966b) *ibid.* 364.

DITCHBURN, R. (1948) *Rev. Opt.* **27**, 4.

DONADIEU. L. (1965) *Onde Electr.* **45**, 958.

DOSSE, J. (1940) *Z. Physik* **115**, 530; (1941a) **117**, 437; (1941b) **117**, 316; (1941c) **117**, 722; (1941d) **118**, 375.

DOSSE, J., and H. O. MÜLLER (1942) *Z. Physik* **119**, 415.

DRECHSLER, M., V. E. COSSLETT and W. C. NIXON (1958) Berlin, 13.

DUCHESNE, M. (1949) *C. R. Acad. Sci. Paris* **228**, 1407; (1953) *Bull. Astr. Paris* **17**, 88.

DÜKER, H. and A. ILLENBERGER (1962) Philadelphia, 1, D-5.

DURANDEAU, P. (1953) *C. R. Acad. Sci. Paris* **236**, 366.

DURANDEAU, P., and C. FERT (1957) *Rev. Opt.* **36**, 205.

DUZER, T. VAN, and G. R. BREWER (1959) *J. Appl. Phys.* **30**, 291.

DYMNIKOV, A. D., T. YA. FISHKOVA and S. YA. YAVOR (1964a) *Zh. Tekh. Fiz.* **35**, 759 (= *Sov. Phys. Techn. Phys.* **10**, 592); (1964b) *Zh. Tekh. Fiz.* **34**, 1171 (= *Sov. Phys. Techn. Phys.* **9**, 1322); (1965a) *Zh. Tekh. Fiz.* **35**, 431; *Sov. Phys. Techn. Phys.* **10**, 340; (1965b) *Zh. Tekh. Fiz.* **35**, 759 (= *Sov. Phys. Tech. Phys.* **10**, 592); (1965c) *Phys. Letters* **15**, 132.

DYMNIKOV, A. D. and S. YA. YAVOR (1963) *Zh. Tekh. Fiz.* **33**, 851 (= *Sov. Phys. Techn. Phys.* **8**, 639).

EHINGER, P. (1954) *C. R. Acad. Sci. Paris* **338**, 879 and 1306.

EHRENBERG, W., and R. E. SIDAY (1949) *Proc. Phys. Soc.* B **62**, 8.

EHRENBERG, W., and W. E. SPEAR (1951) *Proc. Phys. Soc.* B **64**, 67.

EINSTEIN, P. A., and L. JACOB (1948) *Phil. Mag.* **39**, 20.

ELINSON, M. I. (1958) Berlin, 25.

ELLIS, S. G. (1947) *J. Appl. Phys.* **18**, 879.

ENGE, H. A. (1959) *Rev. Sci. Instrum.* **30**, 248; (1961) **32**, 662.

ENGE, H. A. (1964) *Rev. Sci. Instrum.* **35**, 278.

ENGE, H. A. (1967) Deflecting magnets, in *Focusing of Charged Particles* (A. Septier, ed.), Academic Press, New York and London. vol. II, p. 203.

EPSTEIN, D. W. (1936) *Proc. Inst. Radio Engrs.* **24**, 1095.

EVERITT, C. W. F., and K. T. HANSSEN (1956) *Optik* **13**, 385.

EWALD. H. and H. LIEBL (1955) *Z. Naturforsch.* **10a**, 872.

EWALD. H. and H. LIEBL (1957) *Z. Naturforsch.* **12a**, 28.

FELICI, N. J. (1959) *J. Phys. Radium* **20**, 97 A.

FERT, C. (1952) *J. Phys. Radium* **13**, 64 A; (1952) **13**, 83 A; (1954) London, 161; (1956) *C. R. Acad. Sci. Paris* **243**, 1301.

FERT, C., and P. GAUTIER (1951) *C.R. Acad. Sci. Paris* **233**, 148.

FEYNMAN, R. P., R. B. LEIGHTON and M. SANDS (1964) *Lectures on Physics*, vol. II, § 15-5, Addison-Wesley, New York.

FORTESCUE, C. L., and C. W. FARNSWORTH (1913) *Proc. AIEE*, **32**, 757.

FOWLER, H. A., L. MARTON, J. A. SIMPSON and J. A. SUDDETH (1961) *J. Appl. Phys.* **32**, 1153.

FRANCKEN, J. C. (1959/60) *Philips Techn. Rev.* **21**, 11.

GABOR, D. (1942) *Nature* **150**, 650; (1947) **159**, 303; (1948) **161**, 777.

GANS, R. (1937) *Z. Techn. Phys.* **18**, 41.

GARRETT, M. W. (1951) *J. Appl. Phys.* **22**, 1091.

GAUTHIER, P. (1952) *C.R. Acad. Sci. Paris* **235**, 361; (1954) *J. Phys. Radium* **15**, 684.

GAUZIT, M. (1951) *C.R. Acad. Sci. Paris* **233**, 1586; (1953) *Bull. Micr. Appl.* (2) **3**, 37; (1954) *Ann. Phys. Paris* **9**, 683.

GAUZIT, M., and A. SEPTIER (1951) *Bull. Micr. Appl.* (2) **1**, 109.

GERMAIN, C. (1955a) Thèse, Paris; (1955b) *C.R. Acad. Sci. Paris* **240**, 588.

GIANOLA, U. F. (1950) *Proc. Phys. Soc.* B **63**, 1037.

GLASER, W. (1933a) *Z. Physik* **81**, 647; (1933b) **83**, 104; (1940a) **116**, 19; (1940b) **116**, 56; (1941a) **117**, 285; (1941b), **118**, 264; (1943a) **120**, 1; (1943b) **121**, 647; (1948) *Nature* **162**, 455; (1950a) *Proc. Phys. Soc.* B **64**, 114; (1950b) *ZAMP* **1**, 363; (1950c) Paris, communication No. 24; (1950d) *Ann. der Phys.* **7**, 213; (1956) "Elektronen und Ionen-optik" in *Handbuch der Physik (Encyclopædia of Physics)* **33**, 123–395.

GLASER, W., and E. LAMMEL (1941) *Ann. der Physik* **40**, 367; (1943) *Arch. Elektrotechn.* **37**, 347.

GLASER, W., and F. LENZ (1951) *Ann. der Phys.* **9**, 19.

GLASER, W., and ROBL (1951) *ZAMP* **2**, 444.

GLASER, W. and P. SCHISKE (1954) *Optik* **11**, 422 and 445; (1955) *Optik* **12**, 233.

GOBRECHT, R. (1941) *Arch. Elektrotechn.* **35**, 672; (1942) **36**, 484.

GODDARD, L. S. (1944) *Proc. Phys. Soc.* **56**, 372.

GRAY, F. (1939) *Bell. Syst. Techn. J.* **18**, 1.

GRIVET, P. *Electron Microscopy in Metallurgy, Symposium on Metallurgical applications of the Electron Microscope*, London, 16 November 1949; (1950a) *C. R. Acad. Sci. Paris*, **230**, 1152; (1950b) *J. Phys. Radium* **11**, 582; (1951a) **12**, 1; (1951b) *C.R. Acad. Sci. Paris* **233**, 921; (1952a) *J. Phys. Radium* **13**, 1 A; (1952b) *C. R. Acad. Sci. Paris* **234**, 73.

GRIVET, P., and Y. ROCARD (1949) *Rev. Scientifique* **11**, 85.

GRIVET, P., and A. SEPTIER (1960) *Nuclear Instr. and Methods* **6**, 126 and 243.

GUÉNARD, R. (1945) *Ann. Radioél.* **1**, 74.

GUNDERT, E. (1941) *Telefunkenröhre* **19**, 61.

HACHENBERG, O. (1948) *Ann. der Phys.* **2**, 225.

HAHN, E. (1958) *Jenaer Jahrb.* **1**, 184.

HAINE, M. E. and P. A. EINSTEIN (1952) *Brit. J. Appl. Phys.* **3**, 40.

HAINE, M. E., P. A. EINSTEIN and P. H. BORCHERDS (1958) *Brit. J. Appl. Phys.* **9**, 482.

HALL, C. E. (1948) *J. Appl. Phys.* **19**, 198.

HAMPIKIAN, A. (1953) *C. R. Acad. Sci. Paris* **236**, 1864.

HAST, N. (1948) *Nature* **162**, 892.

HAWKES, P. W. (1963) Thesis, Cambridge; (1965a) *Phil. Trans. Roy. Soc. London* **A257**, 479; (1965b) *Optik* **22**, 349; (1965c) *Optik* **22**, 543; (1966) *Springer Tracts in Modern Physics* **42**.

HEIL, O., and J. J. EBERS (1950) *Proc. Inst. Radio. Engrs.* **38**, 645.

HEISE, F. (1949) *Optik* **5**, 479.

HEISE, F. and O. RANG (1949) *Optik* **5**, 201.

HELM, R., K. SPANGENBERG and L. FIELD (1947) *Electr. Comm.* **24**, 101.

HENNEQUIN, J. (1948) *Diplôme d'Études Supérieures*, Paris.

HEPP, G. (1939) *Philips Techn. Rev.* **4**, 225.

HERZOG. R. (1934) *Z. Physik* **89**, 447.

HESSE, M. B. (1949) *Proc. Phys. Soc.* **61**, 233.

HILLIER, J. (1946) *J. Appl. Phys.* **17**, 411.

HILLIER, J., and F. BAKER (1945) *J. Appl. Phys.* **16**, 469.

HILLIER, J., and E. G. RAMBERG (1947) *J. Appl. Phys.* **18**, 48.

HIMPAN, J. (1939) *Telefunkenröhre* **16**, 198; (1942) *Elektrotechn. Z.* **63**, 349.

HINTENBERGER. H. (1948) *Z. Naturforsch.* **3a**, 125 and 669.

HINTENBERGER. H. (1949) *Rev. Sci. Instrum.* **20**, 748.

HINTENBERGER. H. (1951) *Z. Naturforsch.* **6a**, 275.

HOGAN, T. K. (1943) *J. Australian Inst. Engrs.* **15**, 89.

HOLLWAY, D. L. (1955) *Australian J. Phys.* **8**, 74; (1955) *Proc. Inst. Elec. Engrs.* B **103**, 155.

HUBBARD, E. L., and E. L. KELLY (1954) *Rev. Sci. Instrum.* **25**, 737.

HUBER, H. (1949) *Ann. Radioél.* **4**, 26.

HUBERT, P. (1949) *C. R. Acad. Sci. Paris* **228**, 233.

HUGUENIN, L. (1954) *C. R. Acad. Sci. Paris* **239**, 404.

HUTTER, R. G. E. (1945) *J. Appl. Phys.* **16**, 670.

IKEGAMI. H. (1958) *Rev. Sci. Instrum.* **29**, 943.

IVEY, H. F. (1954) *Adv. in Electronics* **6**, 137.

JACOB, L. (1939) *Phil. Mag.* **28**, 81; (1948) *Phil. Mag.* **39**, 400; (1950a) *Proc. Phys. Soc. Lond.* B **63**, 75; (1950b) *J. Appl. Phys.* **21**, 966.

JACOB, L., and T. MULVEY (1949) *Nature* **163**, 525.

JACOB, L., and J. R. SHAH (1951) *J. Appl. Phys.* **22**, 1236.

JOHANNSON, H. (1933) *Ann. der Phys.* **18**, 285; (1934), **21**, 274.

KANAYA. K.. H. KAWAKATSU. H. YAMAZAKI and S. SIBATA (1966) *J. Sci. Instrum.* **43**, 416.

KELLER, R. (1957) C.E.R.N. Sc. 57–30.

KEL'MAN, V. M. and S. YA. YAVOR (1961) *Zh. Tekh. Fiz.* **31**, 1439 (= *Sov. Phys. Techn. Phys.* **6**, 1052); (1963) *Zh. Tekh. Fiz.* **33**, 369 (= *Sov. Phys. Techn. Phys.* **8**, 271).

KEL'MAN, V. M., A. D. DYMNIKOV and L. P. OVSYANNIKOVA (1963) *Bull. Acad. Sci. USSR*, *Phys. Ser.* **27**, 1116.

KERWIN. L. (1949) *Rev. Sci. Instrum.* **20**, 36.

KINDER, E. (1944) *Z. Physik* **122**, 192.

KLEMPERER, O. (1947) *Proc. Phys. Soc.* **59**, 302.

KLEMPERER, O., and Y. KLINGER (1951) *Proc. Phys. Soc.* B **64**, 231.

KLEMPERER, O., and B. J. MAYO (1948) *J. Inst. Electr. Engrs.* **95**, 135.

KLEMPERER, O., and P. WRIGHT (1939) *Proc. Phys. Soc.* B **51**, 296.

KNOLL, M. (1935) *Z. Techn. Phys.* **16**, 467.

KNOLL, M., and E. RUSKA (1932) *Ann. der Phys.* **12**, 607.

KNOLL, M., and R. THEILE (1941) *Forsch. Hochfrequenztechn.* **1**, 487.

KNOLL, M., and H. WEICHART (1938) *Z. Physik* **110**, 223.

KRYLOFF, A. M. (1927) *Mémorial de l'Artillerie Française* **6**, 353.

KUTTA, W. (1901) *Z. Math. Phys.* **46**, 443.

LALLEMAND, A., and M. DUCHESNE (1951) *C. R. Acad. Sci. Paris* **233**, 305.

LANGER, L. M., and F. R. SCOTT (1950) *Rev. Sci. Instrum.* **21**, 522.

LANGMUIR, D. B. (1937) *Proc. Inst. Radio Engrs.* **25**, 977.

LANGMUIR, I., and K. BLODGETT (1923) *Phys. Rev.* **22**, 347; (1924) **24**, 49.

LAPEYRE, R. (1961) *C. R. Acad. Sci. Paris* **252**, 3431.

LAPEYRE, R., and M. LAUDET (1960) *C. R. Acad. Sci. Paris* **251**, 679.

LAPLUME, J. (1947) *Cahiers Phys.* **29-30**, 1 and 55.

LAUDET, M. (1953) *J. Phys. Radium* **14**, 604; (1957) **18**, 73A.

LAW, R. R. (1937) *Proc. Inst. Radio Engrs.* **25**, 954.

LENZ, F. (1950) *Optik* **7**, 243; (1951) *Ann. der Phys.* **9**, 245.

LE POOLE, J. B., and A. C. VAN DORSTEN (1951) *Communication to the St Andrews Congress of the British Microscopical Society.*

LE RUTTE, W. A. (September 1948) *9th Electron Microscope Congress*, Cambridge; Communication B; (1948) *Nature* **161**, 392; (1952) Thesis, Delft.

LIEBL. H. and H. EWALD (1959) *Z. Naturforsch.* **14a**, 84a.

LIEBMANN, G. (1949a) *Proc. Phys. Soc.* B **62**, 213; (1949b) B **62**, 753; (1949c) *Nature* **146**, 149; (1950a) *Brit. J. Appl. Phys.* **1**, 92; (1950b) *Phil. Mag.* **322**, 1143; (1951) *Proc. Phys. Soc.* B **64**, 972; (1952) **65**, 188; (1953) B **66**, 448.

LIEBMANN. G., and L. M. GRAD (1951) *Proc. Phys. Soc.* B **64**, 956.

LIPPERT, W., and W. POHLIT (1952) *Optik* **9**, 456.

LOEB, J. (1947) *Onde électr.* **27**, 27.

LYLE, T. R. (1902) *Phil. Mag.* **3**, 310.

MACNAUGHTON, M. (1952) *Proc. Phys. Soc.* B **65**, 590.

MAGNAN, C., and P. CHANSON (1951) *C. R. Acad. Sci. Paris* **233**, 1436.

MAHL, H. (1940) *Jahrb. AEG Forsch.* **7**, 43; (1942) *Z. Techn. Phys.* **23**, 117.

MAHL, H., and A. PENDZICH (1943) *Z. Techn. Phys.* **24**, 38.

MAHL, H., and A. RECKNAGEL (1944) *Z. Physik* **122**, 60.

MALAVARD, L. in SURUGUE, *Techniques générales du Laboratoire de Physique*, vol. II, Ed. C.N.R.S., Paris, 1940.

MALAVARD, L., and J. PERES (1938) *Bull. SFE*, (5) **8**, 715.

MALOFF, I. G., and D. W. EPSTEIN (1934) *Proc. Inst. Radio Engrs.* **23**, 1386.

MARÉCHAL, A. (1944) *Cahiers Phys.* **26**, 1; (1947) *Rev. Opt.* **26**, 257; (1948) **27**, 73.

MARTON, L. (1946) *Rep. Prog. Phys.* **10**, 204.

MARTON, L., and P. ABELSON (1947) *Science* **106**, 69.

MARTON, L., and R. BOL (1947) *J. Appl. Phys.* **18**, 522.

MARTON, L., M. M. MORGAN, D. C. SCHUBERT, J. R. SHAH and J. A. SIMPSON (1951) *J. Res. Nat. Bur. Stand* **47**, 461.

MARUSE, S., and Y. SAKAKI (1958) *Optik* **15**, 485.

MEADS. P. F. (1966) *Nucl. Instr. and Meth.* **40**, 166.

MECKLENBURG, W. (1942) *Z. Physik* **120**, 21.

MELKICH, A. (1944) Thesis, Berlin, see *Sitzber. Akad. Wiss. Wien Math.-naturw. Kl.* Abt. IIa, Nos. 9–10, 393–471 (1947).

MENDEL, J. T., C. F. QUATE and W. H. YOCOM (1954) *Proc. Inst. Radio Engrs.* **42**, 801.

MENTS, M. VAN, and J. B. LE POOLE (1947) *Appl. Sci. Res.* 1 B, 3.

MEYER, W. E. (1961) *Optik* **18**, 69.

MISES, R. VON (1930) *Z. angew. Math. Mech.* **10**, 81.

MÖLLENSTEDT, G. (1949) *Optik* **5**, 499; (1952) **9**, 473; (1952) *Z. Naturforsch.* **7**, 465; (1956) **13**, 209.

MÖLLENSTEDT, G. and W. BAYH (1962) *Naturwiss.* **49**, 81.

MÖLLENSTEDT, G., and W. HUBIG (1954) *Optik* **11**, 538.

MORTON, G. A. (1946) *Rev. Mod. Phys.* **18**, 362.

MORTON, G. A., and E. G. RAMBERG (1936) *Physics* **7**, 451.

MOSS, H. (1945) *J. Inst. Radio Engrs.* **5**, 10 and 99.

MOTZ, H., and L. KLANFER (1946) *Proc. Phys. Soc.* B **58**, 30.

MÜLLER, M. (1956) *J. Brit. Inst. Radio Engrs.* **16**, 65.

MUSSON-GENON, R. (1947) *Ann. Télécomm.* **2**, 254.

NICOLL, F. H. (1938) *Proc. Phys. Soc.* **50**, 888.

NIJBOER, B. R. A. (1942) Thesis, Groningen.

PACKH, D. C. DE (1947) *Rev. Sci. Instrum.* **18**, 789.

PAUL, W., and M. RAETHER (1955) *Z. Physik* **140**, 262.

PAUL, W., H. P. REINHARD, and V. VON ZAHN (1958) *Z. Physik* **152**, 143.

PEIERLS, E. (1947) *Nature* **158**, 851.

PEIERLS, R. E. and T. H. R. SKYRME (1949) *Phil. Mag.* **40**, 269.

PERES, J. and L. MALAVARD (1938) *Bull. Soc. Fr. Elect.* **8**, 715.

PICQUENDAR, J. E. and O. CAHEN (1960) *Rev. Techn. CFTH* **32**, 59.

PIERCE, J. R., (1939) *J. Appl. Phys.* **10**, 715; (1941) *Proc. Inst. Radio Engrs.* **29**, 28; (1945) *Bell. Syst. Techn. J.* **24**, 305.

PILOD, P. and F. SONIER (1961) *C. R. Acad. Sci. Paris* **253**, 2338.

PINEL, H. (1959) *Ann. Radioél.* **14**, 230; (1960) **15**, 17.

PING KING TIEN (1954) *J. Appl. Phys.* **25**, 1281.

PLASS, G. N. (1942) *J. Appl. Phys.* **13**, 49 and 524.

PLOCKE, M. (1951) *ZAMP* **3**, 441; (1952) **4**, 1.

RAMBERG, E. G. (1939) *J. Opt. Soc. Amer.* **29**, 79; (1942) **13**, 582; (1949) **20**, 183.

RANG, O. (1949) *Optik* **5**, 518.

REBSCH, R. (1940) *Z. Physik* **116**, 729.

REBSCH, R., and O. SCHNEIDER (1937) *Z. Physik* **107**, 138.

RECKNAGEL, A. (1937) *Z. Physik* **104**, 381; (1940) **117**, 67; (1941) **117**, 689; (1943) **120**, 313.

REGENSTREIF, E. (1947) *Ann. Radioél.* **2**, 348; (1950) *C. R. Acad. Sci. Paris* **230**, 1650; (1951a) *Ann. Radioél.* **6**, 51 and 164; (1951b) **6**, 244; (1951c) **6**, 299.

RIDLEY, R. O. (1954) *A.L.R.E.T/M* 103.

RÜDENBERG, R. J. (1948) *J. Frank Inst.* **246**, 311 and 377.

RUSKA, E. (1933) *Z. Physik* **83**, 684; (1934) **89**, 90; (1944) *Arch. Elektrotechn.* **38**, 102.

RUTHEMANN, G. (1941) *Naturwiss.* **29**, 298 and 648; (1942) **30**, 145; (1948) *Ann. der Physik* **2**, 113 and 135.

SAKAKI, Y., and S. MARUSE (1960) Berlin, 9.

SAMUEL, A. L. (1945) *Proc. Inst. Radio Engrs.* **33**, 233.

SANDERS, K. F., and J. G. YATES (1953) *Proc. Inst. Elec. Engrs.* **11**, 167; (1956) C Monograph 195 M.

SANDOR, A. (1941) *Arch. Elektrotechn.* **35**, 217 and 259.

SAUZADE, M. (1958) *C.R. Acad. Sci. Paris* **246**, 272.

SCHAEFER, H., and W. WALCHER (1943) *Z. Physik* **121**, 679.

SCHERZER, O. (1936) *Z. Physik* **101**, 23 and 593; (1947) *Optik* **2**, 114; (1949) **5**, 497; (1950) Paris, communication No. 28.

SCHIEKEL, M. (1952) *Optik* **9**, 145.

SCHIFF, L. (1946) *Phys. Rev.* **70**, 87.

SCHWARZSCHILD, K. (1905) *Abh. Königl. Ges. Wiss. Göttingen* **4**, No. 1–3.

SEELIGER, R. (1949) *Optik* **5**, 490; (1951) **8**, 311; (1953) **10**, 29.

SEMAN, O. I. (1952) *Zh. Tekh. Fiz.* **22**, 1581.

SEPTIER, A. *Diplôme d'études supérieures*, Paris 1948 (see CHARLES, *Ann. Radioél.* 1949, **4**, 1); (1952) *C.R. Acad. Sci. Paris* **235**, 609, 652, 1203 and 1621; (1953a) **236**, 58; (1953b) **237**, 231; (1954a) *J. Phys. Radium* **15**, 573; (1954b) *Ann. Radioél.* **9**, 374; (1955) *C. R. Acad. Sci. Paris* **240**, 1200; (1957) **245**, 1406; (1958) **246**, 1983; (1960) C.E.R.N. 60-6 and 60-39; (1960) *J. Phys. Radium* **21**, 1A; (1961a) *Adv. in Electronics* **14**, 85; (1961b) *C.R. Acad. Sci. Paris* **252**, 2851; (1963) *C.R. Acad. Sci. Paris.* **256**, 2325; (1966) *Adv. in Opt. and Electron Micr.* **1**, 204.

SEPTIER, A., and J. VAN ACKER (1960) *C. R. Acad. Sci. Paris* **251**, 346 and 1750.

SEPTIER, A. and D. DHUICQ (1965) *Jour. Fr. Micr. Electr. Marseille*, Unpublished.

SEPTIER, A., and M. GAUZIT (1950) Paris, communication No. 34.

SEPTIER, A., M. GAUZIT and P. BARUCH (1952) *C. R. Acad. Sci. Paris* **234**, 105.

SHIPLEY, D. W. (1952) *J. Appl. Phys.* **23**, 1310.

SHORTLEY, G., and R. WELLER (1948) *J. Appl. Phys.* **19**, 334.

SHORTLEY, G., R. WELLER, P. DARBY and E. H. GAMBLE (1947) *J. Appl. Phys.* **18**, 116.

SIEGBAHN, K. (1946) *Phil. Mag.* **37**, 162.

SIEGBAHN, K., and H. SLÄTIS (1949) *Phys. Rev.* **75**, 1955.

SOA, E. A. (1959) *Jenaer Jahrb.* **1**, 115.

SOLLER, T., M. A. STARR and E. G. VALLEY (1948) *Cathode ray tube displays*, McGraw-Hill, New York.

SPANGENBERG, K., and M. FIELD (1942) *Proc. Inst. Radio Engrs.* **30**, 138; (1943) *Electrical communication* **21**, 194.

SPEAR, W. E. (1952) *Proc. Phys. Soc.* B **64**, 233.

STEIGERWALD, K. H. (1949) *Optik* **5**, 469.

STÖRMER, C. (1920) *Congrès International des Mathématiques*, Strasbourg, p. 242, published in 1921 by the Bibliothèque Universitaire de Toulouse; (1933) *Ann. der Phys.* **16**, 685.

STRASHKEVICH. A. M. (1964) *Zh. Tekh. Fiz.* **34**, 1401 (= *Sov. Phys. Techn. Phys.* **9**, 1082).

STRUMSKI, C. J., D. I. COOPER, D. M. FRISCHT and R. L. ZIMMERMANN (1954) *Rev. Sci. Instrum.* **25**, 514.

STURROCK, P. (1951 a) *C. R. Acad. Sci. Paris* **233**, 146 and 243; (1951 b) **233**, 401; (1951 c) *Phil. Trans. Roy. Soc. Lond.* A **243**, 387; (1952) A **245**, 155.

SÜSSKIND, S. (1956) *Adv. in Electronics* **8**, 363.

SWIFT, D. (1960) Thesis, Cambridge.

SYMONDS, J. L. (1955) *Rep. Prog. Phys.* **18**, 83.

SYNGE, J. L. (1952) *Rev. Opt.* **31**, 121.

TASMAN, H. A. and A. J. BOERBOOM (1959) *Z. Naturforsch.* **14a**, 121.

TASMAN. H. A.. A. J. BOERBOOM and H. WACHSMUTH (1959) *Z. Naturforsch.* **14a**, 822.

TENG, L. C. (1954) *Rev. Sci. Instrum.* **25**, 264.

THEILE, R., and J. HIMPAN (1940) *Telefunkenröhre* **18**, 50.

TROLAN, J. K., and W. P. DYKE (1958) Berlin, 20.

VAUTHIER, R. (1949) *C. R. Acad. Sci. Paris* **228**, 1113.

VERSTER, J. L. (1961) *Philips Techn. Rev.* **22**, 245.

VINEYARD, G. H. (1952) *J. Appl. Phys.* **23**, 35.

WACHSMUTH. H.. A. J. BOERBOOM and H. A. TASMAN (1959) *Z. Naturforsch.* **14a**, 818.

WAKEFIELD, K. E. (1958) *Report PMS* 23, *Project Matterhorn*, Princeton.

WAKERLING, R. K.. and A. C. HELMHOLZ (1949) *Report T.I.D.* No. 5217, 227.

WALTHER and DREYER (1949) *Naturwiss.* **7**, 199.

WENDT, G. (1943) *Z. Physik* **120**, 720.

WERNER. F. G. and D. R. BRILL (1960) *Phys. Rev. Letters* **4**, 344.

WILDER, T. (1949) *Proc. Inst. Radio Engrs.* **37**, 1182.

WOLFF, P. A. (1953) *Phys. Rev.* **92**, 18.

WOLLNIK. H. (1965) *Nucl. Instr. and Meth.* **34**, 213.

WOLLNIK. H. (1967) Electrostatic prisms, in *Focusing of Charged Particles* (A. Septier, ed.), Academic Press, New York and London, vol. II, p. 163.

WOLLNIK. H. and H. EWALD (1965) *Nucl. Instr. and Meth.* **36**, 93.

YAVOR. S. YA. A. D. DYMNIKOV and L. P. OVSYANNIKOVA (1964) *Nucl. Instr. and Meth.* **26**, 13.

INDEX